Empathy Machines

BLOOMSBURY PODCAST STUDIES

Series Editors:
Martin Spinelli and Lance Dann

Series Editorial Board Members:
Martin Spinelli, University of Sussex, UK
Lance Dann, University of Brighton, UK
Mia Lindgren, University of Tasmania, Australia
Kathleen Collins, John Jay College, USA
Richard Berry, University of Sunderland, UK
Ann Heppermann, Sarah Lawrence College, USA
Tiziano Bonini, Università di Siena, Italy
John Sullivan, Muhlenberg College, USA
Belén Monclús, Universitat Autònoma de Barcelona, Spain

Empathy Machines

This American Life, Podcasting, and the Public Radio Structure of Feeling

JASON LOVIGLIO

BLOOMSBURY ACADEMIC
NEW YORK • LONDON • OXFORD • NEW DELHI • SYDNEY

BLOOMSBURY ACADEMIC
Bloomsbury Publishing Inc, 1359 Broadway, New York, NY 10018, USA
Bloomsbury Publishing Plc, 50 Bedford Square, London, WC1B 3DP, UK
Bloomsbury Publishing Ireland, 29 Earlsfort Terrace, Dublin 2, D02 AY28, Ireland

BLOOMSBURY, BLOOMSBURY ACADEMIC and the Diana logo are
trademarks of Bloomsbury Publishing Plc

First published in the United States of America 2026

Copyright © Jason Loviglio, 2026

For legal purposes the Acknowledgments on pp. vii–ix constitute an
extension of this copyright page.

Cover design: Andrew Walker

All rights reserved. No part of this publication may be: i) reproduced or transmitted in any form, electronic or mechanical, including photocopying, recording or by means of any information storage or retrieval system without prior permission in writing from the publishers; or ii) used or reproduced in any way for the training, development or operation of artificial intelligence (AI) technologies, including generative AI technologies. The rights holders expressly reserve this publication from the text and data mining exception as per Article 4(3) of the Digital Single Market Directive (EU) 2019/790.

Bloomsbury Publishing Inc does not have any control over, or responsibility for, any third-party websites referred to or in this book. All internet addresses given in this book were correct at the time of going to press. The author and publisher regret any inconvenience caused if addresses have changed or sites have ceased to exist, but can accept no responsibility for any such changes.

Library of Congress Cataloging-in-Publication Data
Names: Loviglio, Jason author
Title: Empathy machines : This American life, podcasting and the
public radio structure of feeling / Jason Loviglio.
Description: New York : Bloomsbury Academic, 2026. |
Series: Bloomsbury podcast studies ; 3 |
Includes bibliographical references and index.
Identifiers: LCCN 2025024753 | ISBN 9798765111727 hardback |
ISBN 9798765111680 paperback | ISBN 9798765111703 pdf |
ISBN 9798765111697 epub
Subjects: LCSH: Public radio–United States | Public radio–Social aspects–United States | Public radio–Political aspects–United States | Podcasts–United States | This American life (Radio program) | National Public Radio (U.S.)
Classification: LCC HE8697.95.U6 L68 2026 |
DDC 384.5406/573–dc23/eng/20250624
LC record available at https://lccn.loc.gov/2025024753

ISBN:	HB:	979-8-7651-1172-7
	PB:	979-8-7651-1168-0
	ePDF:	979-8-7651-1170-3
	eBook:	979-8-7651-1169-7

Series: Bloomsbury Podcast Studies, volume 3

Typeset by Integra Software Services Pvt. Ltd.
Printed and bound in the United States of America

To find out more about our authors and books visit www.bloomsbury.com
and sign up for our newsletters.

CONTENTS

Series Preface vi
Acknowledgments vii

Introduction: "Become Empathy Machines!" 1

1   A Feeling Medium 27
2   Voracious Voyagers: NPR Listens 61
3   Feeling Playful: *This American Life* and Narrative Enchantment 95
4   Feeling American 141
5   Feeling Flush: *Planet Money* and the American Dream 175
6   Fellow Feeling: *This American Life* and the Gendered Voice 213
7   Feeling Uncomfortable: The Politics and Aesthetics of Cringe 265

Conclusion: Maybe It's a Feeling? How *This American Life* Long Endured 307

Index 330

SERIES PREFACE

The Bloomsbury Podcast Studies Series sets out to establish Podcast Studies as its own distinct field which spans the Humanities and Social Sciences. It offers granular political, cultural, historical, economic, literary, and data-driven analyses of podcast genres, production practices, institutions and platforms, narratives and semiotics, and national and regional currents by leading scholars and practitioners. With commitments to both accessibility and rigor, the series promotes research and knowledge creation about, through and with podcasting, and offers insights to policy makers, academics, and creatives alike. Its underlying intention is to develop the practical yet sophisticated vocabularies, methodologies and critical tools needed to fully appreciate the depth and dynamism of our newest audio medium and what it contributes to the broader world.

Martin Spinelli & Lance Dann
Series Editors

ACKNOWLEDGMENTS

So many people have talked to me about this project over so many years that I'm tempted to compose a shorter list of people to whom I don't owe a debt of gratitude. But a long book that has been long in the making deserves an acknowledgements page to match.

Big thanks to Martin Spinelli and Lance Dann for believing in this book and including it in their fantastic series. Thanks also to the anonymous reviewers for their excellent feedback and to Katie Gallof Houck and everyone at Bloomsbury for supporting the book from the start.

The people I have had the honor of calling my students over the years have done more than anyone else to help me learn how to communicate ideas about communication and media clearly and simply. The undergraduates in the Media and Communication Studies Department (MCS) of The University of Maryland, Baltimore County, are among the least pretentious and least entitled people I've ever had the pleasure of working with. Too numerous to name here, I give a collective thanks to you for the lessons of clarity you've taught me and for the laughs.

My colleagues on the faculty of MCS have helped me more than they likely realize. Donald Snyder is the most thoughtful and creative collaborator one could hope for; there'd be no MCS without him. Liz Patton's leadership made it possible for me to finish a project that was in danger of stalling. Kristen Anchor, Samirah Hassan, Chung-Wei Huang, Katy Razzano, Bill Shewbridge, Tracy Tinga, and Fan Yang have each taught me something important about teaching and communicating. Big thanks to UMBC colleagues Jessica Berman, Tamara Bhalla, and Christine Mallinson for their years of support and collegiality.

Rachel Buff's steadfast intellectual encouragement and political inspiration have sustained me for my entire career. This book is better for our many conversations. Jennifer Wang's work in radio and podcasting inspired the title for the conclusion of the book and quite a bit more in these pages. Bill Kirkpatrick read large parts of the manuscript and I'm indebted to him for his generosity and tough love, in equal measures. Neil Verma's snapshot of a Post-It Note inspired the book's introduction, and his books and articles have provided an invaluable education in how to think and write about sound. Mike Janssen taught me a lot about writing for a broader audience; any lapses in that regard are in spite of his best efforts.

To speak of radio and podcasting studies as a field or a discipline has always felt a bit aspirational over much of the last twenty-five years. But recently it has begun to sound a bit more apt, even with the usual qualifications about the limits of disciplinarity and the necessarily unstable and overlapping nature of cultural studies scholarship. To the extent that the term has any coherence at all for me, is down to the work of Michele Hilmes, Susan Douglas, and Kate Lacey, the Big Three. In addition, I couldn't have managed much of anything in my career without the scholarship and support of David Hendy, Bill Kirkpatrick, Mia Lindgren, Alex Russo, and Jennifer Hyland Wang. I have likewise benefitted in big and small ways from the scholarship of (and conversations with) colleagues such as Alec Badenoch, Richard Berry, Tiziano Bonini, Kathleen Battles, Dylan Bird, Dolores Inés Casillas, Stacey Copeland, Christopher Cywnar, Christine Ehrick, David Goodman, Brian Fauteux, Dan Marcus, Catherine Martin, Siobhan McHugh, Naomi Mezey, Cynthia Meyers, Matt Mollgaard, Nora Patterson, Janice Peck, Elena Razlogova, Joshua Shepperd, Judy Smith, Susan Smulyan, Martin Spinelli, Andy Stuhl, John L. Sullivan, Neil Verma, Jing Wang, and the late Jonathan Sterne.

So many radio and podcast producers have shared their insights with me over the years that it's daunting to imagine a complete list. In conversations on and off the record, Melissa Block, Jay Kernis, Robert Siegel, and Susan Stamberg taught me so much about NPR that I was forced to rethink some of the foundational assumptions that I brought to this project at its inception. Conversations with Jason DeRose, Aaron Henkin, Andrea Hsu, Sheilah Kast, Jamyla Krempel, Chenjerai Kumanyika, Lawrence Lanahan, Louisa Lim, Mary Rose Madden, Stefanie Mavronis, LaFontaine Oliver, Aimee Pohl, Elissa Nadworny, Nina Totenberg, Amy Scott, Art Silverman,

ACKNOWLEDGMENTS

Lisa Simeone, Scott Simon, Ashley Sterner, Marc Steiner, Flawn Williams, Mary Wiltenburg, and the late Neal Conan helped me understand the public radio structure of feeling.

This book, and so many other much more important things, would not exist without the work of Bill Siemering, whose notion of radio as a "feeling medium" runs through every page of this book.

Thanks also to Ira Glass for chatting with me off the record during two car rides, one of which involved a small fender bender. No radio legends were harmed in the writing of this book.

Big thanks to Rachel Buff, Sheila Doyle, Michael Lorant, Jennifer Loviglio, Gregg Nass, and Felice Shore for moral support. My greatest debt is to my family, Anne Wolf, Benjamin Loviglio-Wolf and Andrew Loviglio-Wolf, whose patience, humor and love sustained me every step of the way. They are the readers I am always writing to and for.

Introduction: "Become Empathy Machines!"

A colleague snapped a picture of a Post-It Note with the words "Become Empathy Machines!" scrawled on it at the Third Coast International Audio Film Festival in Chicago in November 2016 (see Figure 1).[1] It was two days after the stunning election of Donald Trump and one of the radio journalists or producers or storytellers had written this message in response and pasted it to a wall as part of a makeshift Post-It Note mural that sprang up to collect reactions and to identify ways forward in a country that suddenly seemed strange. At the time, I was struck by the instrumentality laid bare in this moment of crisis by the injunction to "become empathy machines." Is that the purpose of journalism? Is that what radio producers should become in an era of rising partisan rancor? Radio had been explicitly understood as an empathy machine since the dawn of *This American Life* in 1996, and if we allow synonyms for empathy, since long before that. Empathy—and the performative versions of it that make for good radio—is a central tenet of modern conceptions of liberalism, one of the values that Trump's first election with its unofficial motto, "Fuck Your Feelings," seemed to rebuke. The nationalist fervor driving the Brexit vote earlier that year and rising authoritarianism in Russia, China, and the Philippines

---

[1] Thanks to Neil Verma for taking this photo and knowing it was the perfect thing to share with me.

FIGURE 1 *"Anonymous Note posted at 2016 Third Coast International Audio Film Festival."*
Photograph by Neil Verma.

suggested a global challenge to traffickers in empathy. Empathy machines, in this context, represented an optimistic doubling down on humanistic values and a specific kind of liberal instrumentality, what Todd Gitlin called the "Administrative point of view."[2]

Authoritarianism and the politics of grievance are nothing new, and, of course, the contemporary politics of grievance on the political right is every bit as emotional in its framing and appeal as the liberal appeal to empathy. It is the explicit embrace of

---

[2]Todd Gitlin, "Media Sociology: The Dominant Paradigm," *Theory and Society* 6, no. 2 (1978): 205–53, http://www.jstor.org/stable/657009.

feeling, empathy in particular, that characterizes the liberal media, a term I use advisedly. I should make clear that "liberalism" in the context of the argument of this book refers both to the classical principle of open markets, freedom of speech, and so on, as well as to the modern US understanding of the term as representing a commitment to diversity, civil rights, and a generally optimistic view of the government's role in addressing social ills.

The imperative to use media to generate fellow feeling is, of course, nothing new, either. The printing press may have been the first empathy machine, although older claims could be made for the alphabet and human speech itself.[3] There's a case to be made for the expressiveness of canine brows in the domestication of dogs, perhaps older still, as a critical development in communication and empathy.[4] The novel, that quintessential medium of modernity, has been hailed for its powers as an empathy machine by authors and readers alike.[5] Film critic Roger Ebert described films as empathy machines with a "liberalizing influence."[6] More recently, virtual reality, with its immersive documentary power, has been similarly praised by journalists and tech evangelists.[7]

---

[3] David Comer Kidd and Emanuele Castano, "Reading Literary Fiction Improves Theory of Mind," *Science* 342, no. 6156 (2013): 377–80; P. Matthijs Bal and Martijn Veltkamp, "How Does Fiction Reading Influence Empathy? An Experimental Investigation on the Role of Emotional Transportation," *PLOS ONE* 8, no. 1 (2013): e55341; Rick Hansen, "How Did Humans Become Empathic?," *Psychology Today*, March 3, 2010, https://www.psychologytoday.com/us/blog/your-wise-brain/201003/how-did-humans-become-empathic.
[4] Juliane Kaminski, Bridget M. Waller, Rui Diogo, and Anne M. Burrows, "Evolution of Facial Muscle Anatomy in Dogs," *Proceedings of the National Academy of the Science of the United States* 116, no. 29 (2019): 14677–81.
[5] Dorothy Woodend, "Books Are Empathy Machines, Says Richard Powers," *The Tyee*, September 22, 2021, https://thetyee.ca/Culture/2021/09/22/Books-Empathy-Machines-Richard-Powers/.
[6] Roger Ebert, "Ebert's Walk of Fame Remarks," *Roger Ebert.com*, 2009, https://www.rogerebert.com/roger-ebert/eberts-walk-of-fame-remarks#:~:text=They're%20not%20only%20entertainment,%20that%20person%20until%20we%20die.
[7] Chris Milk, "How Virtual Reality Can Create the Ultimate Empathy Machine," TED TALK, March 2025, https://www.ted.com/talks/chris_milk_how_virtual_reality_can_create_the_ultimate_empathy_machine. For a discussion of this trend, see B. Belisle and P. Roquet, "Guest Editors' Introduction: Virtual Reality: Immersion and Empathy," *Journal of Visual Culture* 19, no. 1 (2020): 3–10. https://doi.org/10.1177/1470412920906258.

It is no surprise that radio and podcasts have been hailed for this same virtue.[8] The radio-podcast industrial complex represents only the latest moment in the technology of empathy. Understanding the power of contemporary audio, what Michele Hilmes has called "soundwork," to evoke empathy requires an understanding of radio's long history as a "feeling medium."[9] It also requires a bit of a grounding via Sara Ahmed's notion of an "affective economy," her term for describing the mobility of affects as they circulate through subjects rather than residing in them.[10] Finally, it necessitates an exploration into the recent history of liberalism, understood as a tangle of affects as well as a political philosophy. The contemporary case for podcasting and public radio as empathy machines is both part of a long history but it is also very much a product of the specific historical moment, one of the sound effects of liberalism's long death rattle beginning in the 1970s, the first decade of a truly national public radio service in the United States.

In order to do this, we must *listen in* to the different modes of audio across historical eras[11] while also *listening out* for the sounds of alterity[12] that contribute to making radio sound *strange*.[13] My ambition in these pages is to suggest an origin story for the contemporary urgency for more audio empathy machines and perhaps to understand why this appeal may not be sufficient to the

---

[8]Vuyile Madwantsi, "Creating a safe space in sound waves: Podcasts and radio can help promote empathy and reduce stigma," *IOL*, February 5, 2024, https://www.iol.co.za/sunday-tribune/lifestyle/creating-a-safe-space-in-sound-waves-podcasts-and-radio-can-help-promote-empathy-and-reduce-stigma-afb0f8d6-bc5b-4255-b5b3-80b7c6300d7b?utm_source=substack&utm_medium=email.

[9]Michele Hilmes, "The New Materiality of Radio: Sound on Screens," in *Radio's New Wave: Global Sound in the Digital Age*, eds. Jason Loviglio and Michele Hilmes (London and New York: Routledge, 2013), 43–61; William Siemering, "National Public Radio Purposes," *NPR*, 1970. In *Current*, May 17, 2012. https://current.org/2012/05/national-public-radio-purposes/.

[10]Sara Ahmed, "Affective Economies," *Social Text* 22, no. 2 (2004): 117–39.

[11]Susan Douglas, *Listening In: Radio and the American Imagination* (New York: Times Books, 1999).

[12]Kate Lacey, *Listening Publics: The Politics and Experience of Listening in the Media Age* (Cambridge and Oxford: Polity, 2011).

[13]Jason Loviglio and Michele Hilmes, "Introduction: Making Radio Strange," in *Radio's New Wave: Global Sound in the Digital Age*, eds. Jason Loviglio and Michele Hilmes (London and New York: Routledge, 2013), 1–6.

challenges we confront. The appeal of empathy machines, however, has served as a map for the traffic in feelings coursing through public radio and its podcast progeny over the last three decades, as liberalism shifted from a set of institutional commitments to a set of feelings.

By 1930, the first US mass-produced car radio was patented by Motorola, the company's name a portmanteau meant to evoke the coming together of machine and pleasure ("ola" gestured to the Victrola, radio's euphonious ancestor).[14] In Germany, Blaupunkt radios first began to be installed in automobiles around the same time. Among the earliest promotional language for the car radio was the promise of affect management, an electronic mood regulator.[15] The romance of the car-radio assemblage as an empathy machine made for two kinds of transport—physical and emotional—has been a reliable feature of popular culture for nearly a century.[16] And it was an important element of NPR's branding strategy known as "the Driveway Moment," which I will discuss below.

The radio was one of the twentieth-century technologies Raymond Williams identified as central to the novel phenomenon of "mobile privatization," "an at-once mobile and home-centered way of living," pulling in two directions at once, reshaping our understanding of public and private life.[17] Stephen Groening points to the contradictory nature of this pairing of freedom—the promised emotional "transport"—effacing its opposite, "the unfreedom of others" hidden within processes of production with their attendant economic violence. It is important to approach mobile privatization as "a unique modern condition" rather than a technologically specific one. As a "dominant form of subjectivity," then, mobile privatization is not so much a set of capabilities as it

---

[14]H. Raleva, "The naming origin for Motorola," *High Names*, 2013, https://highnames.com/motorola-naming-origin/.
[15]Karin Bijsterveld et al., *Sound and Safe: A History of Listening Behind the Wheel* (Oxford: Oxford University Press, 2014), 103.
[16]A longer version of this argument appears in Jason Loviglio, "The Traffic in Feelings: The Car-Radio Assemblage," in *The Routledge Companion to Radio and Podcast Studies*, eds. Mia Lindgren and Jason Loviglio (New York and London: Routledge, 2022), 226–36.
[17]Raymond Williams, *Television: Technology and Cultural Form*, 2nd ed. (London: Routledge, 1990), 18.

is a problem to be examined.[18] In this way, it stands as an affective record of the early contradictions haunting liberalism as a political and economic structure.

In 2017, one year after the Post-It Note, Julie Snyder, executive producer of *This American Life* (*TAL*), referred to radio as "an empathy machine" in a lengthy podcast interview.[19] Snyder had been in attendance at that 2016 conference. Her talk, given the day after the wall of notes went up, was entitled "Chapter and Verse" and emphasized the importance of "structure" in writing for long-form nonfiction audio works like *Serial*. Perhaps she was the author of the Post-It Note. It is as likely to have been just about anyone else at the gathering; while Snyder was presenting on the main stage, Nikole Hannah-Jones of *The New York Times* and Chana Joffe-Walt of *TAL* jointly presented in a smaller room a session called "Make Them Care: Crafting Narratives About Entrenched Social Problems." By 2016, the notion of empathy as the ethical and aesthetic mission for public radio and public radio-adjacent podcasting had become doctrine.[20]

Only three years later, NPR radio show-turned-podcast, *Invisibilia*, aired an episode titled "The End of Empathy" in which one of the hosts recounted the elaborate implosion of empathy as the show's guiding light, its means and its end. The founding creative team, Elise Spiegel and Lulu Miller and later, Hanna Rosin, it should be said, were all denizens of the same small Third Coast orbit clustered around that same Post-It Note wall in the Hyatt Ballroom in Chicago. New podcasts like *Resistance* (Gimlet), *Not Past It* (Gimlet), *No Compromise* (NPR), *Conflicted* (Evergreen), and *The Argument* (NYT) began to center confrontation and irreconcilability over the emotional labor implied in becoming empathy machines. That same year, on NPR's *Code Switch*, hosts Eugene Demby and Shereen Marisol-Meraji expressed their impatience with the "explanatory comma," the background

---

[18]Stephen Groening, "'An Ugly Phrase for an Unprecedented Condition'. Mobile Privatization, 1974–83," *Key Words: A Journal of Cultural Materialism* 11, no. 59 (2013): 65, https://doi.org/:10.2307/26920341.
[19]Hamish McDonald, "It's a Long Story," *Sydney Opera House*, Podcast, April 17, 2017.
[20]Third Coast International Audio Festival, 2016, Chicago, IL, November 9–11.

information necessary to make sure white people understood the cultural codes of communities of color, the better to empathize.[21] What happened in such a short time?

This story plays out over a much longer period and speaks to a set of enduring tensions at the heart of public radio's relationship to the political and the personal. These tensions, in turn, are wrapped up in the founding contradictions in the creation of public radio in the United States. The strange career of empathy as a public radio shibboleth is just one part of a larger story about public radio's affective economy, its traffic in feelings over the last several decades. Finding a single coherent origin point for radio's status as a "feeling medium" is a mug's game and risks an infinite regress backwards to the first sympathetic vibrations of Tesla's inductive coil.

Instead of starting at the beginning, this book takes up at the beginning of the end as broadcasting's content flows into the newer and distinct medium of podcasting, and then, in remediated form, flows back as a new form of radio. It is at this moment when the role of acoustic affect as a medium for constituting a particular kind of subjectivity becomes easier to discern.[22] The more explicit reckoning with radio and feeling may also be due to the inevitable self-consciousness and reflexivity of the shift of content from one medium to another and back again, a process that involves a double remediation.[23]

By double remediation, I have in mind the formal and tonal transformations radio programs like *This American Life* anticipated before it became a podcast and then helped to import back to other public radio programs, including to some extent even so-called hard news formats like *All Things Considered*. To become podcasting, broadcasting's liveness and immediacy gave way to the thematic richness and the unhurried cadences of conversations

---

[21] Eugene Demby and Shereen Marisol Meraji, "Hold Up! Time for an Explanatory Comma," *Code Switch*, Podcast, December 14, 2016, https://www.npr.org/2016/12/14/504482252/-hold-up-time-for-an-explanatory-comma.
[22] Analiese Richard and Daromir Rudnickyj, "Economies of Affect," *Journal of the Royal Anthropological Institute (N.S.)* 15 (2009): 57–77 © Royal Anthropological Institute.
[23] Jason Loviglio, "*The Daily Dose*: Podcasting and Broadcasting in the Public Interest," in *The Oxford Handbook of Radio Studies*, eds. Michele Hilmes and Andrew Bottomley (Oxford: Oxford University Press, 2024), 441–57.

among friends. A new standard of audio, born on the radio, became associated with podcasts and then taken up by radio shows hoping to become podcast hits.

Radio is no stranger to the process of remediation, self-consciousness, and reinvention, and so it is natural to think about the podcast revolution as just another in a series of technological inflection points. Such a list might include the audion, which enabled broadcasts of the human voice; the higher fidelity of frequency modulation (FM), the revolutionary boost in power and durability wrought by the transistor; the planetary reach of satellites; and the many ways digitalization changed the game of broadcasting. All of this must be understood within the evolution of mobile privatization; the media's power to transport cannot be separated from the history of a particular kind of dominant subjectivity, which in this book I will refer to as "feeling liberal" or more narrowly, as part of the public radio's "structure of feeling." As radio and podcasting remediate each other into new forms, they provide fresh new contexts in which the social and political upheavals of the present moment, and their historical traces, become audible and thus sensible.

There's a built-in process of forgetting in media's history of remediation that helps explain the efflorescence of discourse around podcasting's novel intimacy as if that were not part of its legacy from radio. Daniel Czitrom has pointed to electronic media's "everwhere-ness, all-at-once-ness, and never-ending-ness" as key elements in our amnesia about its history.[24] Even so, podcasting's time-shifting affordances, subscription model, lower capitalization requirements, and earbud-and-smartphone-enabled mobility all make powerful contributions to a new kind of privatized listening. These shifts in the "machinery" of the audio experience from serendipity to recommendation algorithms, from broadcasting's locally confined universality to RSS's global opt-in subscription model, from daypart schedules to time-shifted listening, all represent a further abstraction from the analog to the digital.

The intensification of the focus on the human, the intimate, and empathy associated with these processes can be read as ideological

---

[24]Daniel Czitrom, *Media and the American Mind: From Morse to McLuhan* (Chapel Hill: University of North Carolina Press, 1986), 184.

mystification or as another chapter in the technological sublime. Perhaps we might regard it as an instance of what Lauren Berlant calls "cruel optimism," the desire for "something that is an obstacle to your flourishing."[25] As audio media becomes more personally customizable, collective listening and the notion of an imagined listening public, with all that it represents as a political and social metaphor, recedes into greater abstraction. In this context, the call for more empathy machines makes a kind of compensatory sense. For Williams, this was the central tension in mobile privatization, the feeling of freedom rooted in a complex of market-based constraint, unfreedom, and inequality and isolation.

Radio's moment of reckoning with empathy also coincides with a certain reflexivity in thinking about political and social alignments as neoliberal compromises begin to fracture into seemingly irreconcilable tribal alignments.[26] Podcasting, with its affordance of politically charged privacy,[27] arrived in time to bear witness to the breaking apart of an always tenuous political balancing act in which *feeling liberal* was understood to be either sufficient or at least, a necessary evil.[28] Public Radio, born of Great Society's soaring optimism and subterranean bad faith, has always been an artifact of and the soundtrack for liberalism's massive, impossible compromises.[29] The anecdote I began with, featuring empathy's fall from Post-It Note ubiquity to failed mission, is just one in a series of ironic juxtapositions that can be grouped together as stories

---

[25]Lauren Berlant, *Cruel Optimism* (Durham, NC: Duke University Press, 2011); David E. Nye, *The American Technological Sublime* (Cambridge, MA: MIT Press, 1986), 1.
[26]Louis Menand, "The Rise and Fall of Neoliberalism," *The New Yorker*, June 17, 2023, https://www.newyorker.com/magazine/2023/07/24/the-rise-and-fall-of-neoliberalism; Wendy Brown, *Undoing the Demos: Neoliberalism's Stealth Revolution* (Princeton, NJ: Princeton University Press, 2017).
[27]Michele Hilmes, "But Is It Radio? New Forms and Voices," in *The Routledge Companion to Radio and Podcast Studies*, eds. Mia Lindgren and Jason Loviglio (New York and London: Routledge, 2022), 9–18.
[28]Hilmes, "But Is It Radio?," 9–18.
[29]Jason Loviglio, "Public Radio, *This American Life* and the Neoliberal Turn," in *A Moment of Danger: Critical Studies in the History of U.S. Communication Since World War II*, eds. Janice Peck and Inger Stole (Milwaukee: Marquette University Press, 2011), 283–306.

on a theme: the complex and qualified failures of liberalism and the associated impulses and formal adaptations that have shaped what it means to feel liberal. The limits of empathy in particular, as opposed to compassion or solidarity (or other feelings and commitments), is key here to understanding this history, as we'll see in Chapter 1. The story of *TAL*, its origins and its expansion across media platforms, provides a useful soundtrack for exploring this theme.

I center *This American Life* in this narrative not because it invented something new but because of the multiple ways it served as a nodal point through which new and old things moved. For radio and podcasting's traffic in feelings, *TAL* is Grand Central Station. No other show has been as consistent nor as successful in narrating the emotional experience of the present moment. No other show has been as explicit and didactic about the centrality of feeling in constituting everyday life and the formula for evoking them through narrative.[30] And few other shows have been as relentlessly expansionist in franchising its storytelling brand and affective posture across multiple media outlets and platforms and genres.

TAL also exemplifies the opposite but mutually constituting affordances of public broadcasting and podcasting. It borrowed from and revitalized the public radio traditions of investigative journalism and sonically inventive audio production. It also anticipated the "long tail" business plan and the give-it-away credo of internet content providers. An early adopter of podcasting as a time-shifted delivery mechanism for its broadcast content, it also ushered in appointment listening, a key innovation and disruption in the emerging chaotic attention economy of the twenty-first century. *TAL* bridged the old and the new for decades and now that the new has arrived, it moves from center stage to chorus, still moving through, if no longer directing, the traffic in feelings.

Most of all, *TAL* draws our attention to the role of affect in contemporary soundwork. No show better captures audio media's power to regulate feelings and constitute subjectivity. The show's

---

[30]Mia Lindgren, "Personal narrative journalism and podcasting," *The Radio Journal—International Studies in Broadcast & Audio Media* 14, no. 1 (2016): 23–41, https://doi.org/10.1386/rajo.14.1.23_1.

founder and host, Glass, pioneered an explicitness about this power, building it into the structure of the stories and moving seamlessly into marketing materials and many paratexts across platforms, which is not to say that radio hasn't always been, as Bill Siemering puts it, "a feeling medium." The BBC's premier chronicler, David Hendy, has argued for "rethinking the origins of broadcasting as an emotional project as much as a political one."[31]

There's no better example of the enchantment of radio broadcasts' ability to emotionally regulate the naturally unregulatable than the shipping forecasts. Broadcast for much of the twentieth century on British, Swedish, and Finnish radio, they have become quite literally the stuff of poetry.[32] Like radio's incantatory recitation of the time, weather, and traffic, regular updates impose a narrative and generic order, regulating chaos into a format.[33] For North American ears, the lazy acoustic fuzz of a ballpark on a nighttime baseball broadcast has, for nearly a century, been among the most evocative sounds of summer's stretched-out temporality. Listeners of *Storycorps* can expect to be made to cry during the brief, exquisitely produced dialogues between loved ones. In this period, there's been no shortage of game shows and interviews that can reliably deliver a chuckle or a "gut-punch."[34] In these and other ways, we can think of radio in terms of emotional regimes in which emotional norms were inculcated through rituals, formulae, and discipline.[35]

Even so, *TAL*'s virtuoso narrators welcomed listeners to feel as a way of being in the world and presented storytelling as a way of organizing feelings for ourselves and others. Bringing this off

---

[31] David Hendy, "The Great War and British Broadcasting: Emotional Life in the Creation of the BBC," *New Formations* 82 (2014): 91.
[32] Sanna Nyqvist, "Poetics of the Shipping Forecast," in *Spaces of Longing and Belonging: Territoriality, Ideology and Creative Identity in Literature and Film*, eds. Brigitte le Juez and Bill Richardson (Leiden, The Netherlands: Brill, 2019), 49–63.
[33] Alexander Russo, "Tick Tock Goes the Musical Clock: Time Discipline and Early Morning Radio Programs," in *Radio's New Wave: Global Sound in the Digital Age*, eds. Jason Loviglio and Michele Hilmes (New York and London: Routledge, 2013), 195–208.
[34] Jason Loviglio, "Vox Pop: Network Radio and the Voice of the People," in *Radio's Intimate Public: Network Broadcasting and Mass-Mediated Democracy* (Minneapolis: University of Minnesota Press, 2005), 38–69.
[35] Hendy, "The Great War and British Broadcasting," 96.

required discipline and structure, pattern, and repetitions, all belied by the improvisational house style. No show is more garrulous in its reflexivity when it comes to talking about feelings. Ira Glass put didactic pronouncements on radio's emotional power at the center of the artistic and commercial discourses swirling around the industry, even as he helped to pioneer the cross-media franchising of the brand.

TAL is rightly associated with its founder and host, Ira Glass, but it has been a way station, proving ground, and finishing school for many of the podcast producers, reporters, and executives who now ply their trade across a broad array of commercial and noncommercial podcast shows, networks, and platforms. Ira Glass can, with some justification, claim to have "started this whole thing."[36] His sensibility and his sense of what constitutes a proper story (i.e., its affective power) resonate through more than 860 episodes (and counting) of *TAL* and echo through the work of the show's alumni, including Alix Spiegel of *Invisibilia*; Alex Blumberg, founder of Gimlet Media; writer and podcaster Starlee Kine; and Peter Clowney, who went on to work at American Public Media, SiriusXM, Stitcher and Pushkin.

But it is just as easy to argue that Glass functioned, as his name suggests, as an acoustic mirror for an emerging bundle of affects and aesthetic impulses, now reflecting them towards audio innovations, now bouncing them back upon an audible past. Glass is a central figure in this story because of his virtuoso ability to occupy center stage while constantly deflecting to other voices. Like a matador who makes a trick of appearing and disappearing behind the flourish of a cape, Glass is at his most adept when he is moving between one thing and another.

Empathy Machines explores the history of radio as a feeling medium during an era when feeling liberal shifted and the ground on which liberalism rested shook. It attempts to track these moves without losing focus of the larger lurching changes against which this story of the podcasting/public radio "structure of feeling" emerges, a concept borrowed from Raymond Williams, that I

---

[36]Ira Glass, "Seven Things I've Learned," Keynote Address, Third Coast International Audio Festival, Chicago, IL, November 10, 2017, https://www.thirdcoastfestival.org/feature/seven-things-i-ve-learned.

will define below. *TAL* is the tiny vibrating crystal with which we can tune in the bigger story of modern audio storytelling and the structures of feelings it gave voice to. The book's argument ripples outwards from the texts of the programs to the many paratexts surrounding them (interviews, books, television shows, films, controversies, marketing material, Post-It Notes).

TAL is a massive corpus, too large and multifarious to summarize, and an ambitious brand, sprawling across film, television, CDs, and live shows in addition to radio and podcasting spinoffs and partnerships, like *Planet Money* and *Serial*—far too much to account for in any single book. *Empathy Machines* is not the first; Kristine Johnson wrote a master's thesis on the habits and feelings of *TAL* podcast listeners in 2007; and it certainly will not be the last.[37] Indeed, the biggest ambition I have for this book is that it will inform and provoke further study into the show and to the audio storytelling structures of feeling that course through and around it.

Empathy Machines

"Empathy machine" is a useful metaphor to the extent that it hits our ears as an oxymoron, the opposition between the human and the mechanical, and a productive trope for exploring the role of media, specifically audio media, in fostering human connection, a common, if quixotic, preoccupation. The book aims to better understand the human/nonhuman relationship captured in terms like "empathy machine" (as well as "feeling medium") as historical phenomena specific to the era of mobile privatization and perhaps as partaking in something a bit grander—something about the promises of human technology, both prosthetic and spiritual, stretching back to long before political and technological assemblages like televisions, radios, and telegraphs. Ghosts have long occupied machines, machines have long occupied gardens, just as golems and robots have long occupied the human imagination. A dominant theme

---

[37] Kristine Johnson, *Imagine This: Radio revisited through podcasting*, MA thesis (Fort Worth, TX: Texas Christian University, 2007). For TAL's global influence, see also Mia Lindgren, "'This Australian life': The Americanisation of radio storytelling in Australia," *Australian Journalism Review* 36, no. 2 (2014): 63–75, https://search.informit.org/doi/10.3316/ielapa.912447785683000.

in media history is the role of technology as extensions of human values, such that we no longer hear the metaphors in our arguments. Just as walls and bridges immediately evoke opposite political positions, wavelengths and vibrations become easy shorthand for shared sensibility and human connection.

The idea of radio as an empathy machine stems from a specific history that has been referred to as the technological sublime, an impulse that can be traced back to antiquity but is critical to modern conceptions of democracy, the Enlightenment, and liberalism. The notion that we can invent and communicate our way into equality is central to the way we understand Western history since at least the printing press. The American colonies' high rate of literacy has been suggested as the key to the Revolution's success.[38] And the spread of public libraries, roads, schools, and newspapers was fueled by early republican nationalism, which in turn was bolstered by the "deep horizontal connections" fostered by the daily press.[39] The book, particularly the novel, has been hailed for its powers of inculcating empathy, as mentioned above. Much the same has been said for the theater, which benefits from ancient connections to the democracy (agora) as to the emotional (catharsis).[40]

This history is opposed by an equally compelling one in which the machine represents the antithesis of the values of democratic participation and empathy. From science fiction to folk songs to labor activism, the machine has been a reliable heavy, evoking lust, greed, envy, and other sins. The machine has long served as an all-purpose foil to the human, a point of aspiration and recrimination, a metaphor of human perfectibility and corruption. This tension helps to explain the appeal of the term "empathy machines" as

---

[38]Stephen Earl Bennett, Staci L. Rhine, and Richard S. Flickinger, "Reading's Impact on Democratic Citizenship in America," *Political Behavior* 22, no. 3 (2000): 167–95, http://www.jstor.org/stable/1520046.
[39]Benedict Anderson, *Imagined Communities*, 2nd ed. (London: Verso Books, 2016).
[40]Paul Woodruff, "Sharing Emotions through Theater: The Greek Way," *Philosophy East and West* 66, no. 1 (2016): 146ff. *Gale Literature Resource Center* (accessed July 19, 2024), https://link.gale.com/apps/doc/A439805681/LitRC?u=anon~9af0993&sid=googleScholar&xid=da0273e4.

a way into understanding the public radio structure of feeling. It captures the contradictory currents of feelings and structures swirling around at the twilight of liberalism and opens up ways to think about what comes next as human–nonhuman interfaces and encounters become more complex.

Empathy machine is also a useful trope for thinking about the contradictions inherent in modern liberal governmentality, poised between humanistic principles and cynical instrumentality, democratic vistas and market forces. In this regard, public radio has been the quintessential public institution over the last half century, the soundtrack for the transformation of a liberal consensus through neoliberal shocks and unsustainable contradictions. Tracing the "traffic in feelings" from public radio's early years into its early podcasting era, this book takes seriously the idea that under certain conditions, "affect serves as a medium in which different types of subjects are formed," an insight borrowed from cultural anthropologists Analiese Richard and Daromir Rudnyckyj that I'll explore below. The subjectivity formed through this process, the feeling liberal in neoliberal times, has had to adapt to navigate a shifting aesthetic, social, and political landscape.

Above all, empathy machine is a useful way to think about the very particular relationship to talking about feelings pioneered by *TAL* and carried into many of the most popular, celebrated, and impactful podcasts of the 2010s and beyond. As I'll explore in Chapter 3, *TAL* ran warm and cool in carefully measured proportions, toggling between emotional connection and ironic distancing with an almost mechanical precision. Because Glass hit the live event circuit soon after the show became a national hit, we knew from the start a lot about the instrumentality behind structuring stories to evoke empathy. Glass presented the manufacturing of fellow feeling as both an antidote to the depressing hype machine of mainstream news and entertainment media: an aesthetic reform, and as a balm for the loneliness of contemporary civic and interpersonal life: a social reform.

Did the hollowing out of liberalism by neoliberal privatization create a need for and traffic in empathy? From the perspective of three decades of hindsight, it seems like a prescient and compensatory offering for a culture hurtling towards more electronic stimuli, greater isolation, and (some researchers confirm) less empathy.⁴¹ *TAL*'s preoccupation with empathy and with talking about how to produce it met with rare success in ways that we are now better positioned to explore, even as many of the podcasts that took up this approach have faltered or shifted into new emotional and social paradigms.

The ideas in *Empathy Machines* have developed over the last several years in tandem with a great deal of listening in, listening out, and listening back. I have listened in to the radio and podcasts most readily at hand; the local frequencies and the podcasts atop the recommendation lists targeted to people of my profession and habitus.⁴² Following Susan Douglas, I *listen in* to these signals for the exploratory pleasures of virtual travel, for the sense of connecting to communities of interest. For this project, that has meant tuning in to the frequency of discovery and identification that "evokes a spiritual, almost telepathic contact across space and time" that contemporary well-crafted audio storytelling has mastered.⁴³

Within and beyond that oeuvre, I have also tried to *listen out* in the sense that Kate Lacey has defined it, "as a state of anticipation," part of an already constituted "listening public" not responsive to any specific voice or to a narrowly constituted understanding as what counts as a properly public voice. This requires the difficult work of "listening out for voices that confront and jar as well as those that comfort and support," and which Lacey argues is "an essential technique of democratic political life."⁴⁴ In the context of this project, that has meant more active searching for audio work that might not make it to my ears through the usual digital and analog algorithms. It has also required a posture of being open to new sounds and different voices, listening past the "chapter and verse" of the most successful formulas structuring soundwork.

---

[41] S. H. Konrath, E. H. O'Brien, and C. Hsing, "Changes in Dispositional Empathy in American College Students Over Time: A Meta-Analysis," *Personality and Social Psychology Review* 15, no. 2 (2011): 180–98, https://doi.org/10.1177/1088868310377395.

[42] I have been on a screening committee for The Peabody Awards for radio and podcasts for nearly a decade.

[43] Douglas, *Listening In*, 40.

[44] Kate Lacey, "Listening in the Digital Age," in *Radio's New Wave: Global Sound in the Digital Age*, eds. Jason Loviglio and Michele Hilmes (New York and London: Routledge, 2013), 9–23.

This state of anticipation, this radical openness, is akin, I think, to an understanding of historical analysis as a methodology of "making radio strange." This requires listening for "the irruption of new voices across old borders," which is particularly vital if we are to understand the economic, literary, and ideological tumult in soundwork across broadcasting, podcasting, and streaming.[45]

Finally, I have tried to *listen back* in some of the many senses of the term that Kate Lacey intends. In the simplest sense, this means "listening back to become aware of things that were missed on first hearing."[46] This is deceptively difficult when listening back to radio shows and podcasts from a relatively recent past that can feel continuous with the present moment, but which may provide surprising moments of discontinuity. Listening back, Lacey reminds us, "acts as a reminder that listening happens in real time, in the now, opening an experiential and theoretical gap between the act of listening and the sounds of the past."

Empathy Machines is perhaps, above all, a study of media and affect, specifically the way affects circulate through media, constituting subjectivities, shaping popular discourse, and defining the limits of what is sayable in any given historical moment. It is also an exploration into the way media texts and technologies provide novel or familiar contexts in which to think about feelings and affects and their social power. Finally, it represents an attempt to explore liberalism and neoliberalism as a bundle of affects circulating through and constituted by audio media texts, platforms, and genres. These insights amplify the role of affect studies to understand audio media forms historically and politically.

Structure of Feeling

I have used the term "structure of feeling" in the book's title and above, and now I turn to a brief explanation of how this concept

---

[45]Loviglio and Hilmes, "Introduction: Making Radio Strange," 2.
[46]Kate Lacey, "Listening Back: Materiality, Mediatisation and Method in Radio History," in *The Routledge Companion to Radio and Podcast Studies*, eds. Mia Lindgren and Jason Loviglio (New York and London: Routledge, 2022), 28–38.

helps me to understand the trajectory of empathy and other affects through *TAL* and beyond. The practice of "listening back" reminds me of Raymond Williams's insight that understanding structures of feeling requires attention to "meanings and values as actively lived and felt," which means tracking a moving target, reaching for something still "in solution," vying for but not yet precipitated into a fixed historical solidity. In a Williams-esque moment, Lacey calls attention to how radio's "mediatization of a permanently unfolding perpetual present transformed the modern experience of time and with it the tools and techniques of historiography itself." Williams's great insight was in understanding how difficult, tenuous, and partial this was. Marx and Engels, by contrast, thought that seeing "each Epoch" in its true character was as simple as it was for a "shopkeeper" in "everyday life" to see through a charlatan.[47] The difficult work of understanding one's own historical moment for Williams required paying attention to "social experience … still in process."

This approach privileges "specifically affective elements of consciousness and relationships," a critical pivot away from structuralism and materialism. It is no surprise that Williams developed this concept most thoroughly in his books on literature and film, modern expressive forms through which cultural meanings circulated publicly but which were often experienced as "private, idiosyncratic, and even isolating." Williams's interests were, above all, literary and his attention to "specific feelings, specific rhythms" were crucial to his approach to understanding structures of feeling. In my readings of *TAL* and other audio nonfictional narratives, I hope to bring this same care to the "characteristic elements of impulse, restraint, and tone" that make these stories distinctive and historically specific.[48]

Williams's partial and scattered writings about structures of feeling have been enormously influential, and contemporary scholars have taken up and extended his attention to questions of

---

[47]Karl Marx and Friedrich Engels, *Collected Works*, Vol. 5, trans. Richard Dixon (New York: International Publishers, 1976), 59–62.
[48]Raymond Williams, *Marxism and Literature* (Oxford: Oxford University Press, 1977), 128–35; and *Preface to Film* (London: Film Drama, [1954] 2003), 21–3.

tone and affect. Sara Ahmed's notion of "economies of affect" helps to explain the movement of feelings as a social process.

For Ahmed, feelings take form in their circulation "between bodies and signs" rather than existing prior to this circulation as "private feelings," residing within individuals and awaiting expression through channels of communication. "Emotions *involve* subjects and objects, but without residing positively within them." Ahmed argues that the "non-residence of emotions is what makes them binding," drawing people together in affective economies. Further, the "rippling effect of emotions" binds together "signs and figures and objects" as well, creating narratives and logics all their own. And while Ahmed primarily explores the movement of hate and fear, it can be applied to softer, more complex emotions as well, like empathy. Through this rippling effect, emotions "move sideways," making "sticky" associations between figures.[49]

This is a helpful image to keep in mind as we think through the affective associations rippling across the storytelling of public radio and into podcasting. Thematic unities made explicit, even didactically so, in individual episodes of *TAL*, are recapitulated implicitly on a larger scale across the soundscape of contemporary nonfiction audio storytelling. In this way, the "stories on a theme" structure can be reconsidered as themes on stories; that is, sticky signs, figures, and objects rippling across stories and formats. We can pick up this Ahmadian shift in the movement of empathy from "mission" in Glass's 1998 talk, to failed method in Hanna Rosin's 2017 "The End of Empathy" *Invisibilia* episode. Glass wants to make listeners "have more empathy"; Rosin recognizes that empathy is "the way" stories bind signs, figures, and objects together even when perhaps they shouldn't be. Affects like empathy (and fear) don't pre-exist this traffic in feelings; they are constituted in it. And in the process, "they align individuals with communities."[50]

I use the term "traffic in feelings" because of the way it unites Ahmed's notion of affect's "economic" movement, or circulation, through bodies, and because of the ways it helps to animate William's concept of "mobile privatization" as a central condition (and contradiction) of modern social life. The traffic in feelings is

---

[49] Ahmed, "Affective Economies," 119.
[50] Ahmed, "Affective Economies," 117–39.

also a conceit that helps evoke the historically specific reception practices that are often unspoken but which are at the center of the shift from radio to podcast listening. Radio was, as I will explore in Chapter 1, designed in terms of technology and content to modulate feelings in traffic. The car-radio is an historical assemblage without which it is impossible to understand radio reception in the twentieth century and perhaps especially the rise of public radio to mainstream ubiquity in the 1990s.

The traffic in feelings is a term I hope will help to evoke the layered and contingent readings that follow, attempts to understand meanings still in solution, not yet precipitated into something solid and permanent. Podcasting's digital affordances make listening back newly accessible, transforming our relationship to audio. As I'll try to show in what follows, radio's long relationship with nostalgia and podcasting's self-conscious re-mediation of previous forms makes listening back, in this critical sense, both difficult and necessary.

Sound Affects

Writing about sound is famously difficult. Roland Barthes challenged readers to talk about music without using a single adjective, an exercise in futility, but one with the ambition to understand sound "at the very edge of semantic availability."[51] Barthes's attention to "the grain of the voice" was an attempt to move past indexical meaning to the voice's fleshy substantive embodiment, and perhaps the better to understand its emotional power. Barthes was thinking in terms we might call "associational" and perhaps in its resistance to articulation, akin to the not-yet-precipitated meanings of structures of feeling. Modern sound technologies add to the complexity of understanding sound's emotional impact, especially the sound of the amplified, recorded, and transmitted human voice. R. Murray Schafer and Michael Chion wrote extensively about the emotional, aesthetic, and imaginative power of disembodied voices made possible through recording, broadcasting, and film. In the process, they coined new terminology like "schizophonia" and "acousmetre" as part of a

---

[51]Roland Barthes, *Image-Music-Text* (London: Fontana Press, 1977), 181.

larger struggle to understand the borderless terrain of sound and modern sound technologies.

Susan Douglas has written extensively about the "kindred spirits" haunting radio's "disembodied voice, [which] evokes a spiritual, almost telepathic contact across space and time."[52] Going back as far as the phonograph, Senta Siewert and Carolyn Birdsall have argued, "modern sound technologies have been thematized in terms of the uncanny or insanity."[53] Contemporary scholars of soundwork have inherited a conceptual vocabulary for understanding sounds, especially voices, on radio and podcasts laden with analogies of atavistic spirituality and abnormal psychology. The disembodied voice still carries with it the residue of the uncanny, a word that nicely evokes both ancient and modern sensibilities, ghosts and machines, magic and psychoanalysis. But there are other currents of meaning running through these audio narratives. Radio classics, like *Sorry, Wrong Number* (1941), layer acousmatic horrors one atop the other; but the voices of the women phone operators with their affectless automaton delivery that at last drive the protagonist, Mrs. Stevenson, to madness.

Contemporary scholars like Mack Hagood, Jennifer Stoever, and Christine Ehrick have introduced new ways to understand the resonances among sound, feelings, and the social order. Hagood's study of "orphic media" reminds us of the power of sound technologies to insulate and isolate. Jennifer Stoever identifies "the sonic color line" as the historically layered acoustic forces that shape racial hierarchies and exclusions throughout American history. Ehrick's investigation into "radio and the gendered soundscape" considers the specific social disruptions that the disembodied voice threatens when it is a woman's voice in a patriarchal society, "and from there into larger questions about the way social hierarchies… are reproduced and challenged within the sonic realm."[54]

---

[52]Douglas, *Listening In*, 40.
[53]Senta Siewert and Carolyn Birdsall, "Of Sound Mind: Mental Distress and Sound in the Twentieth-Century Media Culture," *Journal for Media History* 16, no. 1 (2013): 27–45.
[54]Mack Hagood, *Hush: Media and Sonic Self Control* (Durham, NC: Duke University Press, 2019); Jennifer Stoever, *The Sonic Color Line: Race and the Cultural Politics of Listening* (New York: NYU Press, 2016); Christine Ehrick, "Thoughts on the History of Radio and Women's Voices," in *The Routledge Companion to Radio and Podcast Studies*, eds. Mia Lindgren and Jason Lovigliov (New York and London: Routledge, 2022), 78.

These studies and other recent scholarship remind us, even in societies and ages dominated by the visual, of the world-building power of sounds and the systems that record and circulate them. In their close examination of specific social and technological regimes in which sounds circulate, they demonstrate the powerful ways that the traffic in feelings often moves through sound. Self-care, racial and gender animus, and feelings of belonging are bound up in the sonic envelopes, sonic color lines, and soundscapes that help to constitute human experience.

Ahmed uses affect and emotion more or less interchangeably, as indeed many of us do.[55] But there's a distinction worth marking for the purposes of understanding the contingent, hard-to-pin-down, but expansive social power of contemporary radio and podcasting. Affect, as Richard and Rudnyckyj have said, "indexes intersubjective relations" compared to emotion, which connotes "an inner state manifested through outward expression." Quoting Good (2004), they point out that "emotion as an analytical concept still bears the spectre of a psychological individualism," which doesn't as accurately reflect the social processes through which affects can produce subjects and subjectivities, particularly, they argue, during periods of transformation and crisis.[56]

There is also a performative aspect, in the J. L. Austin sense, to affect in this usage: "We take affect to be a form of conduct; a means through which people both conduct themselves and conduct others by structuring possible courses of action."[57] A former NPR correspondent once confided in me about the practice of "mirroring" in which reporters adopt a confessional intimate tone with interview subjects as a way to loosen them up and get "good tape." Robert Smith, of *TAL*-spinoff *Planet Money*, describes his version of this tactic as "modeling big picture ideas," to evoke the kind of responses that make for good radio.[58] This "transitive" quality of affect is not always as consciously instrumental as in this example or in the ones that Richard and Rudnyckyj explore in their

---

[55] Ahmed, "Affective Economies," 117–39.
[56] Byron J. Good, "Rethinking 'emotions' in Southeast Asia," *Ethnos* 69 (2004): 529–33.
[57] Analiese Richard and Daromir Rudnyckyj, "Economies of Affect," *Journal of the Royal Anthropological Institute (N.S.)* 15 (2009): 57–77. © Royal Anthropological Institute.
[58] Jessica Abel, *Out on the Wire: The Storytelling Secrets of the New Masters of Radio* (New York: Broadway Books, 2015), 95.

analysis of corporate strategies for neoliberal transformation of worker behavior and small-donor philanthropy.

Extending Ahmed's argument, Richard and Rudnyckyj argue that just as affect enables circulation within an economy, it is also a medium "within which subjects are formed." "Feeling" on the other hand, takes its meaning very much from the context in which it is used; now a privately felt, pre-existing condition; now a medium through which subjects move. I'll argue, however, that when William Siemering describes public radio as a "feeling medium," he has in mind something closer to affect, that is a social relation rather than a private experience. Radio as a social relation, as a transitive medium for subjects, and at times, as a site for an instrumental approach to generating new subjectivities, is what I have in mind when I talk about empathy machines and the traffic in feelings.

The book is organized chronologically—a history of *This American Life* that is also a history of what liberalism felt and sounded like in the era of neoliberalism. It begins with two chapters designed to situate *TAL* as the heir to a century of broadcast feeling and unsustainable commitments to liberalism. Chapter 1, "A Feeling Medium," explores the theme of radio as a medium of crisis. Its history is often told through the disasters it has covered, and how it shaped our understanding of what disasters are and how we feel about them. It then explores US public radio's very specific historical connection to crisis, a mix of disaster coverage, internal precarity, and emotional immediacy that contributed to its unique mission of providing an "affective education" to its listeners, understood to be "the public," a term so unstable that it too produces its own periodic crises.

Chapter 2, "Voracious Voyagers: NPR Listens," explores the contradictions in public radio's understanding of its audience and the attempts by market researchers to construct an ideal public radio listener using emotional attachments as a proxy for class distinctions. What follows is a set of chapters about five periods in the career of *This American Life*, organized chronologically by different affects, from playfulness to cringe. Following Williams, the focus is on "specifically affective elements of consciousness and relationships," as they are performed across *TAL*'s themed narrative acts.[59] The eras and affects overlap, rippling across each other. Some

---

[59] Raymond Williams, *Marxism and Literature*, 128–35; and *Preface to Film*, 21–3.

are, to this day, still in solution, not yet precipitated into historical solidity.

Chapter 3, "Feeling Playful: *This American Life* and Narrative Enchantment," explores the show as a matrix out of which the public radio structure of feeling developed. Moving back and forth from radio's historical affordances of immediacy and linearity to its self-consciously digital future, the show, which debuted as *Your Radio Playhouse*, revivified radio listening through the play of opposites—empathy and irony, gravitas and whimsy, personal anecdote and grand take-aways. "Feeling Playful," this chapter argues, was *TAL*'s first and most successful affective commitment. These early stories employed the tricks of close-up magic, narrative sleights-of-hand, and disappearing acts to move between feelings and structures, now intimate, now arch. This playfulness involved a rejection of politics in favor of affect and aesthetics in ways that proved unsustainable. It also calls attention to the fatal flaw of public radio empathy; it is often bestowed on the socially marginal by the socially dominant, an implicit hierarchy that structured who spoke, who listened, and who "must be represented" by others.

Chapter 4, "Feeling American: *This American Life* Goes to War," explores *TAL* in its response to the emotional challenges of the twenty-first century, including the aftermath of the September 11 attacks and rapid growth of digital media and web-based content. Empathy and irony had to go to war in two senses—with each other and alongside the rest of the country—as the "American" in *TAL*'s name came into more self-conscious focus. This challenge to the show's playful mood pushed it deeper into the literary structures of storytelling, enabling deeper, darker, and more ambivalent stories of American life. In the process, the show discovered a public appetite for literary long-form nonfiction on the radio, a critical moment in the pre-history of podcasting.

Chapter 5, "Feeling Flush: *Planet Money* and the American Dream," explores *TAL*'s first spinoff show, *Planet Money*, as another critical development in the origin story of podcasting's debt to the public radio structure of feeling. It also provides an analysis of the aggressively neoliberal ideological positioning of *Planet Money*, a collaboration with NPR, during the 2008 mortgage debt crisis and the ensuing Great Recession. This analysis points to the some of the limits of public radio empathy as it ripples across subjects and through stories as part of the public radio structure of feeling.

Chapter 6, "Fellow Feeling: *This American Life* and the Gendered Voice," explores the complicated gender and identity politics of *TAL*'s liberal structure of feeling as they circulate through human voices and vocal performance. The centrality of men's voices, experiences, and feelings in *TAL*, especially in its first two decades, informs its approach to gender, which is post-feminist, and to sexuality, which is an interesting mix of hetero- and homonormativity. Moving beyond gender and sexuality, the chapter examines the role of voice and vocal performance in race and class identity as well, drawing on Bourdieu's notion of distinction to explore public radio's curatorial approach to sound. Vocal performance, in particular, functioned as an implicit form of cultural work in which some voices were embodied by their differences and others were rendered indistinct by their ability to make distinctions.

Chapter 7, "Feeling Uncomfortable: The Politics and Aesthetics of Cringe," investigates the collapse of empathy as the central unit of currency in the affective economy of public radio and its podcast progeny. In close readings of several prominent nonfiction podcasts, this chapter analyzes the ways that soundwork producers respond to the impossibility of telling stories of fleeting Driveway Moments of empathy. Instead, they stage public scenes of disappointment and discomfort in which expectations for moments of public and private comity meet intractable contradictions that have long haunted the empathy machines of audio narrative. These scenes of discomfort are echoed in the economic and political turmoil in the audio industries themselves, adding another layer of difficulty to the challenge of audio storytelling. Across these awkward public scenes, however, we can hear the faint strains of a new civic impulse; fantasies of public life, however doomed, provide in these stories a new animating spirit for the public radio structure of feeling.

The conclusion, "Maybe It's a Feeling? How *This American Life* Long Endured," attempts to explain how the show has survived the chaos, retrenchments, and scandals that wreaked havoc on the rest of the industry that it did so much to inspire. Recapping the history of the show's movement across affects and eras, the conclusion returns to the notion of *TAL*'s constitutional playfulness and to a resilient civic impulse as old as radio and as current as the latest social media trend. If podcasting's loudest claim of difference from radio was its "freedom" from constraints of time, format, and technological overhead, the public radio structure of feeling discovered, perhaps due to its platform agnosticism, a different kind of freedom in its commitments to something called the public. In a short afterword, I examine the emerging rhetorical power of liberalism as it animates the decidedly illiberal far-right and masculinist audiosphere. This disturbing phenomenon, and the political power it represents, suggest an increased urgency into greater clarity about what it means to feel and sound liberal.

CHAPTER ONE

A Feeling Medium

Radio in a State

The history of radio is often told through a series of crises. The wireless telegraph operator aboard the *SS Californian* hung up his earphones and went to his bunk moments before the distress signal came in from the *Titanic*. Legend has it that a young David Sarnoff, future head of Radio Corporation of America, singlehandedly managed the wireless distress calls from his office in the Wanamaker's department store. The story, though apocryphal, speaks to the enduring mystique of radio's symbiotic relationship to crisis, a throughline of disasters simultaneously declared and managed via radio broadcast. Wireless transmissions in the United States were heavily relied upon by the Department of the Navy, an indication of the early understanding of radio's relationship to national security, both as menace and utility.[1] Early wireless experimenters wrought havoc by sending out fake distress calls, adding an additional layer of uneasiness to the early history of radio. It's easy enough to tell the story of radio from 1912 forward through exceptional moments of drama and crisis: Walter Winchell's breathless coverage of the 1932 Lindbergh Baby kidnapping and subsequent murder trial has been called the first media event in the

---

[1] Eric Barnouw, "US Navy regulates early wireless," *A Tower in Babel*, Vol. 1 (New York: Oxford University Press, 1996), 7–39.

era of modern celebrity journalism.² Herbert Morrison's anguished cry, "Oh, the humanity!" as he witnessed the *Hindenburg*'s fiery explosion over New Jersey in 1937 is probably the first electronic media catchphrase of the twentieth century.

The imagined community of Benedict Anderson's newspaper-reading public became, in the simultaneous reception of the radio news bulletin, a community drawn together "*in a state*," in a multilayered sense of the term coined by Butler and Spivak: affectively *and* politically; in crisis *and* as a geo-political entity.³ FDR's massively popular, highly choreographed, and sparingly used fireside chats were among the first and most effective uses of the performative and regulatory power of radio to create and manage the crisis in the very utterance of it. Roosevelt's Banking Crisis address in 1933 is credited with simultaneously defining and defusing a nationwide panic. It was also the first broadcast to be (retroactively) categorized as a "Fireside Chat," consolidating Roosevelt's status as the "Radio President."⁴ Roosevelt's Fireside Chats in May, 1940 and December, 1941 likewise declared states of emergency and states of war, even as they conveyed, in a medium-specific way, a sense of confidence about their resolution.⁵

Scholars have parsed the prosodic elements of Roosevelt's vocal performance (i.e., cadence, pace, pitch) to better explain the affective power of these radio events.⁶ One listener responding to a 1933 address put it like this: "It was just like having a good friend

---

²Neal Gabler, *Winchell: Gossip, Power and the Culture of Celebrity* (New York: Vintage, 1995), 207–13.
³Credit for this useful pun goes to Judith Butler and Gayatri Chakravorty Spivak's *Who Sings the Nation-State? Language, Politics, Belonging* (London, New York, Calcutta: Seagull Books, 2007), 3–4.
⁴Russell D. Buhite and David W. Levy, eds., *FDR's Fireside Chats* (Norman, OK: University of Oklahoma Press, 1992), 11; Edward D. Miller, *Emergency Broadcasting and 1930s American Radio* (Philadelphia: Temple University Press, 2002), 78–9; Jason Loviglio, *Radio's Intimate Public: Network Broadcasting and Mass-Mediated Democracy* (Minneapolis: University of Minnesota Press, 2005), 18–20.
⁵I presented a version of this argument in Loviglio, *Radio's Intimate Public*, 1–37.
⁶Earnest Brandenburg and Waldo W. Braden, "Franklin D. Roosevelt's Voice and Pronunciation," *Quarterly Journal of Speech* 38, no. 1 (February 1952): 23–4.

sit down and talk your troubles over with you."⁷ It was Roosevelt's jaunty calm, Morrison's unhinged panic, Winchell's rat-a-tat cadence that made each broadcast effective and affective. From the start, as David Hendy has pointed out with reference to the origins of the BBC, radio's story has been impossible to understand without attending to "a whole messy hinterland of feelings and moods" on both sides of the microphone.⁸

Newscasts about the assassinations of the 1960s, President Nixon's 1974 resignation, up through the morning of September 11, 2001, helped to frame and answer the question, "where were you when you learned that ___ happened?" The question brings together historical events and moments of reception, a tribute to the revolutionary transformation that radio broadcasting's "liveness" brought to world events and human experiences of them. The question "where were you?" implies the thrilling idea that, wherever you are, thanks to radio, "you are there," which was the title of a popular CBS program from the late 1940s that reenacted iconic historical moments.⁹ This sense of liveness—of immediacy—is part of the bundle of affects associated with "mobile privatization," "an at-once mobile and home-centered way of living," which Raymond Williams understood to be a powerful mechanism for reinforcing the status quo in which public matters became swathed in layers of domesticity.¹⁰

Radio helped write its own history of ersatz crises, as with the 1938 *War of the Worlds* broadcast, a hoax that generated a little panic and a lot of radio audience research.¹¹ It was neither the first nor the last radio hoax, nor were hoaxes, wars, and disasters the only

---

⁷J. E. Baudo, Brooklyn NY, to Franklin D. Roosevelt, March 13, 1933. Personal File 200 (PPF 200). Franklin D. Roosevelt Library. Hyde Park, NY.
⁸David Hendy, "The Great War and British Broadcasting: Emotional Life in the Creation of the BBC," *New Formations* 82 (2014): 82–99, muse.jhu.edu/article/558912.
⁹Martin Grams, *Radio Drama: American Programs 1932–1962* (North Carolina: McFarland & Company, Inc, 2000).
¹⁰Raymond Williams, *Television: Technology and Cultural Form*, 2nd ed. (London: Routledge, 1990), 18.
¹¹Hadley Cantril, *The Invasion from Mars* (Princeton: Princeton University Press, 1982).

crises that radio mediated.[12] Edward Miller has shown how radio newscasts, starting in the 1930s, first declared and then regulated states of emergency on a daily basis, a key development in the grammar of electronic journalism and a key function of radio's role as an ideological state apparatus.[13] Kathleen Battles demonstrated how the police "radio car" functioned in popular culture as an extension of this logic of technological surveillance by the state.[14] Evoking and assuaging moments of crisis has been a durable and unique affordance of the newscast ever since. An alertness to danger and an openness to feeling reassured, a version of what Michael Schudson has called "monitorial citizenship,"[15] became part of the structure of feeling evoked by national broadcasting's first decades.

I return to the pun, borrowed from Butler and Spivak, that radio brought citizens together "in a state"[16] because of how nicely it gestures to the link between politics and affect, because of what it can tell us about radio's historical relationship to evoking and assuaging political and emotional states of unrest. It can help us "rethink the origins of broadcasting as an emotional project as much as a political one."[17] If democratic feeling could be promulgated via radio, so could fears about political catastrophes. While Roosevelt embodied radio's prosthetic extension of the democratic polity, radio producers like Orson Welles dramatized radio's nightmare potential as a machine of mass panic and political violence. Welles's "allegories of fascism and anti-fascism," like *The Fall of the City* (1937, written by Archibald MacLeish, but starring Welles), *War*

---

[12]Kate Lacey, "Assassination, insurrection and alien invasion: Interwar wireless scares in cross-national comparison," in *War of the Worlds to Social Media: Mediated Communication in Times of Crisis*, eds. Joy Hayes, Joy, Kathleen Battles, and Wendy Hilton-Morrow (New York: Peter Lang, 2013), https://hdl.handle.net/10779/uos.23389721.v1; Kathleen Battles and Joy Elizabeth Hayes, "The Enduring Significance of the War of the Worlds as Broadcast Event," in *The Routledge Companion to Radio and Podcast Studies*, eds. Mia Lindgren and Jason Loviglio (New York and London: Routledge, 2022), 217–25.

[13]Miller, *Emergency Broadcasting*.

[14]Kathleen Battles, *Calling All Cars: Radio Dragnets and the Technology of Policing* (Minneapolis: University of Minnesota Press, 2010).

[15]Michael Schudson, *The Good Citizen: A History of American Civic Life* (New York: Free Press, 1998), 16.

[16]Butler and Spivak, *Who Sings the Nation-State*, 4.

[17]Hendy, "The Great War and British Broadcasting," 82–99.

of the Worlds (1938), and even his voice work on *The Shadow* (1938) drew upon the very mechanisms of radio broadcasting to warn about its acousmatic power to sow fear and to reap political disaster.[18]

The Fall of the City, a modernist take on a Greek tragedy, complete with chorus, is composed of moments of affective rippling—now fear, now joy, now despair—through a massive crowd in "the great square of the city." As rumors of an unknown and unseen conqueror bear down on the city, priests and ministers and ghosts and citizens shout impassioned speeches on the virtues and dangers of freedom. The impact of these voices, like those of the radio broadcast itself, is immediate and social—affective. A priest's impromptu sermon sparks a spontaneous dance through the crowd: "A current of people of coiling and curling through people," is how Welles narrates it in the voice of the announcer, a proxy for a radio newscaster, simultaneously inside and outside the diegesis.[19] Moments like this demonstrate Sara Ahmed's argument that "emotions are not simply 'within' or 'without' but that they create the very effect of the surfaces or boundaries of bodies and worlds."[20] The announcer, moved by the beauty of the scene, describes "a circling of people through people like water through water," an apt account of this rippling motion of affect, of the power of affect to constitute social formations as it moves through them.

Early audience research suggests that listeners understood the radio, and their own listening, to be part of a shared vigilance. "We have so much faith in broadcasting," one listener told pioneering radio researcher Hadley Cantril in his study of the *War of the World*'s panic. "In a crisis it has to reach all people. That's what radio is here for."[21] US broadcasting's regulatory language borrowed from nineteenth-century utilities law, placing the "public interest, convenience, and necessity" at the center of its considerations. This meant that even as commercial networks gobbled up bandwidth in the early network era, radio, its audience, and its mission

---

[18]Michael Denning, *The Cultural Front: The Laboring of American Culture in the Twentieth Century* (New York: Verso, 1997).
[19]Archibald MacLeish, *The Fall of the City* (Columbia Broadcasting System, 1937).
[20]Sara Ahmed, "Affective Economies," *Social Text* 22, no. 2 (2014): 117.
[21]Cantril, *Invasion from Mars*, 3.

was understood to be synonymous with "the public." Lacey has argued that the "listening public," shaped by recorded and broadcast sound technologies, created "a condition of plurality and intersubjectivity" for early twentieth-century listeners necessary for modern democratic life.[22]

Listeners also counted upon radio's public reach to provide moments of affective power from its national address and personal touch, as when a broadcaster lost his composure in the face of a tragedy or when a president tempers a talk on banking with a baseball analogy.[23] A "media event" was, above all, a moment of shared emotion, a "rippling" of affective current through the ether.[24] During the Second World War era, human interest shows achieved such moments when they surprised servicemen with prizes during live on-location broadcasts at military bases. The reliable formula of patriotism, consumerism, and emotional outbursts became a central part of an explicit wartime broadcast partnership between the Office of War Information and the commercial networks during the Second World War. Massively popular daytime serials were known for emotional excess and slow plotting, but they also kept up a brisk schedule of national defense talking points, pushing war bonds one week and Red Cross donations the next.[25]

Nostalgia

While radio's history can be told along a timeline of national crisis, the daily experience of listening has also been associated with "a persistent sense of spiritual longing and loss" and the emotional compensations of "camaraderie and mutuality coming from the sky above."[26] As the BBC put it in 1947, "radio has always been a

---

[22]Kate Lacey, *Listening Publics: The Politics and Experience of Listening in the Media Age* (London: Polity Press, 2013), 5, 8.
[23]Loviglio, *Radio's Intimate Public*, 38–69.
[24]Ahmed, "Affective Economies," 119.
[25]Gerd Horten, *Radio Goes to War: The Cultural Politics of Propaganda During World War II* (Oakland, CA: University of California Press, 2003).
[26]Susan Douglas, *Listening In: Radio and the American Imagination* (New York: Times Books, 1999), 40.

medium well suited to the nostalgic."²⁷ Broadcasting's most powerful programming insight has always been affective—nostalgia for that old home music for listeners on the move. Radio's history of music programming is inseparable from the history of US im/migration and industrialization in the twentieth century. Radio makers, from old-time musicians to conservative talk radio hosts, have been explicit in their exploitation of nostalgia as the central emotional gratification of the medium. Radio formats featuring plain-spoken members of the listening public brought the added dimension of "fellow feeling" to the experience of nostalgia. Often presented as emissaries from the residual pre-modern world, the voices of the people provided the "gut-punch" of authenticity and the balm of empathy to slickly produced and commercially profitable audience-participation shows, perhaps the most enduring format in electronic media.

As Martians began to destroy New Jersey in Welles's version of *War of the Worlds*, listeners in and out of diegesis of the radio play listened to the familiar strains of "Stardust," by 1938 a bona fide "oldie." In between sinking ocean liners and exploding zeppelins, radio offered the sounds of home, starting from before the network era.²⁸ Early hits like *The National Barn Dance* (1924) brought the reassuring sounds of that old-time music to millions of migrants to urban industrial centers. Local ethnic and foreign-language stations in urban industrial centers featured the music (and language) of home for immigrants from Mexico and all across Europe.²⁹

The cultural politics of the Depression-era New Deal drew on the notion of democratic feeling in the new mass media of musical recordings, motion pictures, and radio. Roosevelt's political and policy successes are hard to separate from his mastery of media,

---

[27]Quoted in Kathryn McDonald, "Scripting the radio interview: Desert Island Discs," *Radio Journal: International Studies in Broadcast and Audio Media* 18, no. 2 (2020): 180.

[28]Susan Douglas, *Inventing American Broadcasting, 1899–1922* (Baltimore: Johns Hopkins University Press, 1989).

[29]Clifford J. Doerksen, *American Babel: Rogue Radio Broadcasters of the Jazz Age* (Philadelphia: University of Pennsylvania Press, 2002); Gene Fowler and Bill Crawford, *Border Radio: Quacks, Yodelers, Pitchmen, Psychics, and Other Amazing Broadcasters of the American Airwaves*, revised ed. (Austin: University of Texas Press, 2002).

chiefly radio, as has been well documented.[30] Near the end of his first fireside chat in March 1933, Roosevelt was able to "catch the note of confidence from all over the country," as if his words and their emotional impact were communicating back and forth in real time. From the start, he made explicit the importance of the radio apparatus to the feeling of national unity necessary to take on challenges like the banking crisis—rhetorically dividing and then uniting bankers, savers, and hoarders—and later, the looming war, uniting isolationists and anti-fascists. When he proclaims at the conclusion of his first Chat that "We have provided the machinery to restore our financial system and it is up to you to support it and make it work," he appears to use "machinery" to link the bank reforms, the radio apparatus, and the affective current rippling through the country.[31] As novelist H. G. Wells wrote, Roosevelt was "a ganglion for reception, expression, transmission, combination and realization."[32]

Susan Douglas has captured the combined emotional and political power of early network radio in her analysis of the compensatory appeal of the "linguistic slapstick" of Depression-era radio comedies, particularly the way that democratic values sympathetic to the feelings of "the forgotten man" made their way into the silliest of premises, such that the underdog often got the upper hand, even when it was only through misdirection, malapropism, or small moments of verbal anarchy.[33] Audience participation shows likewise featured the untutored voices of the man-in-the-street chatting with the smooth-voiced host. This juxtaposition is also one of network broadcasting's earliest impulses; just as soon as there was such a thing as a professional radio voice, interest in and affection for the voices of "real people" became an irresistible, and often quite economical, programming

---

[30]Horten, *Radio Goes to War*; Loviglio, *Radio's Intimate Public*; William Leuchtenberg, *The FDR Years: On Roosevelt and His Legacy* (New York: Columbia University Press, 1995).
[31]Buhite and Levy, eds., *FDR's Fireside Chats*, 12.
[32]H. G. Wells, *Experiment in Autobiography: Discoveries and Conclusions of a Very Ordinary Brain* (New York: Little, Brown, 1985).
[33]Douglas, *Listening In*, 100–23.

option. For young people, including recently arrived immigrants, that came of age in the middle decades of the twentieth century; radio jingles, catchphrases, and hit songs provided a lingua franca, which in turn became the basis for a shared repository of nostalgia and generational identity.[34]

Radio's status as a feeling medium is also evident in the history of its installation in the automobile. From the start, long before it made sense aesthetically or technologically, the radio was promoted as a tool of emotional regulation and transportation for drivers. Clunky, delicate, and impractical, the first radio set was installed in a "wireless car" at the 1904 St. Louis Expo by inventor Lee Deforest.[35] As radio became more mobile with the invention of the transistor, the car was increasingly designed and experienced as domestic space. Amid the postwar sprawl, the car radio was advertised as a trusty mood regulator when navigating stressful traffic.[36] The postwar era also brought new formats with a new focus on youth and local audiences and an intensification of the car-radio's power to transport and explore new relationships and identities.[37] The 1964 FCC rule that FM stations broadcast unique programming, rather than simply duplicating the content of a sister AM frequency, made space for a new kind of expressiveness on the air just in time for rock's concept albums, urban and college-based countercultures, with densely settled reception zones, and the coming together of the personal and the political. By the 1990s, noise-cancelling sensors built into automobiles brought acoustic cocooning to a new level of sophistication. Silence was not the goal so much as affective autonomy. It was important that "drivers feel themselves to be in control of their acoustic environment."[38]

---

[34]Douglas, *Listening In*, 100–23; Renato Rosaldo, "Imperialist Nostalgia," *Representations* no. 26 (1989): 108.
[35]M. B. Schiffer, *The Portable Radio in American Life* (Tucson: University of Arizona Press, 1991), 23.
[36]Karin Bijsterveld, Eefje Cleophas, and Stefan Krebs, *Sound and Safe: A History of Listening Behind the Wheel* (Oxford: Oxford University Press, 2013), 103.
[37]Douglas, *Listening In*, 219–55.
[38]Bijsterveld et al., *Sound and Safe*, 5.

Public Radio and Crisis

US radio in the 1960s presented a model of mobile privatization and smooth broadcast flow that was ideal for commercial penetration by advertisers and record labels, but it left much to be desired for critics who pushed for a truly national public service instead of the thin patchwork of underfunded community stations housed mostly in colleges around the country. Criticisms of commercial broadcasting in this era invoked the language of crisis, as in FCC Chair Newt Minnow's famously apocalyptic description of television as a "vast wasteland."[39] Calls for better stewardship of the public airwaves, more educational programming, and greater diversity of interests served on TV and radio shaped the years-long process culminating in the 1967 passage of the Public Broadcasting Act, which created the Corporation for Public Broadcasting (CPB), and eventually, National Public Radio.[40] This is another chapter in US broadcast history that can be told as a series of crises right up to the present moment. Indeed the Carnegie Commission on Educational Television, formed by Congress in 1965, spoke "urgently" in the Cold War rhetoric of the era, of a need for "an instrument for the free communication of ideas in a free society."[41]

Like Moses in the bullrushes, National Public Radio was nearly cast away before it had a chance to begin. The Public Broadcasting Act of 1967 was always primarily about the creation of a public television service; public radio was an afterthought and was left out altogether in the final draft. The words "and radio" had to be surreptitiously added with Scotch tape in the text of the "Public Television Bill" the night before the House vote by a shadowy group of radio partisans who pulled it off through "subterfuge

---

[39]Newton N. Minow, "Television and the Public Interest," address to the National Association of Broadcasters, Washington, D.C., May 9, 1961.
[40]Shepperd has explored the long history of activism that made possible the kind of commitments and momentum leading up to the 1960s commissions, reports, and speeches that made something like the creation of the CPB, what he calls, "the victory of public broadcasting," possible. Josh Shepperd, *Shadow of the New Deal: The Victory of Public Broadcasting* (Champaign: University of Illinois Press, 2023).
[41]Carnegie Commission on Educational Television, Summary, 1967. Current Publishing Committee and National Public Broadcasting Archives.

and dumb luck."[42] Even so, the legislation directed the lion's share of the funding to public television.[43] Since then, the lore of public radio's narrow escape has grown and with it, an anxious sense of being at odds with its own mission and the world it is charged with reporting on. As Tom McCourt has put it,

> the system has been riven from the outset by a contradictory mission: to create a single national identity while giving voice to those excluded by the marketplace ... the embers of self-immolation smolder constantly.[44]

The hyperbolic, almost comic, sense of precarity captured in this passage echoes through a great deal of the literature on the history of public broadcasting. As the editors at *Broadcasting* observed in 1973, "crisis... is something that public broadcasting has never been able to escape."[45]

As we will see below, the contradictory mission had its roots in the longer history of US broadcasting's ambivalent relationship to the public and the marketplace.[46] And in some of the inevitable structural limitations: "puny resources," a mission to provide an alternative to the status quo, and vast ambitions "to cover events in a first-rate fashion" in the words of "founding mother," Susan Stamberg.[47] As Josh Shepperd has shown, in the United States, at least, "public media started at a disadvantage."[48] Debates about

---

[42] Jack Mitchell, *Listener Supported: The Culture and History of Public Radio* (Westport, CT: Praeger, 2005), 36–7, 59; See also, Michael McCauley, *NPR: The Trials and Triumphs of National Public Radio* (New York: Columbia University Press, 2005), 21–2.
[43] Mitchell, *Listener Supported*, 8; McCauley, *NPR*, 18.
[44] McCourt, Tom, *Conflicting Communication Interests in America: The Case of National Public Radio* (Westport, CT: Praeger, 1998), 88.
[45] *Broadcasting*, November 12, 1973, 26–32.
[46] See also Laurie Ouellette, *Viewers Like You: How Public TV Failed the American People* (New York: Columbia University Press, 2002), which tells a similar story of the besetting contradictions in the creation of public television.
[47] Susan Stamberg, "Introduction: In the Beginning, There Was Sound but No Chairs," in *This Is NPR: The First Forty Years* eds. Cokie Roberts et al. (San Francisco: Chronicle Books, 2001), 15; "NPR's Frischknect: Don't Look Back," *Broadcasting*, January 17, 1977, 65.
[48] Shepperd, *Shadow of the New Deal*, 170.

inclusion and exclusion, fairness and bias, shared identity and niche marketing, were central to public radio's history of crisis. Fears of "self-immolation" shaped a history of political self-consciousness and caution.[49]

Such tensions lie at the heart of the liberalism of the postwar era—soaring ambitions of the Great Society and entrenched, atavistic inequalities; idealistic Cold War rhetoric of human freedom and brutal Cold War proxy wars; liberalism as empathy and liberalism as technocratic administration. Communication, James Carey has observed, works as transmission for the purposes of social control and as ritual, for the purposes of social connection.[50] Public radio operated in both modes simultaneously, transmitting and negotiating crisis and constituting rituals of fellow-feeling among listeners, producers, and on-air talent.

Even for a medium with a history of symbiotic relationship to crisis and disaster, NPR's record is striking: The station's flagship news program, *All Things Considered* (*ATC*) was born coincidentally, but consequentially, on May Day of 1972 in the midst of the teargas and chaos of a massive anti-war demonstration in Washington, DC. Reporter Jeff Kamen interviewed protesters, questioned motorcycle cops, and opined that "Today in the nation's capital, it is a crime to be young and to have long hair."[51] This broadcast, and its place in NPR lore, helped to define an enduring, if misleading, reputation for anti-establishment bias, which persists to this day, part of the political instability at the heart of the crisis/growth paradox.[52]

Morning Edition, the network's second and eventually premier daily news program, began on the first morning of the Iranian

---

[49]Mitchell, *Listener Supported*, 59–77.
[50]James Carey, *Communication as Culture: Essays on Media and Society* (Boston: Unwin Hyman, 1985), 14–23.
[51]http://www.npr.org/programs/atc/atc30/timeline/index.html, accessed July 9, 2012; see also Douglas's account of this debut broadcast in *Listening In*, 321–2.
[52]It is beyond the scope of this chapter to fully engage the arguments about NPR's perceived liberal bias. However, this matter has been taken up elsewhere. Jeffrey A. Dvorkin, "NPR In 1988: 'News That Soothes,'" *NPR*; Tim Groseclose and Jeffrey Milyo, "A Measure of Media Bias," *The Quarterly Journal of Economics* 120, no. 4 (2005): 1191–1237; Norman Solomon, "NPR and the Fallow Triumph of Public Radio," *Alternet*, April 11, accessed on February 2, 2012.

hostage crisis in 1979, quite by chance (unlike ABC's *Nightline*, which began in direct response to the unfolding hostage crisis). NPR's overall audience skyrocketed during the first Gulf War in 1990–91 and kept growing, with predictable growth spurts in times of national crisis. Nicols Fox, writing for the *American Journalism Review* puts it baldly: "Public Radio owes a lot to the firm of Cheney, Schwarzkopf & Powell. The Gulf War—like Watergate, Iran-Contra and Tiananmen Square—brought the news-hungry to a halt at the NPR signal."[53] By the early 1990s, NPR staff were boasting of their "air superiority" over the television networks in their coverage of the Gulf War, a pun that reinforced the network's symbiosis with crisis and that reinforced left-wing charges that it had become "National Pentagon Radio" in its uncritical foreign policy coverage.[54]

The size of the network's audience spiked again after the attacks of September 11, 2001, resulting in enduring gains for the network and helping to justify the addition of new foreign bureaus and correspondents in more of the world's hot spots. The attacks also helped to spin off the network's Sonic Memorial Project, the intimate interviews show *Story Corps*, and to revive the popularity of first-person nonfiction narratives, particularly stories of trauma and resilience, in audio formats.[55] Unlike CNN and other electronic news outlets, which required fresh infusions of shock and awe to get its ratings bumps, NPR tended to hold a good chunk of its new audience post-crisis.

NPR grew in audience, revenue, and national stature during the 1990s and early 2000s, a period when other electronic and print news outlets were in free-fall. In an era of genuine crisis for traditional journalism and in the tumult of its own impossible situation, NPR evolved and thrived. Even so, concern that NPR was "too liberal" for its public, both from within and without the

---

[53]Nicols Fox, "NPR Grows Up," *American Journalism Review* (1991): 30–6, https://www.cjr.org/special_report/trump-bump-sopan-deb-katy-tur-sam-sanders-prachi-gupta-fahrenthold.php/.
[54]Mitchell, *Listener Supported*, 180–1; Scott Sherman, "The Good, Gray NPR," *The Nation*, May 23, 2005, https://www.thenation.com/article/archive/good-gray-npr/.
[55]Elisia L. Cohen, "One Nation Under Radio: Digital and Public Memory after September 11," *New Media & Society* 6, no. 5 (2004): 591–610.

organization, was a constant and very productive anxiety right up to the election of Donald Trump in 2016 and again in the months before his reelection, culminating in its defunding in 2025.[56] NPR pioneer Jack Mitchell fretted that "our (anti-war) attitude probably did not seem reasonable and responsible to most Americans, who overwhelmingly re-elected Richard Nixon as president over George McGovern in 1972, eighteen months after *All Things Considered* took to the air."[57] In the late 1970s, Nina Totenberg railed against her colleagues who seemed to be "crusading for a cause" rather than reporting the news.[58] In 1983, Robert Siegel, then head of news, worried if "we are really describing the country that voted for Ronald Reagan."[59]

As John Durham Peters has shown, liberalism's openness to its illiberal critics is its greatest and most self-defeating feature.[60] The pivot away from its liberal origins became a defining ritual, a reflexive response to each new crisis. Liberalism, always a difficult term to pin down, functioned in this ritual as a floating signifier, unmoored to any fixed meaning. Liberalism was public radio's besetting sin and greatest asset, its origin story that no amount of atonement or pivoting or maturing could seem to redeem. And yet each fresh crisis brought new chances for redemption, for reinvention. Just as neoliberalism fed upon liberalism's liver all day long only to see it regenerated overnight, so NPR endured round after round of "maturation" from its naïve origins in liberalism, only to encounter new crises and new market-based opportunities. It was precisely this liberal approach to moving rightwards that powered public radio's promethean atonement and explosive growth in audience, market share, and revenue.

---

[56]Anna Tauzin, "Perceived Orientation of NPR Programming," *NPR Audience and Insight: NPR.com*, https://www.slideshare.net/slideshow/perceived-orientation-of-npr-programming-1693424/1693424; Uri Berliner, "I've Been at NPR for 25 Years: Here's How We Lost America's Trust," *The Free Press*, https://www.thefp.com/p/npr-editor-how-npr-lost-americas-trust.
[57]Mitchell, *Listener Supported*, 73.
[58]Mitchell, *Listener Supported*, 78.
[59]Laurence Zuckerman, "Has Success Spoiled NPR?" *Mother Jones*, June/July, Qtd. in Mitchell, *Listener Supported*, 167.
[60]John Durham Peters, *Courting the Abyss: Free Speech and The Liberal Tradition* (Chicago: The University of Chicago Press, 2005).

When thinking about the symbiotic relationship between growth, revenue, and reputation on the one hand, and periodic existential crisis on the other, consider the foundational role of the seasonal pledge drives that every local NPR member station conducts multiple times a year. On-air appeals from the familiar voices of local and national hosts regularly ask listeners to imagine their daily routine if their NPR member station cannot pay its bills and ceases to broadcast. This ritual began in 1983 in response to a budget crisis that very nearly bankrupted the network. The so-called "Drive to Survive" brought in millions over the course of three days and saved the network while providing Congress a rationale for continued underfunding of public broadcasting.[61] The mendicant posture of public radio was seen by many as an indictment of the federal government's weak commitment to public broadcasting and more broadly, of the inherent limits of public institutions under liberal capitalism.[62]

Because US public radio was only partially funded by the government, it operated between the competing pressures to raise money through, among other strategies, fundraising, corporate sponsorship, and licensing deals while still retaining the cultural cachet of alternative media that appeals to its valued imagined audience of educated, upper middle-class listeners. As Eleanor Patterson has argued, *TAL*'s formation as a public radio franchise was shaped by the decentralized structure and funding policies of US public media and the industrial structure of US public radio industries encouraged programs like *TAL* to pursue media franchising. *This American Life*, Patterson argues, was recognizable as a public radio brand through its discursive opposition to commercial media.[63]

---

[61]Mitchell, *Listener Supported*, 109.
[62]Jessica Clark, and Patricia Aufderheide, "A New Vision for Public Media," in *Media and Social Justice*, eds. Sue Curry Jansen, Jefferson Pooley, and Lora Taub-Pervizpour (Cham: Springer, 2011), 55–67; Louisa Lincoln and Victor Pickard, "Reimagining American public media: A key infrastructure for local journalism," *Journalism* 26, no. 2 (2024): https://doi.org/10.1177/14648849241248018; Robert McChesney, *Rich Media, Poor Democracy: Communication Politics in Dubious Times* (New York: The New Press, 2016).
[63]Eleanor Patterson, "This American Franchise: *This American Life*, public radio franchising, and the cultural work of legitimating economic hybridity," *Media, Culture, & Society* 38, no. 30 (2016): 450–61.

Consider also the now-defunct two-year cycle of Congressional funding for the CPB and the constant threats from Republican lawmakers to "zero out" NPR's budget. Such threats, often made by lawmakers from rural jurisdictions ill-served by commercial radio and heavily reliant on local NPR member stations for news, weather, and public affairs programming, were until 2025, mostly disingenuous, designed to raise money and support from their conservative base.[64] In addition to short-term political benefits for politicians, such threats were enormously productive for NPR, rallying liberal defenders, spurring listener donations, and generating talking points for reformers keen to pivot the network away from liberal politics and towards heavier reliance on corporate underwriting as the main source of revenue. The success of these defunding efforts in 2025, part of a broader dismantling of the federal government's commitment to funding education, public health, foreign aid, and other intistutions of the liberal state, seemed both inevitable and shocking after decades of proleptic crisis discourse.

As early as 1998, Audience Research Analysis, Inc., invoked Congressional threats, "as a ringing wakeup call" in its recommendation to increase the network's focus on the well-heeled "highly educated" portion of listeners.[65] Referencing a mid-1990s GOP threat to eliminate NPR's budget, the report made explicit this productive tension: "That crisis unleashed enormous creative energy throughout the public radio system."[66] A decade later, amidst the shocks of the financial crisis that led to the Great Recession, then-president Vivian Schiller echoed this logic: "I'm here to tell you today, and I will continue to say this... until I'm blue in the face[—]this is a crisis we will not waste."[67]

This is, in a small way, the logic of the "shock doctrine," an explicit approach to economic reforms laid out in great detail by the neoliberal economists at the University of Chicago in the 1970s in

---

[64]David Margolick, "National Public Rodeo," *Vanity Fair*, January 17, 2012.
[65]David Giovanonni, Leslie Peters, and Jay Youngclaus, *Audience 98: Public Service, Public Support*, ed. Leslie Peters. Audience Research Analysis.
[66]*Audience 98*, Report: 144.
[67]"NPR downsizing is a 'crisis we will not waste'," editorial, *Current*, April 9, 2009, https://current.org/2009/04/npr-downsizing-is-a-crisis-we-will-not-waste/.

their capacity as advisors to Chilean dictator Pinochet. Since then, it has migrated into mainstream economic thinking, fueling heady talk of "disruptors" bringing "creative destruction" to "stagnant" sectors of the economy. As Naomi Klein has argued, such verbiage often obscures sharply anti-democratic policies, deep cuts in social spending, and massive profit opportunities for financial institutions at the expense of public institutions.

The logic of lurching financial crises, followed by market-friendly reforms, shaped the development of the network for fifty years, paralleled by a nagging internal monologue about its liberal bias requiring constant compensatory editorial activism. Crucially, "disaster capitalism" is effective precisely because of its emotional force—its power to shock, and more, to "socialize" emotional upset into new affects and new kinds of subjectivity, a tactic that has been taken up in less violent corporate contexts.[68] Public radio was both the best and worst public institution to practice these shocks on, as its founding philosophy, its guiding "purpose," was neither partisan nor political but deeply committed to the idea that "listeners should feel."[69]

"Listeners Should Feel"

In its early years, *All Things Considered* was celebrated for its improvisational vibe, its immediacy, and its capacity for both gravitas and whimsy, twin features the network has striven to keep in balance ever since. It was also hailed for its therapeutic qualities, its capacity to offer succor to Americans by dint of radio's intimate address, and because of its adherence to the network's first program director, William Siemering's commitment to "celebrat[ing] the human experience." Charles Kuralt's memory of listening in 1971 to the program while in a country store in rural Virginia captures the intended therapeutic appeal:

---

[68]Naomi Klein, *The Shock Doctrine: The Rise of Disaster Capitalism* (New York: Picador, 2007), 131–70; Analiese Richard and Daromir Rudnickyj, "Economies of Affect," *Journal of the Royal Anthropological Institute (N.S.)* 15 (2009): 57 © Royal Anthropological Institute.
[69]William Siemering, "National Public Radio Purposes," *NPR*, 1970. In *Current*, May 17, 2012. https://current.org/2012/05/national-public-radio-purposes/.

I bet that man and his daughter in the store where I bought the cheese and apples listen to *All Things Considered* not only because it informs them but also because it makes them feel better. It makes me feel better too.[70]

While its debut broadcast, including hair-raising coverage of the massive anti-war demonstration, channelled gravitas amidst crisis, the story leaned into interpersonal understanding rather than sensationalism. First-person accounts of people from varied backgrounds, in their own words, became a hallmark of the show's reportage. As Geneva Overholser has pointed out, "the very first words on the very first episode of *All Things Considered* were those of a Black nurse talking about her heroin addiction."[71] NPR's flagship news program staked its claim to an odd pairing of mandates: to respond to crisis and to make people "feel better." These competing impulses formed the warp and woof of the public radio structure of feeling over the next five decades.

NPR's founding "Purposes," penned in 1970 by founding program director William Siemering, was a mission statement for a medium of feeling. "Listeners," it insists, "should feel that the time spent with NPR was among their most rewarding in media contact." The first three words of this statement, "listeners should feel," exemplify the editorial philosophy of the network for most of the last half century. The editorial attitude it promised, in the utopian idiom appropriate for a mission statement and for the era, "would be that of inquiry, curiosity, concern for the quality of life, critical, problem-solving, and life loving."[72]

It is a manifesto that crackles with the emotional immediacy characteristic of the social movements of the period. NPR would regard its audience as "curious, complex individuals who are looking for some understanding, meaning and joy in the human

---

[70]Charles Kuralt, "Preface," in *Every Night at Five: Susan Stamberg's All Things Considered Book*, ed. Susan Stamberg (New York: Pantheon, 1982), xi.
[71]Geneva Overholser, "NPR Offers News and Companionship," in *What Good Is Journalism? How Reporters and Editors are Saving America's Way*, eds. George Kennedy and Daryl R. Moen (Columbia: University of Missouri Press, 2005), 40.
[72]Siemering, "National Public Radio Purposes."

experience." The language sounds dated despite and because of its universalizing idealism. The network's founders enlisted feeling in the service of liberal democracy, an institution simultaneously in crisis and enjoying a kind of rejuvenation, through the cultural and political movements and the Great Society programs of the 1960s.

Perhaps it is more precise to say that, for the fledgling NPR, feeling was the mode contemporary liberal democracy required. "The formulation of public policy," "NPR Purposes," argued in language borrowed from political science, depended upon a new emphasis on "the influence of personal motives, ambitions, and emotions." Moreover, America needed an "affective education" in civics and "the cultivation of competence in the emotional and interpersonal."[73] Siemering saw NPR as a key part of this effort, naming as one of its chief goals, to "increase the pleasure of living in a pluralistic society." In this version of utopian liberalism, pluralism was the given; pleasure was the ambition. The document reads like a word cloud of feeling terminology: "joy" or versions of it occur four times, along with references to "pleasure," "hate," "envy," "love," and "pride." The statement concedes that "hate" and "envy" are inevitable in human experience but twice identifies "apathy" as the new network's true nemesis.

> "National Public Radio will serve the individual," it will promote personal growth; it will regard the individual differences among men with respect and joy rather than derision and hate; it will celebrate the human experience as infinitely varied rather than vacuous and banal; it will encourage a sense of active constructive participation, rather than apathetic helplessness.[74]

The high-flown idealism of this rhetoric, combined with the paltry funding from the CPB and the political upheaval of the early 1970s, can, in retrospect, seem like obvious markers of an unsustainable and doomed project.[75] The original mission to be an alternative

---

[73] Siemering, "National Public Radio Purposes."
[74] Mitchell, *Listener Supported*, 55.
[75] *Broadcasting*, November 12, 1973, 26–32.

service, a complement to mainstream journalism, inevitably clashed with the ambitions of the journalists who built the network into an increasingly professional and influential operation.[76] "The vox populi became the voice of the best professionals," Jack Mitchell, *All Things Considered*'s first producer, conceded with rueful pride.[77] In the dominant narrative, NPR's programming evolved to tell the stories that were exploding around it, creating newer, larger audiences that required the network to reconceptualize its listenership and to recalibrate its mixture of amateurish immediacy with its growing professionalism.[78] The emphasis on the individual, a shared value for both Cold War Liberalism and the counterculture that rose up in opposition to it, provided another layer of built-in contradiction.

The history of the network can be approached as a battle over competing understandings of the concept of public life—a universalizing utopianism evoking a broad and evolving human rights vision of humanity or a market-friendly spirit of competition and consumer choice. In both versions, the logic of inculcating empathy listener by listener, story by story, compelled program directors, underwriters, producers, and audience alike. Within the newsroom, a similar détente prevailed among the old guard, attached to the amateur spirit of experimentalism and the new guard of hard-nosed journalists with mainstream ambitions: both sides agreed on the value of storytelling, the personal, and an "affective education." Linda Wertheimer, one of the latter, conceded that the power of emotional proximity was the network's secret weapon. Despite her inside-the-beltway bona fides, she was committed to "keeping our voice small and intimate, a voice in the car on the ride home, on the radio in the kitchen."[79]

---

[76]See Siemering on NPR's alternative, complementary mission, "NPR's Frischknect: Don't Look Back." On the tension between old and new guards at NPR in the 1980s, see McCauley, *NPR*, 82–3; and Mitchell, *Listener Supported*, 90–1; Scott Sherman, "Good, Gray NPR," 34–40.
[77]Mitchell, *Listener Supported*.
[78]See for example the insider histories written by Mitchell, *Listener Supported*, and McCauley, *NPR*.
[79]Werthheimer, *Listening to America*, 16.

In some ways, NPR's increased mainstream success amidst an increasingly conservative era provided ideal conditions for maintaining this tonal compromise. Barbara Ehrenreich has argued that the individualism of the generation that came of age in the 1970s turned into something "meaner" and less optimistic as it assumed institutional power in the 1980s.[80] Susan Stamberg put it slightly differently, stating "the general level of caring declined" in the new decade and that "Special interest groups expressed the emotions of public life the Reagan-Bush years."[81]

This commitment to a small and intimate voice required both art and science, and a shared approach from the technical, marketing, and editorial staff at NPR. A 1971 press release announcing the debut of *All Things Considered*, promised "a relaxed pace, unique for broadcasting [that would]... allow form to organically follow function."[82] At the technical level, NPR quickly developed a reputation for pristine sound quality, starting in the 1980s, when it began using the state-of-the-art studios, microphones, sound engineering and satellite uplinks. NPR learned early on from its commercial counterparts that radio should be engineered to be heard while in the car, its audience an imagined community of commuters.[83] Sound engineers working for car companies long understood the interiors of cars to be "acoustic cocoons," in which the drivers' affective state had to be managed. NPR adopted the Neumann U-87 microphone, as its "house standard." On this microphone, the bass roll-off is typically switched on, to eliminate the lower frequencies of the human voice because, according to Fox:

> Most listeners to NPR are in cars, there's rumbling sounds of car engine and the road, etc. and they want their voices to be higher—above all that sound—and thus clearer and easier to hear, cleaner.[84]

---

[80]Barbara Ehrenreich, *Fear of Falling: The Inner Life of the Middle Class* (New York: Pantheon Books, 1989).
[81]Susan Stamberg, *Talk: NPR's Susan Stamberg Considers All Things* (New York: Perigree Books, 1993), 6.
[82]ATC Original Press Kit., 1971. Susan Stamberg Papers, Library of American Broadcasting Series 2: 1–5. Box 5.
[83]Bijsterveld et al., *Sound and Safe*, 104.
[84]Fox, "NPR Grows Up," 30–6.

Some of NPR's sound engineers epoxied the switches to bass roll-off to thwart hosts trying to sound more bassy and "authoritative," an indication of how intense the commitment was to consistency and to the in-the-car sound.[85] The importance of "clean, clear sound" as part of this process came up frequently in interviews with NPR sound engineers. Network studios avoided right angles and solid walls to cut down on reverb, and computers and CPUs were kept away from the microphone to prevent ambient buzzing sounds. The meticulous editing of interviews has become "a masterful art" for many programs, another distinctive feature of public radio's relationship to voice. Digitally removing pauses, "ums," and coughs, makes interviewees "sound far more articulate." Mixed with previously recorded ambient sound, or "situational sound effects," interviews done in the field can be made to feel more "lush," evocative, and immediate.

NPR reporter Alex Chadwick, a role model for the young Ira Glass, calls the resulting product, "a higher reality, a cleaner, more articulate reality."[86] Experts also point to the relatively high sound volume of the NPR broadcast (relative to commercial stations playing music or talk shows), giving a "far wider dynamic range" and "much more nuance." Because there are no commercials, there is no need to constantly turn the volume up and down to accommodate varying levels, creating a more even, pleasant, and engaging experience. Engineers "take great pain not to overly saturate the sound." Unlike commercial radio, NPR uses no volume compression when it sends its signal via satellite to affiliates around the country. "We keep it as pure as possible."[87] Purity, clarity, and cleanliness represent public radio's technical and aesthetic distance from the commercial dial, with its loud commercials, shrill announcers, and voice-tracked DJs and tinny compression.

The theme of purity and authenticity runs through NPR's relationship to vocal performance training as well. NPR, which has trained many of public radio's most well-known voices, has

---

[85] Adam Ragusea, "A top audio engineer explains NPR's signature sound," *Current*, June 5, 2015, https://current.org/2015/06/a-top-audio-engineer-explains-nprs-signature-sound/.

[86] Adam Ragusea, "Why you're doing audio levels wrong, and why it really does matter," *Current*, July 14, 2014, http://current.org/2014/07/why-youre-doing-audio-levels-wrong-and-why-it-really-does-matter/.

[87] Ragusea, "Why you're doing audio levels wrong."

long relied on voice coaches to help hosts and reporters bring out "the most emotive, evocative and distinctive qualities" in their voices. "Host Whisperer" David Candow, among others, worked with hosts, anchors, and reporters to improve their pitch, pace, volume, and rhythm. Getting on-air talent to "sound more like themselves" was the mantra for most of these coaches. Ira Glass echoed this injunction, urging his colleagues to speak in the natural, conversational styles that make for better storytelling, greater empathy, and a more intimate connection with listeners.[88] Avoiding the "cookie-cutter" sound of other radio performers, public radio offers listeners an "artisanal" approach to voice. For Candow, improvements in sound technology have opened the field of radio to a wider range of voices. "Now cars are quieter, and most car radios produce CD-quality sound. Ergo, radio can tolerate a wider range of voices."[89] These sonic effects also helped to distance NPR from the emotional upset associated with driving in traffic, the tumultuous news of the day, and the theatrical outrage of talk radio on the AM dial. Even during periods of internal crisis, NPR managed to sound to its millions of listeners "like a sea of tranquility."[90]

These formal efforts at clarity and tranquility were matched by an editorial approach to content that suffused a great deal of the "small and intimate" storytelling approach to journalism that the network became known for. Listening to NPR's voices and stories, as Kuralt observed, made people "feel good," perhaps in more ways than one. Across multiple accounts from listeners, audience researchers, on-air talent, and producers, emotional authenticity was cited as key to NPR's success in making the news a calming daily ritual, rather than a source of anxiety.

At the same time, it was an experience in a shared community somehow both broad and select, democratic and distinctive. Kuralt's experience of witnessing a shopkeeper and his daughter listening to *All Things Considered* sets in motion a fantasy of national reception: "listeners... in buses wandering down back

---

[88]Paul Farhi, "When This Guy Talks, NPR Listens," *Washington Post*, August 31, 2008, https://www.washingtonpost.com/wp-dyn/content/article/2008/08/29/AR2008082900683.html.
[89]Farhi, "When This Guy Talks."
[90]Margolick, "National Public Rodeo." Douglas made explicit the sonic and emotional contrast between NPR and AM conservative talk radio in *Listening In*, 284–327.

roads, in country stores in Virginia, in California sports cars and New Mexico bunkhouses and in the cuddy cabins of lobster boats off the coast of Maine—so many Americans who otherwise have so little in common sharing this one experience." These listeners, he implies, become, in the act of listening, extensions of the program's qualities: "curious, thoughtful and humane."[91]

Reporters and hosts like Susan Stamberg leaned into these themes as the network developed its distinctive voice in the 1970s and 1980s. "I tend to look for novels in news events—for conflicts, relationships, and feelings... I want to know how events affect ordinary people and I would rather talk to them than officials."[92] This confession frames her account of moderating a call-in program for President Jimmy Carter, an unprecedented journalistic feat, suggesting how carefully Stamberg adjusted the tension between emotional immediacy and professional gravitas. "My voice is as soft as the belly of a kitten," she recalled of her interview with notorious Watergate figure John Ehrlichman. "But I'm asking killer questions."[93]

Stamberg was the first woman to anchor a nightly national newscast in US history, and her voice, full of the music of curiosity, was the network's defining soundmark for much of its history. Stamberg, thirty-one years old when she was hired, brought youthful exuberance to the job that was a departure from newscasting's baritones with their supposedly neutral midwestern accents. Stamberg's voice was "nasal, quizzical, and unashamedly female," as Lisa Phillips put it. It came, she said, "with a hometown—New York—and an ethnicity—Jewish." Curating distinctive voices "rich with the rhythms and accents of their regions" was another explicit way in which *All Things Considered* initially sought to sonically mark its difference from what had come before, according to Stamberg.[94]

NPR's commitment to many voices included those who brought regional, as well as gender, diversity to the airwaves. Occasional commentators Baxter Black, a cowboy poet from Texas; Vertamae

---

[91]Kuralt, "Preface," ii.
[92]Stamberg, *Talk*, 56.
[93]Stamberg, *Talk*, 56.
[94]Lisa Phillips, *Public Radio: Behind the Voices* (New York: Vanguard Press, 2006), 3.

Grosvenor, a culinary anthropologist born in the Gullah community of North Carolina; and Kim Williams, a naturalist, checked in during the late 1970s and early 1980s with field reports from their corners of the country. Andrei Codrescu, a Romanian American artist living in New Orleans, began to bring his thickly accented English and droll humor to NPR in 1983. Putting these folks on air seemed to address the network's vision of speaking in "many voices and many dialects." The intent, Mitchell wrote, was explicitly democratic, to be "representative of the nation. That meant white, black, Hispanic, Asian and as many women as men."[95]

Picking Stamberg was a controversial choice in 1971. Early feedback on Stamberg from station managers around the country wasn't encouraging. "She sounded too New York, too Jewish, too off-putting," according to Mitchell.[96] Her success stemmed not from the particularities of her identity but from the lushness of the personal that suffused her voice and interviews. Many of the listeners who wrote in to the network about Stamberg seemed to agree that her vocal performance had a powerful and immediate impact on their experience of the news. "I especially appreciate the emotion you inject into people you talk to and subjects you investigate, wrote one woman in 1976."[97] Such sentiments are common in the fan mail for Stamberg over her long career. "I was just beginning to zone out when my ears perked up at the sound of your voice," Edwin Baker wrote to Stamberg in June 1996. "Your voice has that effect on me... what stands out in every venue is your enthusiasm, curiosity, passion, concern, zeal, and interest."[98]

Like many listeners, and like the audience dial-operated meters introduced in that decade, Baker charts the moment by moment shifts in his attention during the broadcast flow. "Whenever I hear Susan penetrating my early morning fog," wrote Arlie Anderson in 1999, "I bounce awake instantly because I don't want to miss a word... Her voice, her musical speech patterns, and her laugh

---

[95] Mitchell, *Listener Supported*, 48.
[96] Mitchell, *Listener Supported*, 69.
[97] Howard E. Mead, Chicago, IL, September 19, 1976, *Susan Stamberg Papers*. Series 6, box 49. Fan Mail, 1972–7. Library of American Broadcasting, College Park, MD.
[98] Edwin R. Baker, June 1, 1996, *Susan Stamberg Papers*. Series 7, box 51. Fan Mail. Library of American Broadcasting, College Park, MD.

bring joy into my life."⁹⁹ Such encomiums must have gratified NPR producers, who had released a glossary of terms as part of a training handbook in the 1980s, that included this definition of "Good Tape": "Material that causes the ears to perk up, the eyes to open, the back to tingle, and the voice to emit a cry of joy."¹⁰⁰

Another common theme was the shared emotional response between Stamberg and listeners and among listeners. The "warmth and bubbling enthusiasm with which you approach each interview," wrote Sharon Shipley in 1976, "is so apparent to us listeners."¹⁰¹ Like Roosevelt, Shipley understands feeling as broadly shared, "bubbling" or "rippling" as Ahmed puts it, across the social bodies. "All who hear [you] feel the emotional drive and power with which you ignite the airwaves," agreed Wallace D. Peterson in 1998, who also marveled at Stamberg's access to the "sotto voce inner-truth we all occasionally hear" and confirming for her "that you have hit your audience dead center in their emotions."¹⁰²

The implicit and at times explicit message was that the job of delivering the news was at least as much about managing the bubbling, rippling emotional effects of world events as they crashed over listeners. "When our minds overload with empathy and hurt... how can we cope?" asked RH in 1994. "When you're not there," RH continued, referring to the fact that by the mid-1990s, Stamberg was only an occasional presence at the anchor's desk, "the disasters always seem worse and the triumphs less triumphant."¹⁰³

Appreciation for Stamberg's voice as an affective feature of listening to the news can be traced back to her heyday as *All Things Considered* anchor in the 1970s when the very fact of a woman's voice delivering the nightly news represented a milestone for

---

⁹⁹Arlie Anderson, Rochester, June 18, 1999, *Susan Stamberg Papers*. Series 7, box 51. Folder 6. Fan Mail. Library of American Broadcasting, College Park, MD.
¹⁰⁰ATC—Handbook May 1989, *Susan Stamberg Papers*. Series 2, Subseries 3, Box 5. Library of American Broadcasting, College Park, MD.
¹⁰¹Sharon Shipley, Maryvale Missouri, September 10, 1976, *Susan Stamberg Papers*. Series 6. Box 49. Fan Mail, 1972–7. Library of American Broadcasting, College Park, MD.
¹⁰²Wallace D. Peterson, Santa Monica, CA, May 15, 1998, *Susan Stamberg Papers*. Series 7, box 51. Folder 6. Fan Mail. Library of American Broadcasting, College Park, MD.
¹⁰³RH Downers, Grove IL, Fall 1994, *Susan Stamberg Papers*. Series 7, box 51. Fan Mail. Library of American Broadcasting, College Park, MD.

women's rights. Lauren Matho, writing in 1974, credited Stamberg with sustaining her on long car commutes, through "the chauvinism of law school" and the news of the day: "I would hear your voice and be consoled."[104] "I've never yet tuned in to *All Things Considered* even on a terrible news day," Betsy Frier wrote to her in 1976, "when the world was literally falling apart, that you didn't make me feel better."[105] Many fan letters to Stamberg confirm the research on the car radio as an emotional regulator, an empathy machine: "Tuesday's program had me at one moment needing to pull off to the side of the road to contain my tears," wrote Bob Ilchik in 2002, "and then not even halfway to my destination it had me laughing and grinning." Ilchik was moved first by Stamberg's account of Palestinian "misery and humiliation," then by her "teasing levity" when discussing the Bush Administration's color-coded terror alert system.[106]

Indeed, at times the news itself was less important than the sound of her voice: "Saturday morning, I got into my car, NPR came on the radio and I heard your voice," Cynthia Williams wrote in 2006 on the day of her diagnosis with cancer, "I don't remember the subject of your report, but somehow hearing your voice made me feel just a few steps closer to normal. Your wonderful voice had been part of so many of my mornings that hearing you allowed me to believe that life could go on, would go on, despite my diagnosis."[107] It is important to recall that as early as 1979, a young Ira Glass was working behind the scene, cutting tape, for Stamberg among others. The emotional power attested to by Stamberg's audience in her fan mail lay in the emotional approachability in her voice, an attribute that Glass has claimed for his own show and while also crediting his NPR mentors.

---

[104]Lauren Matho, Portland, OR, April 30, 1974, *Susan Stamberg Papers*. Series 6. Box 49. Fan Mail, 1972–7. Library of American Broadcasting, College Park, MD.
[105]Betsy Frier, correspondence. December 1976, *Susan Stamberg Papers*. Series 6. Box 49. Fan Mail, 1972–7. Library of American Broadcasting, College Park, MD.
[106]Bob Ilchik, Tucson, AZ, March 13, 2002 (email), *Susan Stamberg Papers*. Series 7. Box 51. Fan Mail, 2002–9. Library of American Broadcasting, College Park, MD.
[107]Cynthia Williams, Fruitland, MD, October 5, 2006, *Susan Stamberg Papers*. Series 7. Box 51. Fan Mail, 2002–9. Library of American Broadcasting, College Park, MD.

Driveway Moments

With this history in mind, consider the "driveway moment," NPR's bespoke term for the archetypal scene of public radio reception in which the listener-as-motorist is imagined to be spellbound by the act of listening at the end of the commute home. This ritualized reception, a moment of non-productive pleasure, and the sense of time stolen from economic and domestic roles, gradually became a potent symbol of public radio's artistic merit and its restorative spiritual powers starting in the 1990s when many listeners were documenting such moments in fan mail to Stamberg.

It was during this time that *Your Radio Playhouse*, the original title of *This American Life*, first hit the airwaves, bringing back to the airwaves a newfound commitment to appointment listening, rooted in long-form storytelling and designed to produce arresting moments for motorists and other radio listeners. Glass often exhorted listeners to drive when listening to his show: "Radio is for driving," he proclaimed frequently with gnostic assurance.[108] The history of the driveway moment doesn't directly acknowledge the role of *This American Life* in producing this kind of listening, in part because it wasn't produced or distributed by NPR. (It was, however, carried on NPR member stations around the country starting in 1996.) It does acknowledge it, more obliquely, by bestowing the honorific of "driveway moment," to several stories and storytellers associated with the program.

The development of the driveway moment from unspoken theme to overdetermined commodity is an apt model of Raymond Williams's concept of the "structure of feeling" as a vivid but mostly unacknowledged cultural force, "taken to be private, idiosyncratic, and even isolating," though "in analysis, it has its emergent, connecting, and dominant characteristics."[109] Lingering in the driveway, prolonging the liminal space/time of driving between work and home, the archetypal NPR listeners are caught up in the matrix of narrative and feeling, solitude and community, mobility and stasis. Heeding the imperative to "only connect"

---

[108]Liv Combe, "Off the Cuff With Ira Glass," *The Oberlin Review*, September 23, 2011, https://oberlinreview.org/3437/news/off-the-cuff-with-ira-glass/.
[109]Raymond Williams, *Marxism and Literature* (Oxford: Oxford University Press, 1977), 132.

by sitting alone in a car, moments of empathy in sonic cocoons epitomize the ambivalent publicness at the heart of the public radio structure of feeling. By the mid-1990s, public radio had become the quintessential cultural institution of the professional managerial class, "societally conscious, highly educated" and increasingly, acutely self-conscious; the perfect moment for a new kind of radio storytelling.

By the early 2000s, it had become a marketing concept, useful for selling collections of audio features on CD and buzzy shorthand useful for pledge drives, those seasonal appeals for listener donations to local NPR affiliates, likely as familiar to Americans as they are foreign to radio listeners across the rest of the globe. Public broadcasting, it should be remembered, means something different in the United States than in most countries: more a public/private partnership drawing funding from three main sources: The first was an anemic and routinely threatened federal commitment funneled through the now-defunct Corporation for Public Broadcasting (CPB), an ostensibly neutral filter separating Congress from the creative and editorial functions of public radio and television alike. The second was comprised of a variety of corporate underwriting deals, in which genteelly enunciated "brought to you by" on-air credits constituted good-will advertisements for consumer brands, philanthropic endowments, and industry trade groups. The third source was "listeners like you" who were urged in on-air pledge drives to chip in with membership donations to support their local NPR affiliates, who in turn purchased programming from networks, chiefly NPR, but also the PRX and the BBC.

The driveway moment is another chapter in the longer history of the car radio as emotional transportation discussed above. The pleasure of listening to radio voices in a parked car was noted long before NPR seized upon the term. Saul Bellow waxed nostalgic about the cars pulled over, bumper to bumper, in the 1940s along Chicago's Midway to listen to one of President Roosevelt's Fireside Chats, an image I borrowed to evoke a national radio public made up of intimate moments of shared reception in an earlier book.[110] Contemporary accounts have likewise emphasized the car

---

[110]Loviglio, *Radio's Intimate Public*, 1.

radio's potential for emotional transportation through music, a compensatory gift that exceeded the literal mobility of a moving car:

> We would go out to the garage—winter or summer, it didn't matter—open up the garage doors and turn on the car radio… We listened to the "travelin' songs"… We "rode" in the car… We still got home after dark, but we'd never been anywhere but out to the garage.[111]

The advances in automobile interiors, along with low reverberation asphalt, noise-canceling sensors, and high-fidelity speakers helped to make the car an ideal listening space, but these technologies alone cannot explain the pleasures attested to in these accounts. A listener in one study reported listening to his radio in his parked car at home, not for its sonic isolation, but for its publicness: "I wind the windows down so I can hear what's going on."[112]

For public radio, the driveway moment invited a kind of listening worthy of virtuosic sound production and intentional rather than distracted listening of news and human-interest stories exhaustively reported, carefully recorded, and lovingly edited and engineered to take full advantage of the audio medium. The network's reputation for meticulous attention to sound quality, their "secret sauce," was much admired (and lampooned), and resulted in an unmistakable sonic identity that can be quickly recognized when turning the radio dial.[113]

While these efforts contributed to NPR's distinctive "stationality,"[114] they also can be understood within the broad framework of the "soundscape of modernity," in its preoccupation with artificially dampening of reverberations and "technical mastery" of the sonic environment.[115] NPR's reporters, hosts and

---

[111]Philip Jeter, "Enter the Forties: Riding on Fumes and Traveling with Talk and Tunes," in *Moving Sounds: A Cultural History of the Car Radio*, eds. P. Johnson and I. Punnett (New York Peter Lang, 2018), 66.
[112]Michael Bull, "Soundscapes of the Car: A Critical Study of Automobile Habitation," in *Auditory Culture Reader*, eds. Michael Bull and Les Back (Oxford: Berg, 2003), 357.
[113]Ragusea, "A top audio engineer explains."
[114]Golo Föllmer, "From stationality to radio aesthetics," in *Sound as Popular Culture: A Research Companion*, eds. J. G. Papenburg and H. Schulze (Boston, MA: MIT Press, 2016), 306.
[115]Emily Thompson, *The Soundscape of Modernity: Architectural Acoustics and The Culture of Listening in America, 1900–1933* (Cambridge, MA: MIT Press, 2002).

anchors were instructed to speak very close to the microphone (but not too close) and, above all, to be themselves, although sometimes that required corrections in how they pronounced their own names.[116] Driveway moments, claimed longtime NPR producer Jonathan Kern, "are not born, they're created."[117]

Testimonials to the moving tug of radio narrative, even after the car has stopped moving, have been featured in the marketing strategies of other national broadcasters.[118] The driveway moment also anticipated the kind of reception practices that we've come to associate with podcasting: solitary, intensive, targeted rather than broadly public, and foregrounded as a primary activity, rather than as mere accompaniment to travel. Featured voices on the NPR web page dedicated to the "special series" of driveway moments include favorite *This American Life* alumni as well as Robert Krulwich of *Radiolab*, The Kitchen Sisters, Jay Allison of *This I Believe*, and Adam Davidson of *Planet Money*.

The *Driveway Moments* CD collections, first issued in 2003, promised "literate, intelligent, mirthful, and moving" stories. The stories, presented by comedians, radio diarists, and singer-songwriters, were advertised as quirky, spellbinding, and above all, emotional. The first collection included David Sedaris's *Santaland Diaries*, along with an interview with animal scientist Temple Grandin about her work in designing ethical slaughterhouses. Composer Charlie Barnett waxed nostalgic in another about a cross-country hitchhiking trip in 1974 over a plinking banjo and fiddle score. The tagline for the series, "Radio stories that won't let you go," was a play on the concept of a parked car and a promise of the emotional tug of narrative.

Over seventy stories, gathered from 2003 to 2009, an era in which NPR began to reposition itself from a premiere broadcast journalism network to a transmedia brand, are gathered under

---

[116]Mark, Jerkowitz, "Jane Christo: The aloof, devoted, intense, insensitive, shy, tyrannical perfectionist behind WBUR," *The Boston Globe*, December 4, 1997, D1-D5.
[117]Liane Hansen, "Crafting Driveway Moments," *Weekend Edition Sunday*, NPR, July 20, 2008, https://www.npr.org/2008/07/20/92716706/crafting-radios-driveway-moments.
[118]Andrea Gallacher, "Radio Feature: Dead or Alive?," in *The Routledge Companion to Radio and Podcast Studies*, eds. Mia Lindgren and Jason Loviglio (New York and London: Routledge, 2022), 245.

the Driveway Moment moniker; importantly, this designation was conferred after the initial broadcasts, to retroactively mark the stories as commodities that can be listened to multiple times, given as gifts, and collected. Many of the stories centered on the nuclear family and the gothic mysteries attending their dissolution and decomposition, also as we will see, a dominant theme for early *This American Life* episodes. These collected driveway moments represented the network's self-conscious effort to identify and market the network's affective power on the listeners, movement through time and space, simultaneously arresting and moving. You can hear in these stories the hard work of imagining the soundscape of a new century coming into being, not with a roar but in the dulcet tones of empathic storytellers.

And, because the period from 2003 to 2009 coincided with the worst years of the US wars in Iraq and Afghanistan, many of the stories relate small moments of human connection amidst the horrors of war. One story, told by a former prisoner, of the love he and his fellow inmates developed for a stray cat that wandered into the prison yard, makes the point that "caring makes us human," a sentiment very much in line with the promises for enriched emotional life, and the promise of empathy in particular, at the heart of public radio and its podcast progeny. By 2014, NPR's new hit podcast *Invisibilia*, co-hosted by *TAL* alum Alix Spiegel, hailed new listeners with the tagline "Listen. Feel. Different," an injunction that seemed very much in the spirit of Siemering's Founding Purposes document of nearly a half century earlier. That same year, Meta's purchase of Oculus VR for $2 billion was hailed by Mark Zuckerberg as the fulfillment of "a dream—you're actually present in another place with other people."[119]

Public Radio and Branding Feeling

The year 2003 was also the year of the Joan Kroc gift to the network. Kroc, heiress to the McDonald's fortune, gave over

---

[119]Mark Zuckerberg, quoted in Brooke Belisle, and Paul Roquet, "Guest Editors' Introduction: Virtual reality: immersion and empathy," *Journal of Visual Culture* 19, no. 1 (2020): 3–10, https://doi.org/10.1177/1470412920906258.

$200 million to NPR, a game-changing bequest for the network that had always lived in a precarious relationship to finances. The gift made headlines partly due to its size but also because no two organizations at the turn of the century more iconically evoked opposite demographic groups and cultural meanings than McDonald's and NPR. In some ways, this realization also helped to clarify the concept that NPR was, in fact, "a brand," the first in a new wave of audio narrative brands, including *This American Life,* an important step in its transformation from radio show to podcast to franchise.[120] What seemed like a comically unlikely pairing of McDonald's and public radio in 2003 became a new commonsense understanding of the influence of narrative forms on a generation of audio storytelling by 2025 when scholars Sharon and Johns averred, "*This American Life* is the McDonald's of podcasting."[121]

Feeling as a balm to individual as well as collective hurts was a reliable current running through fifty years of public radio's most successful innovations. Such stories were designed to make listeners "feel good" in more than one sense of the word. In 1998, Glass declared public radio's mission: "to tell us stories that help us empathize." One early *This American Life* fan described listening as utterly pleasurable, "it feels like you're doing something good, staying in tune with the world, in the tiniest way possible."[122] In 2004, McCauley described NPR as a balm to listeners whose "idealistic dreams of community were shattered in the 1970s."[123] A 1988 audience research study commissioned by NPR had already confirmed this insight, using "psychographics" to argue for a "unified" programming sound "targeted" to the psychological predilections of their most coveted listeners.[124]

---

[120]Eleanor Patterson, "This American Franchise," 458.
[121]Tzil Sharon and Nicholas John, "Talking the talk: A Conversational Cross-Cultural Analysis of a Podcast Story Told to Three Different Audiences," in *Podcast Studies: Practice into Theory*, eds. Lori Beckstead and Dario Llinares (Waterloo: Willfrid Laurier University Press, 2025), 163–4.
[122]Kristine Johnson, *Imagine This: Radio Revisited Through Podcasting*, MA Thesis (Fort Worth, TX: Texas Christian University, 2007), 36–7.
[123]McCauley, *NPR*, 2.
[124]S. Thomas, J. Thomas, and Theresa R. Clifford, "Issues and Implications," *Audience 88* (Washington, DC: Corporation for Public Broadcasting, 1988), 26–7.

CHAPTER TWO

Voracious Voyagers: NPR Listens

Among the oldest tropes in radio history is the question: *Is anyone out there?* Because of the vicissitudes of weather, the physical properties of radio waves skimming across the ionosphere, the amateur quality of receivers and transmitters, and mostly because the signals were *cast broadly*, it was impossible to know in radio's early experimental period how many listeners were out there or where they were. The very first amateur broadcasters, known as DXers, took to their homemade sets in the 1910s with appeals for some kind of response from their far-flung listeners. Listeners, who likened tuning in to these amateur signals to "fishing," sent postcards to DXers letting them know the date, time, and location of the broadcasts they had "pulled in." Audience research was radio's earliest impulse and its very first format.[1]

A century later, despite the revolutions in telecommunications, computing, and market research, reliable information about radio's audience, and that of streaming and podcasting for that matter, still seemed aspirational.[2] Improved technology for tracking listeners had been outmatched by an explosion of ways to listen. Podcasts provided firmer numbers for downloads only, but not for actual

---

[1] Susan Douglas, *Inventing American Broadcasting, 1899–1922* (Baltimore: Johns Hopkins University Press, 1989).
[2] Brian Barletta, "The Measurement Excuse," *Sounds Profitable: The Business of Podcasting*, Podcast, March 21, 2023.

listening. Personal People Meters (PPMs) for audio signals couldn't always distinguish intentional listening from the incidental kind and they didn't really work with headphones, which largely defeated the purpose. Remarkably, into the twenty-first century, audience researchers still called people at random on the phone or asked a sample of people to fill out daily listening diaries, methods that date back to the 1930s.[3] Radio historians, steeped in the ethereal romance of the medium, may be forgiven for taking some comfort in the persistence of the phantom quality of its audience.

NPR's engagement with audience research began early and gathered strength in the mid-1980s and early 1990s, as the network adjusted to its increased reliance on corporate underwriting and listeners' donations after the 1983 fiscal crisis.[4] Faith in market-based research was met with scepticism by many journalists and station program managers, who subscribed to a more intuitive sense of the relationship between public radio and its listeners.[5] This struggle, between "speaker-centered" and "listener-focused," pitted two approaches to public radio that drew strength from the same ideological sources—liberal notions of public service—and expressed themselves in the idiom of emotional connection as the key to successful public radio. In the end, both sides in this debate arrived at the same basic conclusion: public radio was for listeners pretty much like the people who made public radio (i.e., educated, curious, and affectively liberal).

Public radio producers researched their audiences since the on-air light first lit up at NPR in 1971, and here too, satisfying answers have often proved elusive or temporary. But these efforts do tell us quite a bit about what sort of audiences have been imagined, desired, and fretted about. Listening for a public radio audience has proven to be noisy work, producing its own formats, anxieties, and fantasies. If the commercial radio broadcasters' besetting question since 1920 was, "Who's out there?" then public radio producers' parallel question was, "Who is the public?" Questions about the

---

[3]Hugh Malcolm Beville, *Audience Ratings: Radio, Television, and Cable* (Hillsdale, NJ: Lawrence Earlbaum Associates Publishers, 1988), 4; David Giovanonni, Leslie Peters, and Jay Youngclaus, *Audience 98: Public Service, Public Support*, Leslie Peters, ed. Audience Research Analysis, 165–6.
[4]Jack Mitchell, *Listener Supported: The Culture and History of Public Radio* (New York: Praeger, 2005), 120–1, 133–4, 146–7.
[5]Mitchell, *Listener Supported*, 121.

"phantom public" have haunted democratic theory for more than a century and the nature of radio's service to "the public interest, convenience, and necessity," has been the subject of contentious debate since at least 1927.[6]

The history of audience research has been animated by similar questions and provided answers that were accurate enough to drive programming and investment decisions, yet provisional and hesitant enough to require regular reiterations and ever-newer models. In a 2006 audience research report for NPR, researchers insisted that success in programming and fundraising required a "consistently defined listener."[7] Of course, the audience and market research industry that grew up in and around NPR, with its ever-evolving methods and psychographic categories, meant that definitions of listeners and methods of measuring and describing them would take different forms.

For public radio, finding its public meant negotiating tensions and contradictions at the heart of the country's complicated relationship to the notion of the public good. "There is no one universal definition of public radio," bemoaned one industry insider in 1979, a full decade into its modern era. "It almost does not exist as such."[8] This uncertainty helps to explain public radio's history of crisis, detailed above, but it also speaks to a historical legacy of confusion that predates the creation of NPR and the CPB. The research and arguments for the need for public broadcasting that led to the 1967 Public Broadcasting Act acknowledged that a purely commercial system failed to adequately serve the public interest. Such arguments are as old as broadcasting, with roots in the 1920s, as Robert McChesney has documented.[9] They took on

---

[6]Walter Lippmann, *The Phantom Public* (New Jersey: Transaction Publishers, 1925); John Dewey, *The Public and Its Problems* (New York: Holt, 1927); Bruce Robbins, *The Phantom Public Sphere* (Minneapolis: University of Minnesota Press, 1993); Robert W. McChesney, *Telecommunications, Mass Media, and Democracy: The Battle for the Control of U.S. Broadcasting, 1928–1935* (London: Oxford University Press, 1995); Josh Shepperd makes the case that the struggle for noncommercial radio and the birth of modern US media studies were closely intertwined, part of a shared struggle to "increase democratic participation," Josh Shepperd, *Shadow of the New Deal: The Victory of Public Broadcasting* (Champaign: University of Illinois Press, 2023), 174.

[7]*Audience 2010* Report 6 May 9, 2006, 12

[8]Larry Josephson, "Why Radio?" *Public Telecommunications Review* 7, no. 2 (March/April 1979): 6–18.

[9]McChesney, *Telecommunications, Mass Media, and Democracy*.

new life in the 1960s, after decades of commercial broadcasting created what FCC chairman Newt Minow characterized as a "vast wasteland." If commercial broadcasting could not adequately serve the public good, what specifically would a public broadcasting service contribute?

The answers were themselves contradictory and left the public broadcasting system with an impossible, and as we've recently seen, ultimately doomed task. Public television and radio would serve "minority" audiences of very different sorts: the well-educated elites turned off by low- and middle-brow commercial dreck, racial minorities whose lives and experiences were completely excluded, and children for whom quality educational programming was rare. Finding a coherent identity for underfunded, single-channel national radio and TV systems would be difficult, if not impossible.

In Jack Mitchell's history-cum-memoir, *Listener Supported*, the idea of a truly catholic schedule of public-serving programming is gently mocked as a temporary and youthful folly: "NPR thought we could reach many minorities who'd be patient to listen to each other," he noted dryly.[10] As soon as public broadcasting was understood to serve only small fractions of the larger public, it lost a truly public constituency and thus the kind of coherent mission and robust political constituency enjoyed by more broadly public institutions. With this tenuous tether to the public good and the federal funds that undergirded it, public radio would have to find new forms of cultural authority and new forms of revenue through its audience.

What emerged from this besetting contradiction was a decades-long struggle to straddle almost impossible divides between different minority audiences and between the universalist meanings inherent in the concept of public and market-based approaches to audience. Finding the sweet spot between the highly educated elite and racial minorities was not possible so long as both were considered equally important. NPR and many of its member stations chose to court the former through empathetic representations of the latter. "We needed a service that focused on the societally conscious," Mitchell noted, "not a service that was societally conscious."[11]

---

[10]Mitchell, *Listener Supported*, 55.
[11]Mitchell, *Listener Supported*, 184.

In order to speak coherently and earnestly *to* social elites *about* socially marginalized people, questions of race, class, and ethnicity would have to be obscured in favor of a new vocabulary of "values, interests, and beliefs."[12] Questions of politics would have to be reframed in the idiom of feelings in general, and empathy in particular. Building a science of audience measurement and fundraising within the framework of public service required more than simply borrowing market research methods of commercial broadcasters; new instruments, more subtle and penetrating, had to be devised to help public radio find its public.

In this way, NPR and the CPB turned to psychographics, an emerging field dedicated to the study and exploitation of taste cultures and to recasting the meaning of the word "public" from civics to affects. Audience research activities of NPR redefined the meaning of "public service" from an imperative about its reach, accessibility, and utility to more market-friendly notions of quality, emotional impact, and customer satisfaction. Public service, in this definition, could be measured in donations, time spent listening (TSL), and in the more amorphous metrics of affective investment. Listeners would have to feel their way into this exclusive new public. If the public no longer referred to a theoretically universal audience, perhaps it could be understood as a set of aesthetic and genre qualities, found in the style and substance of the programming itself and echoed in the emotional connection on the other end of a million radio receivers.

But what were those qualities associated with this new public? More than three decades of public radio research emphatically provide a single answer: education. Listening to public radio was highly correlated to high levels of educational attainment. This answer served a multitude of purposes, not the least of which was obscuring the race, class, generational, and regional hierarchies overlapping with education level. It also served as a flattering portrait of an audience whose close identification with public radio paid off twice: once through voluntary donations and again through underwriting dollars from corporate sponsors eager to get their names in the ears of a coveted demographic. The highly

---

[12]S. Thomas, J. Thomas, and Theresa R. Clifford, "Issues and Implications," *Audience 88*, Washington, DC: Corporation for Public Broadcasting, 1988.

educated audience reflected its status back on the network and its programming, leveraging state and federal funds.

Research and anecdotal data provided public radio producers with additional answers to the question of who was listening. Public radio attracted listeners who were not only highly educated, but also "inner-directed and societally conscious," per one study. The surfeit of descriptions for listeners stemmed not from their ubiquity but instead from the uncomfortable realities of public radio's narrow demographic niche and its impossible political relationship to sustained, disinterested federal (that is, public) funding. Of course, the audience research industry was well compensated for updating the research periodically with catchy new formulations for capturing some of the enduring contradictions facing public radio. Meanwhile, the political economy of neoliberalism was becoming more entrenched, choking off alternative ways to imagine national institutions and public life as Lisa Duggan and Wendy Brown have demonstrated.[13]

The notion that public radio needed sophisticated audience research was a challenge to a more intuitive approach to making radio, one that assumed an easy equivalence between listeners and producers, the public and the makers of public radio. This speaker-based approach, as Stavistsky has called it, developed over decades in the local, community-based educational radio stations scattered around the country. It drew inspiration from the personal affective appeal created by rock-and-roll DJs throughout the 1950s and by the politics of amateurism and intuition of the 1960s counterculture. The notion of radio as a "tribal drum," per Marshall McLuhan, resonated for a generation of radio producers for whom the mass media seemed to enact a "retribalization" of human civilization. Improvisation, not research, was the preferred method in this approach.[14]

The public in public radio, in this residual model, referred to listeners and producers alike. For public radio producers, anchors, and hosts, this sense of an essential equivalence between speaker

---

[13]Lisa Duggan, *The Twilight of Equality? Neoliberalism, Cultural Politics, and the Attack on Democracy* (Boston, MA: Beacon Press, 2004); Wendy Brown, *Undoing the Demos: Neoliberalism's Stealth Revolution* (New York: Zone Books, 2025).
[14]Susan Douglas, *Listening In: Radio and the American Imagination* (New York: Times Books, 1999), 25.

and listener was deeply ingrained in the culture. As founding *All Things Considered* producer Jack Mitchell said from the start, "the listeners we attracted were pretty much like us."[15] "NPR news," another former public radio producer, McCauley stated, was "made by people like me for people like me."[16] NPR's voice consultant David Candow enjoined reporters to imagine they're speaking "to a trusted and respected friend on the air."[17,18] Linda Wertheimer described NPR's debut as "the very beginning of a conversation with this country," and saw her role as "listening to America," as much as talking to it.[19] Cokie Roberts described her career at NPR as "forty years of... connection and friendship with listeners."[20]

The notion that public radio was made by and for the same sort of people, highly educated and socially minded, proved to be a powerful and useful one through the years of political and financial turmoil and periods of vertiginous audience growth. And it provided a handy frame for interpreting the often confusing and even contradictory data produced by audience researchers. As public radio sought to chart a new path in between the market rationality of commercial sponsorship on the one hand and the public service model of full government support on the other, it had to imagine a new kind of public too and a new way to talk to it. It also had to figure out how to listen to it.

Pressure was exerted on the "speaker-centered" approach to programming by the growing ubiquity for market research across all sectors of broadcasting, along with the uncomfortable awareness that the upper-middle-class Americans on either side of the public radio signal were not representative of the entire US public. The nagging concern that public radio was out of touch with the public that it was supposed to serve proved an equally powerful force on programming. Robert Siegel's concern that NPR failed to "describe

---

[15]Mitchell, *Listener Supported*, 145.
[16]Mitchell, *Listener Supported*, 114.
[17]Farhi, "When This Guy Talks."
[18]Mitchell, *Listener Supported*, 145.
[19]Linda Wertheimer, *Listening to America* (New York and Boston: Houghton Mifflin, 1995), xvii.
[20]Cokie Roberts et al., *This Is NPR: The First Forty Years* (San Francisco, CA: Chronicle Books, 2010), 9.

the country that voted for Ronald Reagan," mentioned earlier, was only one of many instances in which this form of self-doubt animated the editorial processes within public radio institutions.

The idea that listeners and producers were the same sort of people, and the spasms of self-doubt that call this idea into question, map nicely onto Barbara Ehrenreich's account of the "retreat from liberalism" carried out by the professional middle class (PMC), including importantly, those who worked in the media industries. In *Fear of Falling: The Inner Life of the Middle Class*, Ehrenreich makes the case that this rightward turn was "an episode in the life of the middle class, a change of mind, a shift in consciousness" begun in the 1970s, a shift that has been described elsewhere as the birth of neoliberalism.[21] It follows from a moment of class consciousness in which the PMC understands itself as a kind of "elite" class, but thanks to its lack of inherited wealth, the perceived power of liberal economic policies, and the rigors of the "meritocratic" capitalist system, an imperiled one. "This emerging self-image," argued Ehrenreich, "led to the adoption of the kind of outlook appropriate to an elite, which is a conservative outlook, and ultimately indifferent to a non-elite majority."[22] The concern that the liberalism of America's media and government was out of step with the "real America" proved a powerful psychological and rhetorical lever to prise open a new kind of politics.

Projecting its own insecurity onto the mythic "working class" became a way to dress up a new-found illiberal selfishness in the unpretentious garb of Middle Americans. In other words, an emergent elite class, newly conscious of itself and including a cadre of professional communicators, squinted hard enough to see its own reflection in a mass of average Americans on whose behalf it claimed to speak. In place of the "libidinal" spirit of 1960s liberalism rose a conservative superego, an impulsive policing of generosity that betrayed a "softening" of American and individual resolve. This displacement became a way for the affluent, highly educated PMC to project a certain ambivalence about its own "softness" onto the liberalism that it was retreating from, according to Ehrenreich.

---

[21]David Harvey, *A Brief History of Neoliberalism* (London: Oxford University Press, 2007).
[22]Barbara Ehrenreich, *Fear of Falling: The Inner Life of the Middle Class* (New York: Pantheon Books, 1989), 11.

Meanwhile, in a parallel displacement, public radio professionals strained to hear their own voices echoed in those of their listeners, setting up a productive tension with a vast unserved American public just out of reach. Public radio's rightward pivot, which I have argued is a habitual impulse and defining ritual, was born in the historical moment Ehrenreich describes of the PMC's birth of self-awareness as an imperiled elite beginning in the 1970s and 1980s.

As political winds shifted back and forth over the next several decades, public radio continued to understand its relationship to its audience and to its claims of publicness as similarly imperiled. By aligning itself with an audience defined by its elite educational status, public radio institutions like NPR and programs like *This American Life*, were doomed to repeat a pattern governed by a fear of falling: to preserve itself, public radio must be simultaneously an elite service and one that represents an imagined American public lurking in the blue-collar shadows of its conscience. In what follows, I will provide a history of public radio's ambivalent search for its own audience against the backdrop of Ehrenreich's analysis of how *feeling liberal* drove the PMC's rightward turn politically and inward turn socially.

"Premeditated Elitism"

Because of the structural contradictions built into public radio, the idea of a truly representative public audience was almost immediately abandoned in favor of reaching out to a more narrowly defined community along the lines of education, culture, and social class. Indeed, the Carnegie Commission on Educational Television, mentioned in the previous chapter, already identified the inevitability of a "minority" audience for public broadcasting before the Public Broadcasting Act was drafted.[23] The experiences with programming, fundraising, and audience measurement of the handful of powerful university-owned (e.g., WHA in Madison, WI) or municipal-owned (e.g., WNYC in New York) "educational" stations suggest that as

---

[23] Carnegie Commission on Educational Television, "Public Television: A Program for Action," Carnegie Commission on Educational Television, 1967.

early as 1952, noncommercial stations dedicated to public service "did not attempt to serve all of the people all of the time, but rather to serve some of the people especially well all of the time."[24]

Public broadcasting, coming into nation-wide distribution fifty years into the commercial broadcasting era could only be compensatory and alternative, an implicit critique of the mass-mediated public constructed by the for-profit networks on radio and television. If the "public" in "public radio" was going to mean anything, it was going to have to be a sensibility, a structure of feeling for an imagined community of like-minded people who understood themselves to be distinct. This meant representing an educational elite and racial, national, and other minorities. This dual constituency formed the public of public media, imperfectly and asymmetrically, informing research and storytelling, underwriting and empathy, ever since. Negotiating this uneven and unstable dual constituency proved to be highly productive, generating audience research, industry histories, and insider memoirs, which celebrated and fretted about the limits of the public radio audience in equal measure.

The search for public radio's audience, like the PMC's search for a mythical "middle America," was at one level an emotional journey. In both accounts, the dominant mode of perception was described in *feeling* rather than *cognitive* terms. For Ehrenreich, a mass-mediated "fear of falling" structured the rightward pivot of society in the 1970s. For McCauley, emotional succor for "shattered dreams" lay at the heart of listeners' affection for public radio. Audience research, a business of numbers, would have to find an emotional register to communicate its findings and to argue for its own necessity.

Public radio's search for its audience evolved over the years from simple numbers, to demographics to "utiligraphics"[25] to "psychographics"[26] to confessional yarns. Yet, there has been a circular quality to these efforts underscoring the supplicant quality of radio's primal question: "Is anyone out there?" Throughout these iterations, the notion of public radio's public remained vexed. In

---

[24]A. G. Stavitsky, "'Guys in Suits with Charts': Audience research in U.S. public radio," *Journal of Broadcasting & Electronic Media* 39, no. 2 (1995): 177–89, https://doi.org/10.1080/08838159509364297.
[25]Stavitsky, "Guys in Suits with Charts."
[26]Stavitsky, "Guys in Suits with Charts."

each era, the search for listeners provided novel ways to understand dueling notions of what kind of public it served. Along the way, public radio audience research gave voice to the phantom public haunting radio broadcasting since its inception.

Throughout the 1970s, local NPR affiliates and smaller stations aspiring to that status felt growing pressure to use audience research to guide programming. The arrival of Tom Church, formerly of Arbitron, to CPB's audience research department in 1976 was a watershed moment in the shift to "audience-centered" programming. Under his leadership, audience research activities expanded across the industry, an approach that was in direct conflict with the philosophy of many station managers across the country who saw the relationship between their stations and the public they served in more intuitive, local terms. As members of a community, they reasoned, public radio stations were well positioned to serve that community in its diverse and particular ways. Or, as one station employee put it at a 1978 conference after sitting through a presentation by Church: "Arbitron is bullshit!" But the forces inside the CPB pushing for better audience data, "guys in suits with charts," in Alan Stavitsky's memorable phrase, were relentless. Their Office of Communication Research funded eight seminars across the United States between 1978 and 1981 on "Public Radio and the Ratings," to bring station managers on board. They also provided a forum for research proponents to argue that conducting research did not in itself compromise a public station's mission.

It was, according to Stavitsky, a kind of war of attrition with one side grinding down the resistance of the other.[27] At odds were two different conceptions of public radio's mission and mandate. In 1976, Larry Josephson called public radio "a monastery of liberal humanism in the dark age of mercantilism... a conservatory of diverse minority cultures... and the methadone clinic of the 'news junkie.'"[28] In his defense of public radio's inherent localism, he captured the romantic conception of public radio exceptionalism that was under siege by a growing CPB commitment to the nationalizing force of audience research:

---

[27]Stavitsky, "Guys in Suits with Charts," 5.
[28]Josephson, "Why Radio?" 6–18.

In northern Minnesota, public radio is reflected off the Iron Range, amplifying the populist voice of community: miners, mavericks and timberwolves (KAXE).... In Ramah, New Mexico, public radio speaks Navajo (KTDB); in Santa Rosa, California, Spanish (KBBF); in Warrenton, North Carolina, it speaks with a rural black accent (WVSP); in Washington, D.C., in the jazz idiom of the urban black experience (WPFW), as well as in the Appalachian twang of bluegrass (WAMU) and in the "high church" unction of classical music.

(WETA-FM)

An argument made in poetry, and a riposte to the turgid prose of audience surveys. The notion that audience research and other market-driven methods were inadequate to the task of capturing public radio's distinctiveness, lyricism, and multivocality stems from William Siemering's Founding Purposes document, which sought the many voices and dialects of the nation, and which encouraged the participation of local member stations in the creation of the national network sound.

CPB's investment in audience research methods benefitted from developments in consumer psychographic research like the VALS (Values and Lifestyles) methodology, introduced in 1978. Grounded in the social psychology of Reisman and Maslow, VALS proposed a novel form of market segmentation. Research focused on psychographic types tended to obscure some of the more obvious and uncomfortable class, race, age, and regional distinctions that vexed public radio's claim to publicness. This approach sprung from the same intellectual traditions as Herbert Gans's liberal, class-conscious work from earlier in the decade. VALS represented an attempt to apply the insights of cultural and class differences to the science of marketing.[29]

By the early 1980s, NPR was buying survey data directly from national consumer research firms like Simons Market Research. Audience research was beginning to drive the network's approach to programming even as its potential as a tool for underwriting increased. In the 1980s, the two objectives became strongly linked. Per Stavitsky, "The network's audience research operation became

---

[29]Ralph Engelman, *Public Radio and Television in America: A Political History* (Thousand Oaks: SAGE Publications), 116.

analogous to a commercial station's sales department."³⁰ By the time of NPR's nearly fatal financial crisis of 1983 and ensuing "drive to survive" fundraising campaign, the network understood that its future financial health depended upon robust audience research as a means of attracting underwriting dollars.

The FCC may have understood this as well; by 1984, they softened the regulations against "enhanced underwriting," which meant that corporations and foundations could purchase more than a mere mention of their names on the public airwaves. Now they could also specify what they made, did, or advocated. On the strength of these changes, and under Douglas J. Bennet's leadership, NPR saw its revenue from corporate contributions increase almost fivefold between 1983 and 1988.³¹

But most listeners still didn't donate money. As NPR lurched from crisis to crisis and gained in audience share, the need for a stable donor pool increased. David Giovannoni's landmark 1985 audience study, nicknamed "The Cheap 90," for the approximately 90 percent of listeners who don't donate (actually closer to 83%), described these listeners as "a major source of potential revenue" but ultimately argued for jettisoning them in favor of more listeners with the characteristics of those who already donated. While presented as a source for objective data, the report also demonstrated a philosophy about the inevitability of a business model in which public radio would be "weaned" off government funding, as part of a natural maturation process. It also provided a set of strategies for changing how listeners understood public radio as a crucial first step in changing their behavior. The Cheap 90 was a handbook for redefining "public" to mean "listener supported," marking a crucial shift away from tax-based funding to something closer to a subscriber model.

Getting listeners to understand that public radio was dependent upon listener support was simply a matter of communicating a fact about the emerging business model of the industry. But by suggesting that this shift was part of an inevitable process, it also meant persuading listeners to internalize the market-based logic which held that nothing should be publicly funded permanently. The

---

³⁰Stavitsky, "Guys in Suits with Charts," 8.
³¹Mitchell, *Listener Supported*, 127.

study's "utiligraphic" emphasis found that the longer one listened, the more likely one was to donate. Changes to programming could induce longer spells of listening. This required a narrow audience focus, rather than the expansive one of Siemering's original vision. Here, too, business strategies were being repackaged as a commonsense way to improve public service.

Other donor traits seemed harder to inculcate. The study found that donors tended to be older, affluent, white, and male. Giovannoni argued that such demographic traits were merely incidental to educational level, which just so happened to be skewed by age, race, income, and gender. Highly educated listeners donated more often. "To the extent that individual programs and formats... attract better educated listeners with higher incomes... they are components of broader station programming strategies which can increase audience support."[32] In this way, the Cheap 90 swept away the race and class and gender implications of its proposed strategy to focus narrowly on those who shared the traits of donors.

Part of the problem, in other words, was that the cheap 83 percent were simply the wrong kind of listeners. Getting and keeping the right kind of listeners meant catering to those who gave or who were easily persuaded to give, a narrower slice of the audience. Increasing time spent listening (TSL) required appealing ever more narrowly to the most desirable type of listener, or in the words of legendary public radio executive Jane Christo: "super-serve your super-core."[33]

For Giovannoni, increasing TSL meant creating programming that "enticed" listeners to incorporate public radio into daily routines and even to interrupt those routines: "Whenever a listener stays in his car after parking to hear the end of a piece, public radio is doing what is required to "turn listeners into contributors." With this in mind, we can regard audience research as the incubator out of which the driveway moment, as aesthetic ideal and as marketing concept, was born. One of the concluding suggestions to station managers was simply to "attract better educated listeners with

---

[32]David Giovannoni, *Public Radio Listeners: Supporters and Non-Supporters* (Washington DC: CPB, 1985).
[33]Ashley Sterner, personal communication, Baltimore, MD, 2015.

higher incomes." Giovannoni acknowledges with equanimity this tension at the conclusion of the study:

> while public broadcasts can accept their service to better educated listeners with higher incomes as a fortuitous result of their quality programming, many will find the strategy of "premeditated elitism" unacceptable.[34]

Within a few years, thanks to the influence of the VALS approach, NPR had hit upon a simple, powerful message: "Public radio's listeners are different from other listeners in their demographics, values, and lifestyles."[35]

"The fundamental proposition of AUDIENCE 88," the 1988 study that took two years to complete and set the tone for the next fifteen years, was "know your audience." It exuded a new confidence, promising "a detailed portrait of the public radio audience" that could help to "pierce the veil between the broadcaster and the listener, and to capture the clearest possible picture of the people who welcome public radio into their lives." Shifting away from lamenting the "cheap" majority who didn't contribute, it "discovered" an idealized public radio audience, "a special kind of listener at the heart of the public radio," waiting to be counted and served. It also shifted away from the complicated diplomacy required by the insight that public radio listeners spent more time with commercial competitors. Public radio listeners were, quite simply, "special."

The *Audience 88* study asserted a sharper delineation between public radio listeners and everyone else and argued for a "strategy of unified programming appeal" based on an "explicit decision to focus on a particular group of listeners ... their needs and interests." It also encouraged an integrated approach to audience research based on "in-depth reports on Underwriting, Advertising & Promotion, Programming, and Membership." In these ways, the study made the case that public radio should be approached within the framework of the commercial broadcasting with which it shared

---

[34]Giovannoni, *Public Radio Listeners*.
[35]Thomas, Thomas, and Clifford, "Issues and Implications."

bandwidth. Instead of positioning public radio as an alternative to that system, *Audience 88* declared that the commercial market for radio listeners dictated that it must aggressively court a "special niche" audience within it.

In making the case for public radio listeners' difference, *Audience 88* reads like a primer written for Ehrenreich's PMC. Public radio listeners are special, an elite by dint of education, occupation, culture, and values, with "education... at top of the list." This audience, "significantly better educated than the U.S. population as a whole," required a different sort of programming. The study maps the audience onto a hierarchy of education, described as a "continuum," which predicts how "public radio-like" they are. "The further people pursue their education," gush the authors Thomas and Clifford, "the more likely they are to pursue public radio." "Over 70 percent of public radio's heavy core listeners have graduated college, and nearly half (46 percent) went on to graduate school. This educational attainment correlates highly with income and profession."[36]

Using VALS profile typologies, *Audience 88* found that the label "Inner-Directed, Societally Conscious" was "an extraordinarily powerful predictor of public radio use." This contradictory pairing of descriptors captures the mystique at the heart of public radio's conception of itself and its audience. This double-consciousness characterizes Ehrenreich's new class: inner directed toward its sense of itself as an emergent elite and anxiously aware of its position relative to the larger social world in which it must exercise its dominance. NPR's heavy core listeners constituted an educational elite, high-achieving, highly self-actualized, and socially aware, but not in ways that might endanger their own status. Descriptions invoking these and similar qualities, rather than rigid demographic categories like race, class, region and gender, formed the "new vocabulary" that Thomas and Clifford celebrated. By this time, Stavitsky argued, audience research at NPR became "analogous to a commercial station's sales department," interested less in developing insights for programming than for underwriting and listener support.[37]

---

[36]Thomas, Thomas, and Clifford, "Issues and Implications."
[37]Stavitsky, "Guys in Suits with Charts," 8.

The study also argues for diversity across lines of race and political orientation. Despite the risible claim to racial diversity (the study admits that 91 percent of its audience is white), the study argues forcefully that *anyone* could be a public radio listener. This elision of mainstream demographics within an elite taste culture represented the difficult dance that public radio researchers and consultants felt compelled to perform on behalf of its niche audience. Focusing on the core, while acknowledging a broad diversity among all listeners, enabled the authors of *Audience 88* to navigate the tricky problem of public radio's publicness.

Emphasizing heavy core listeners and their traits was presented as a rational way to grow the audience and perhaps a marketing strategy for attracting underwriters to a coveted market. Finally, in its desire to grow the heavy core, it demonstrated the aspirational nature of these traits. Unlike demographic traits like race and ethnicity, theoretically anyone could acquire the coveted educational and social traits associated with the most public-radio-like. "In sum, while public radio serves millions of Americans from all walks of life, it speaks in an especially compelling way to a certain kind of listener."[38]

As the audience research model became more dominant, the language used to describe listeners would require endless innovation, to preserve the mystique of a social bloc simultaneously democratic and elite, public and selective. As the residual "speaker-centered" philosophy of programming continued to lose ground to a "listener-focused" approach, the intuitive spirit of the former found a new home in improvisational style, and a newfound delight in anecdote and serendipity in storytelling. It also found a welcome home in the old notion that public radio's listeners and producers were essentially alike, an idea reinforced by vocal performance styles that conveyed conversational intimacy.

Ten years later, with *Audience 98*, public radio audience research achieved a new level of sophistication and commitment to the "listener-focused" approach to programming. Drawing on survey data from nearly eight thousand listeners, Arbitron diary entries, and a new set of psychographic profiles from the VALS2 program, *Audience 98* promised to be "the most comprehensive

---

[38]Thomas, Thomas, and Clifford, "Issues and Implications."

picture of public radio's audience taken to date, recording over 200 characteristics for ... public radio listeners."[39] *Audience 98* was the most confident expression of the philosophy that Giovannoni defensively referred to as "premeditated elitism" thirteen years earlier in the "Cheap 90" study. That confidence extended not only to the principle that a highly educated audience was the proper quarry for public radio's programming, marketing, and fundraising efforts but also that such an audience was not just served by public radio but also created by it. Going forward, development professionals would be called on to earn more from listeners per listener-hour.

When attempting to metaphorically map out public radio music listeners in an imaginary town, a later study adopted its own class- and race-inflected neighborhoods with euphemistic terms like "uptown jazz" vs. "downtown jazz." The town had "leading citizens" who listened to "upstairs classical." It also had "Strugglers and Believers," two of the "less coveted" members of the public radio community, who "lived on the outskirts of town."[40] In this way, the audience research literature demonstrates the continuity between explicitly exclusionary rhetoric and what Ahmed calls "the 'ordinary' work of reproducing the nation... through the repetition with a difference, of some *sticky words* and language."[41] Class and race hierarchies "stick" to the demographic and psychographic categories in ways that "create effects."

Elsewhere, the authors of the *Audience 98* study fall back upon the argument of self-selection: like seeks like, one of public radio's oldest conceits. "We're an industry of highly educated, values driven professionals who rely on the support of highly educated, values-driven people."[42] In this equation of radio producers and radio listeners, the demographic makeup of the audience, the putative purpose of this lengthy study, can be reduced to the notion, common among those program directors most hostile to audience research in the 1970s and 1980s, that public radio producers saw

---

[39]Giovanonni, *Public Radio Listeners*; Leslie Peters and Jay Youngclaus, *Audience 98: Public Service, Public Support*, ed. Leslie Peters, Audience Research Analysis, 1998.
[40]Peters and Youngclaus, *Audience 98*, 76.
[41]Sara Ahmed, "Affective Economies," *Social Text* 22, no. 2 (2014): 21.
[42]Peters and Youngclaus, *Audience 98*, 26.

their listeners reflected back to them, like Narcissus gazing into a pool. Later, sounding a bit less patient, the authors suggest that public radio has an attitude that most Americans simply do not share. That's not good or bad. "It's just how radio works."[43]

Color-blindness, however, represents the grand overarching narrative of the report. Public radio "transcends color through its very indifference to it." Embedded in this notion that ignoring racial difference is the best way to achieve a universal audience is a set of assumptions about the racial neutrality of institutions dominated by white people. Embedded in these assumptions is the idea that white institutions' exclusivity is based on merit rather than bias.

> When we say we want a different audience, *we're really saying we want an audience that isn't so highly educated...* We may even compromise deeply held ideals and highly esteemed standards.[44]

Audience 98 is mostly sanguine about the implications of "premeditated elitism," and to the extent that it is defensive about catering to the rich, it only reinforces the idea that its desired audience, like Ehrereich's PMC, are affluent, but not securely so. And yes, those listeners must have money to spare, but this is not rich people's radio. Most gifts come from people whose annual income is modest to moderately upper middle class.[45] While arguing against wealth as a significant variable in listeners' likelihood of becoming "givers,"[46] to use the vaguely Randian term employed here, the authors of the report finally concede that 40 percent of each donation is influenced by household income.[47]

The report emphasizes the importance of "a sense of community" for the coveted educational elites. Citing "some fancy statistical footwork," it argued for the idea that "personal importance" includes the idea of community bonds. Giovannoni calls for more work on this connection, moving it "from poetic rhetoric to further, serious research."[48] The abrupt elision of the category "personal

---

[43]Peters and Youngclaus, *Audience 98*, 35.
[44]Peters and Youngclaus, *Audience 98*, 41.
[45]Peters and Youngclaus, *Audience 98*, 113.
[46]Peters and Youngclaus, *Audience 98*, 135.
[47]Peters and Youngclaus, *Audience 98*, 126.
[48]Peters and Youngclaus, *Audience 98*.

importance," which has throughout the report been linked to individualistic psychographic profiles, with a desire for community bonds speaks to the profound ambivalence at the heart of the public radio structure of feeling.

This seeming contradiction is echoed in McCauley's 2006 description of NPR listeners shattered by the failure of communitarian ideals of the 1960s turning to public radio for succor and an imagined community. These same listeners, he also argues, are not interested in the social commitments of specific political movements, only the feeling of connectedness. In this way, *Audience 98* manages to negotiate the tension of feeling liberal in neoliberal times. Insisting on finding evidence for the implicit personal desire for community and for the promise of a "portable" community, Giovannoni echoes Ahmed's insight that feelings and communities are mutually constituted. "Emotions work by sticking figures together… a sticking that creates the very effect of a collective." The report's insight that coveted listeners seek virtual community that "travels with them," points to the ways the industry had already begun to imagine the audiences that would move so eagerly into podcasts in the next two decades.

Meanwhile, a new public radio show was gaining in popularity at a rate not seen since *Car Talk* went national in 1987. By 1998, *This American Life* had found a way to harmonize the divide between market research and intuition. Host Ira Glass, an NPR veteran, took to the work of making fund-drive pitches, becoming the "medium's most ardent fund-raiser" with the same lovingly handcrafted approach he brought to his weekly radio show.[49] Long before his 2015 proclamation that public radio was "ready for capitalism," Glass distinguished himself as unusually gifted at the tricky work of asking listeners for contributions in support of local programming.[50]

---

[49]Marshall Sella, "The Glow at the End of the Dial; Ira Glass Is, Um (Pause, Delete) … Listening: The Perfectly Edited World of His 'American Life,'" *The New York Times*, April 11, 1999, https://www.nytimes.com/1999/04/11/magazine/glow-end-dial-ira-glass-um-pause-delete-listening-perfectly-edited-world-his.html.

[50]Felicia Greiff, "Ira Glass: Public Radio is Ready for Capitalism," *Ad Age*, April 30, 2015, https://adage.com/article/special-report-tv-upfront/ira-glass-public-radio-ready-capitalism/298332.

He did it by leaning into the very contradictions around class and audience diversity that had proved so vexing and divisive at NPR. Listeners were, like Glass, "middle class people with normal jobs and responsibilities." Urging such listeners to "walk past" the rational self-interest common to their milieu, Glass made an emotional appeal. It meant searching beyond "reason," which he describes as "standing in a corner, looking all cool and smoking a cigarette" to embrace an emotional logic that, in surpassing reason, hewed to "a larger truth" about what "feels right" that was ultimately an even surer guarantor of middle-class status.[51]

Twenty-First-Century Trajectories

As I explored in Chapter 1, the first decade of the twenty-first century was a period of tumult and innovation, for NPR and for the world it covered. It was a period in which the network came into its own as a major multi-media presence and shed the last vestiges of its alternative legacy. NPR's weekly listeners grew by 58 percent between 2000 and 2010.[52] As NPR grew in listeners, and expanded through multi-platform initiatives and foreign bureaus, mainstream newspapers, magazines, and broadcast news haemorrhaged market share and journalists. Like previous crises, the events of September 11 and the subsequent US military responses provided fresh evidence of the network's "air superiority" over rivals. The $200 million bequest of Joan Kroc in 2003 provided NPR with "a literal foundation" that it had previously lacked, and which positioned it to invest heavily in the transition to multi-platform journalism.[53]

Even so, it didn't take long for NPR to produce a new psychographic model of its audience and to find data pointing to a new crisis. This time, the emphasis was on generational differences, a category dismissed as one of the "usual suspects" in *Audience 98*. In 2006, *Reinvigorating Public Radio's Public Service & Public Support* broke listeners and potential listeners into five generational

---

[51]"*This American Life*-Evergreen Fundraising Spots," PRX, https://beta.prx.org/stories/129659.
[52]Meredith Heard, "Generational Differences in Internet Usage," *Go Figure*, Blog, April 11, 2011, https://www.npr.org/sections/gofigure/2011/04/11/135312499/generational-differences-in-internet-usage.
[53]Robert Siegel, Personal communication, September 29, 2017.

categories: "The Matures, The Silents, The Boomers, and "Gens X & Y."[54] Gone too was the triumphal confidence that characterized the previous study. *Reinvigorating* begins on a note of alarm. "Public radio is no longer a growth industry. Public radio's national loss of audience momentum is real." Listener donations "showed signs of softening."[55] Subsequent reports echoed the alarm: "A Troubling Loss of Momentum," "A Disturbing Downturn in Loyalty," and "Losing our Grip." Public radio's stars, proclaimed to be "in alignment" in the 1980s, may "have unaligned" in recent years.[56]

The report also shifted the blame for revenue shortfalls from "the Cheap 90" (i.e., listeners) to program directors and other radio staff who failed to heed "the inescapable laws of public service economics."[57] "It's the invisible hand that now guides the money into and through and out of your station," Giovannoni insisted, not a shared investment in supporting public broadcasting. Among recommendations, he called for greater investments in fundraising staff, the better to "mine" and "extract" money from listeners and underwriters. A series of reports titled *Audience 2010* that came out in 2006 on behalf of The Radio Consortium, echoed the anxiety about lost market share. *Audience 2010* struck a grimmer tone than previous big audience research reports. It warned of a coming downward spiral in the public service economy absent a robust response.

Instead of marshalling political resources for greater public investment in public media, a strategy used by corporate sectors like agriculture and the fossil fuel industry, market research firm Walrus Research, author of *Audience 2010*, argued for doubling down on the privatization of public radio revenue. The only way to stop the "historic loss of momentum, would be for public stations "to earn more from listeners per listener-hour."[58] The report encourages

---

[54]Key Findings from 2006 PRPD Session. http://www.walrusresearch.com/images/Key_Findings_for2006_PRPD_Session.pdf.
[55]http://www.walrusresearch.com/reports/Audience%202010%20-%20Interim%20Report%201.pdf.
[56]http://www.walrusresearch.com/reports/Audience%202010%20-%20Interim%20Report%201.pdf; http://www.walrusresearch.com/reports/Audience%202010%20-%20Interim%20Report%206.pdf.
[57]*Audience 2010*, May 9, 2006. Sec 6.
[58]*Audience 2010*, May 9, 2006. Sec 6.

increases in funding to support "earned" revenue. And since only "listener-sensitive" revenue (i.e., funds donated by listeners and paid by corporate underwriters) count as "earned," attempts to increase federal funding, grants, or more support from the station's licensee organization were considered "unearned," and somehow not worth the investment.

During times of meteoric growth in audience and revenues and during periods of stagnation, the answer was always the same: shift the burden of revenue from the public tax base to "listener-sensitive income." The report calls for an intensification of the market rationality that previous research had prescribed and which the public radio industry had, by 2006, largely adopted. It was no longer sufficient to straddle public service and commercial imperatives. The mechanism for this intensification would be doubling down on radio's power as a feeling medium.

Program directors must now "recommit to making each minute of programming a more powerful, compelling, appropriate, listenable, listened to, personally important, and highly valued service."[59] Listeners should feel good. The report ends with a call to all stations, to "strengthen loyalty" by dumping programs that don't serve the "cume," in other words, those target listeners who already do the most listening.[60] Pushing back on "calls for younger and more racially- and educationally-diverse audiences," the report warns that such attempts are misguided because programs for listeners "outside the public radio mainstream," simply "don't belong" on most stations. Coveted high education and racially homogenous listeners will "flee" in the face of such efforts, an impact felt not just during a specific show: "their impact ripples throughout the day and across the week."[61] Listeners should feel "at home" in their own neighborhood.

The passage reprises the neighborhood/community metaphor of *Audience 98*, making clear that public radio's neighborhoods were vulnerable to white flight and economic fallout to public life as a consequence.[62] It also evoked Ahmed's account of the rippling

---

[59]*Audience 2010*, May 9, 2006. Sec 6, 13.
[60]*Audience 2010*, May 9, 2006. Sec 6, 23.
[61]*Audience 2010*, May 9, 2006. Sec 6, 22.
[62]*Audience 2010*, May 9, 2006. Sec 6, 14–15.

quality of racial animus across sticky associations embedded in "normal discourse," the qualities that make the cume a coherent grouping of listeners come into being through "the accumulation of affective values."[63] Sticky feelings like loyalty and fear "increase in affective value as an effect of the movement between signs: the more they circulate, the more affective they become, and the more they appear to 'contain' affect."[64] In this way, "each minute of programming" represents a crucial moment of exchange in the traffic in feelings.[65]

By 2010, the network had embraced a fully multiplatform world. And it did this in part by making explicit the network's market-orientation and by fully engaging audiences online. One of the biggest initiatives of this period was NPR Listens, "an audience advisory panel," launched in 2007 by Lori Kaplan, NPR's director of Audience Insight and Research. Reaching out to the most dedicated audience members through social media platforms like Facebook and Twitter, NPR Listens acquired thirty-thousand volunteer audience panelists by 2010. Members of NPR Listens, regardless of qualities, were entitled to discounts at "The NPR Shop," an online source for branded swag and Driveway Moments CDs.

In 2010, NPR turned to SmithGeiger for another psychographic study of its audience, this time to link listener types specifically to consumer brands. The key psychographic categories the study found that NPR should be targeting were "Voracious Voyagers, Team Captains, and Dutiful Aggregators"; "Low Opportunity Categories," include "The Strugglers" and "The Traditionalists," according to the study.[66] These categories, products of the same VALS methodology used in previous studies, described not just NPR listeners but people in general (i.e., American consumers of

---

[63] Ahmed "Affective Economies," 120.
[64] Ahmed, "Affective Economies," 120.
[65] Eric Nuzum, "Treat Every Moment of Every Episode Like It Is the Last Chance You'll Ever Have to Do This Thing That We All Love So Much," *The Audio Insurgent*, April 3, 2023. Remarkably similar sense of urgency to the 2010 recommendation—"every moment." For a recent example of the brisk traffic in racial animus in contemporary media, see also, https://www.nytimes.com/interactive/2022/04/30/us/tucker-carlson-tonight.html.
[66] SmithGeiger, *NPR Audience Segmentation and Growth Opportunities*, September 2010.

news media, to be more precise). The categories were created based on a forty-two-minute quantitative online survey of 3,700 people.

Once they placed them all into these "key audience segments," SmithGeiger followed up with in-home ethnographic interviews with 13 Voracious Voyagers, Team Captains, and Dutiful Aggregators. The Strugglers (who are poor) and the Traditionalists (who are incurious) were not interviewed as they were thought to provide "a lower return on investment." NPR programming, it should be noted, was at this time still featuring the stories about the poor and giving lengthy and largely sympathetic coverage to the traditionalist conservatives who made up the rank and file at the "Tea Party" demonstrations taking place across the country.

"Voracious Voyagers" on the other hand, represented the very beating heart of the cume, a psychographic grouping that epitomized the swirling currents of enthusiasm, ambivalence, and uneasiness at the heart of the public radio structure of feeling. Voracious Voyagers "are passionate about their hobbies." They bring a "liberal, scientific filter for the world and a penchant for diving deeply into whatever they do." Unlike the incurious Traditionalists, they "embrace technology, explore culture, and eschew organized religion. They approach life with a sense of adventure and curiosity." Voracious Voyagers say things like "I can't wait to travel the world"; "I want to love my work and be financially stable"; "I love to get out of the house and meet up with my friends or check out a new band"; "I keep several web browsers open at work." Team Captains were described as equally exuberant, if a bit more mainstream. They are "characterized by a sense of optimism, self-confidence, and enthusiasm for life." Team Captains say things like "I'll go first!"; "When I read an interesting article online, I share it with all my friends on Facebook"; "I can get everything I want to know in one place on Yahoo!"; and "I'm leading my ideal life."

Dutiful Aggregators, perhaps less confident and exuberant, were no less industrious. They are known for "prodigious media consumption, an abiding desire to understand all sides of an issue, and a need to take advantage of all that life offers." They seemed to acknowledge the enormous amount of work necessary to attaining their goals. However, they also seemed more capable of uncertainty. "They are well-informed, with a broad array of media sources consumed in a typical week. Members of this group see life in shades of gray, with moderated and often conflicting opinion sets." They say things like "Education is important to me"; "I make it a priority to continuously learn new things"; I don't believe in absolutes"; and "Of course I check a variety of news sources."

Each of these groups present their own set of promises and challenges. Voracious Voyagers are evangelists for NPR; they love it. In their enthusiasm and liberalism, they may be to blame for spreading NPR's reputation for left-wing bias. They're "slightly repelled" by attempts to increase accessibility to other types of listeners and, the study's authors imply, slightly repellent to the rest of the imagined community. Dutiful Aggregators listen to NPR the most of any group but tend also to be very critical of the network's lack of "accessibility" to a wider range of listeners; included in this critique is a problem with accessibility in "tone." Or, as *Current* put it in its summary: "Study sees growth if NPR loosens up, sounds less elite."[67] "Winning over a Team Captain means winning over a leader," the study crows.

To better serve all three, especially the Team Captains, who the study implied were especially coveted in part because most of them didn't listen at all, would require some change in tone. To survive and thrive, NPR must pivot away from public commitments to commercial ones, from the perception of a liberal bias to an embrace of a more conservative values, from secularism to greater respect for the religious values of those wary Team Captains and Dutiful Aggregators. SmithGeiger's study reveals the extent to which NPR now saw itself as a brand among brands, a corporate entity whose audience's attachments to it could be understood entirely through the psychographic categories of consumer culture.

Audience research like SmithGeiger's has at least two audiences of its own: the program directors and producers who ostensibly want to better understand the audience and the corporate underwriters and marketers who want the same thing, though for different reasons. The translation of listeners into consumer-based psychographic types made it far easier to understand NPR as a brand among brands. When asked what sorts of brands they most wanted NPR to emulate, Voracious Voyagers selected The

---

[67]Karen Everheart, "Study Sees Growth if NPR Loosens Up, Sounds Less Elite," *Current*, September 20, 2010, https://current.org/2010/09/study-sees-growth-if-npr-loosens-up-sounds-less-elite/.

Discovery Channel, Target and "independent booksellers." For Team Captains, it was similar, only with Marriott hotels instead of bookstores. Dutiful Aggregators were drawn to Wal-Mart. It's not immediately apparent how these insights can be easily incorporated into programming ideas. But it's a bit easier to imagine how this information can be leveraged into underwriting pitches, though you can hear the yowls of indignation about Walmart coming from the Voracious Voyagers.

This study, titled "NPR Audience Segmentation and Growth Opportunities," urged the network to steer away from many of the things it had come to represent while holding on to everything that has made it unique. This ritual and paradoxical pivoting away from itself had become a key part of the NPR's reflexive response to crisis, as seen by its wobbly self-defense when threatened with federal defunding. We can see this same impulse in the study's tortured logic and hectoring tone on how to be more accessible to non-white listeners. "It is critical to note that a dismissive attitude toward religion and spirituality is likely to alienate a portion of this [African American] audience."[68] Placing secular, science-loving listeners on one side and African American listeners on the other, this passage reinforces the racial imaginary that audience researchers had developed over two decades of reports for NPR. Conceding that the network is dismissive of religion and spirituality it frames as inevitable the conflict between Voracious Voyagers, with their Darwin bumper stickers on their Subarus on the one hand, and the promise of a truly accessible public radio service on the other.

Audience Research as Farce

In 2013, in yet another weird pivot away from itself, NPR created a web feature that seemed to make a travesty of the Sisyphean labor of audience research and blurring of lifestyle brands with the network's targeted demographics and psychographics. Doubling down on the quirkiness of listener profiles, and the absurdity of being all things to all people, the network launched a web-based

---

[68]SmithGeiger, *NPR Audience Segmentation.*

campaign called "Interesting People, Interesting Radio." Part survey, part slot machine, part Facebook quiz, the site sent up the reductive psychographic branding approach of matching sociocultural details of listeners with specific programs. A "hang-gliding accountant from San Diego," for instance might, favor *Morning Edition*. A "scuba-diving barista from Fort Worth," on the other hand, might prefer *On the Media*. And "a book club film buff from Indianapolis," might prefer *Travel with Rick Steves*.

The explicitly unscientific matching game poked fun at the customizing algorithms that were, at that very moment, enabling rising corporate behemoths like Amazon and Netflix to reconstruct the relationship between audiences and content. The "interesting listeners" feature was an implicit acknowledgment perhaps that the advice from *Audience 2010*, making sure that "each element (of programming) appeals to a consistently-defined listener regardless of its source of production" made no sense given the constantly changing psychographic definitions of listeners.[69]

The "Interesting People, Interesting Radio" feature also pointed to the ways that, after decades of searching for "Who's out there?," NPR was, at some level, acknowledging the extent to which audience research was a commodity fetish and a marketing ploy, an attempt to build a brand around a set of values and sensibilities most likely to win underwriting support and listener donations. Together, all the studies and market reports seem to have achieved an unlikely outcome of bringing together the two fiercely opposed schools of thought regarding how to think about audience. In the late 1970s and early 1980s, public radio program directors and producers chafed against the emergent power of quantitative audience research methods and priorities. They felt strongly that, as members of a local community and as radio professionals, they knew what their listeners wanted. Their interests and demographics were continuous with those of their listeners. Over three decades of audience and market research, of demographics and psychographics, blogs and charts and focus groups, the logic of this intuitive equivalence between NPR and its audience seems to have held, even strengthened.

---

[69]*Audience 2010*, Report 6, May 9, 2006, 12.

The impulse to bring the weird reflexivity of audience research into program content led to other novel web-features. *Go Figure* was a short-lived blog developed by the NPR Audience Insight & Research team, as "an experiment in research transparency."[70] In practice, it seemed more like a vanity project by self-proclaimed "research nerds" who, had they submitted to SmithGeiger's psychographic study, would have surely been classified as Voracious Voyagers: enthusiastic to the point of being evangelical. Between 2010 and 2012, the blog featured weekly updates on audience research in a chatty, first-person style apparently meant to convey transparency. It also had the effect of reducing the distance between NPR staff and its listeners, a much-touted affordance of interactive digital media. And as we've seen again and again, the similarity between listeners and producers has been an enduring conceit of public radio. *Go Figure*'s informal, enthusiastic, and self-reflective researchers presented themselves as ideal types for the kinds of psychographic types they were in the process of creating. "How I Became a Research Nerd" was a recurring feature on *Go Figure*. Described as "a fun look into the lives of the researchers behind Audience Insight & Research," it was also a revealing performance of the psychographic profile of an ideal type of NPR audience member.[71]

Careful to distinguish themselves from the "editorial" side of the network, the research nerds identified themselves as part of the highly educated "lifelong learners" that comprised their ideal audience. *Go Figure* turned the spotlight onto behind-the-scenes NPR staff members, making them celebrities to demonstrate the ways they were just like us. Further, they invited us to think about the gathering of personal information by a media organization as a game we have been invited to play, not as objects of surveillance, but as fellow data nerds. "Not all of my questions lead to something useful," Jamie Helgren, an intern in Audience Insight & Research conceded in her 2012 blog posting. "But they're fun to research anyway. For example, how many NPR listeners own cats, drink gin,

---

[70]Lori Kaplan, "A (Sort of) Farewell," *Go Figure*, Blog, August 17, 2010, https://www.npr.org/sections/gofigure/2012/08/17/159005508/a-sort-of-farewell.
[71]Jessica Ruiz, "How I Became a Research Nerd: Jessica Ruiz," *Go Figure*, Blog, July 21, 2011, https://www.npr.org/sections/gofigure/2011/08/02/138155083/how-i-became-a-research-nerd-jessica-ruiz.

AND listened to 80s pop music in the last six months? Answer: not many. You know who you are."[72]

Go Figure's reflexivity also demonstrated the ways in which audience research understood consumption habits as a cipher for psychographic identity. A few days before Christmas 2011, research nerd Lori Kaplan juxtaposed her own family's desire for a holiday ski trip to the travel patterns of NPR listeners. She took heart in discovering that listeners, like her, "are more likely to take vacations—skiing, beach, national park," than other Americans. "Given the audience's general interest in the world and typically higher levels of disposable income, this finding is not shocking."

By May 2011, the blog had stopped regular postings. Kaplan suggested gamely that it was a victim of its own success. She also acknowledged that as much of their research data was "proprietary," the blog's goal of transparency had become impossible. But Kaplan assured her readers that it wasn't a real farewell and linked to other pages on the NPR website where insider news could be found. Both of those pages were defunct or had been re-routed to the "about" page by the end of that year. But Kaplan wasn't wrong that by 2012 the self-reflexive spirit of audience and marketing research had already come to suffuse the network's other platforms of audience engagement. The goal of bringing audience research, marketing research, and brand identity out from "behind the curtain" had largely been successful.

Conclusion: Crisis and Opportunity

The network's tropism for crisis could make these programmatic gestures of transparency awkward at times. 2011 proved to be the network's "*annus horibilius*."[73] In the space of a few weeks, NPR

> fired its most conspicuous, popular black voice, Juan Williams, raising questions about its commitment to free speech in the

---

[72] Jamie Helgren, "Trivial Pursuit: NPR Listener Edition," *Go Figure*, Blog, January 31, 2012, https://www.npr.org/sections/gofigure/2012/01/31/146097703/trivial-pursuit-npr-listener-edition.

[73] David Margolick, "National Public Rodeo," *Vanity Fair*, January 17, 2012, https://www.vanityfair.com/news/business/2012/01/National-Public-Rodeo.

process. Then it essentially fired the woman who had fired him. Then it fired the woman who had fired the woman who fired him, along with its chief fund-raiser. All this had been embarrassingly public and poorly explained, and from an outfit whose business is explication.[74]

It seemed a circular firing squad. Reports of a "simmering battle" between journalists and executives about the direction of the network spread. Had the network changed or had the culture? It seemed the liberal center could not hold amidst the "partisan sniping of the sound-bite era." More than this, the Williams firing put the network in the crosshairs of its own impossible positions on racial inclusiveness, along several axes. Williams had admitted (on Fox News, no less) that he was frightened of Muslim-appearing passengers when he boarded airplanes. NPR's insufficiently critical posture on the war on terror (according to liberal critics) and its poor track record on hiring and retaining African American staff meant that the aftermath of the Williams debacle was a perfect storm. For the newly emboldened anti-tax Tea Party, the calls to defund the network sounded increasingly sincere and bloodthirsty. The traffic in feelings was brisk but feeling good was now a harder sell; roiling and rippling through the current were other, less comfortable affects.

This long history of self-conscious audience research and the spasms of self-inflicted crisis cast a somewhat unflattering light on NPR and suggests a manipulative rather than an authentic relationship to radio's emotional power. As I stated in the introduction, the term "empathy machine" juxtaposes instrumentality and kindness without resolution, though many who use the term seem not to notice the tension.

This chapter and the one before it sought to ground the public radio structure of feeling—with its genius for fellow feeling, its clumsy self-regard, its class chauvinism, and its virtuousic storytelling—in the historical conditions of its development. Public radio's mission in the United States was always nearly impossible; its improbable success for almost sixty years suggests the power of productive tensions. NPR and the public radio programs that

---

[74]Margolick, "National Public Rodeo."

listeners found on member stations at the end of the twentieth century and the start of the twenty-first, reached listeners precisely because they articulated "the messy hinterland of emotions" that characterized an era.

Between late 2005, just before the launch of *Audience 2010*, and the debacles of 2011, NPR had been rather quietly developing a batch of podcasts aimed at new channels, niche audiences, and carefully crafted for new audio environments. Largely inspired by *This American Life*, this new initiative looked for smaller stories, each designed around discrete moments of feeling, the better to move its listeners who were dispersing across platforms and time-shifted reception practices. Much of the early podcast content was re-packaged broadcast stories, stripped of program-specific theme music and re-imagined as stand-alone material. But a good deal of it grew out of "alt.NPR," an experimental incubator for podcasts, drawing on new voices, and targeting narrower audience slices.

Some of the programs featured lifestyle or hobby topics, dedicated to upmarket listeners interested in food, wine, and poetry. Others, like *Radio Juventud*, which targeted Spanish-language youth, seemed to be reaching out to audiences outside the coveted demographic that researchers were asking the network to prioritize over all others. *Brini Maxwell's Hints for Gracious Living*, hosted by a drag queen, offered an irreverent take on lifestyle program that fell outside the ambit of the brand identities and psychographic models of Voracious Voyagers.

Love+Radio and *Benjamen Walker's Theory of Everything*, two bona-fide podcast hits that helped establish the independent, non-profit Radiotopia network, got their start on alt.NPR. The former, hosted by Nick van der Kolk, has been widely praised as a ground-breaking sonic innovation that "screwed around with the boundaries of what one usually expects when listening to a radio story or even a simple interview." These hidden origins give weight to the idea that the podcast sound that came to dominate the commercial and non-commercial platforms and networks in the 2010s originated in public radio. Indeed, Radiotopia was not the only independent podcast network to emerge from public radio origins, as we'll explore in future chapters.

The years of research and development on the technology and creative sides that went into the podcasts of alt.NPR represent an enormous and largely uncredited gift to the independent podcast industry that exploded in the 2010s. It is curious that the experimentalism with new sounds and niche audiences ran parallel to an era in which the hyper-market discipline of audience research firms, with their gospel of super-serving the super core, held such sway at the network.

It is a strange irony that this undercurrent of new voices was likely only possible thanks to the largesse of Joan Kroc, whose McDonald's fortune bolstered a network that had teetered on the edge of insolvency for decades prior. In terms of distribution, alt. NPR's podcasts represented "a revolution," in the words of Eric Nuzum, director of programming and acquisitions during this period. The challenge was how to keep the same "ethos, feel, and vibe" that listeners had come to expect from on-air programming.[75] Fortunately for Nuzum and others at alt.NPR, other public radio shows were reaching a significant and growing portion of their listeners via podcasts by 2010. Chief among them was *This American Life*, which since its debut podcast in 2006, was rarely out of the iTunes Top 5 most downloaded podcasts for the next decade.

The history of public radio's search for its own audience tells us more about the network's priorities and anxieties than it does about the listeners themselves. However, this distinction seemed to melt away in the first decade of the new century, as a general understanding of "the NPR listener" seemed to take hold in the popular imagination. The consensus among those opposed to audience research and those in favor of it never wavered from the idea that those who made public radio and those who tuned in were essentially the same sort of people. In the logic of global brand identities, the only constant was managing impressions and responding to crises. Public radio's adaptation to new online platforms and to new corporate branding opportunities didn't abandon Siemering's injunction that listeners should feel. On the contrary, it centered feeling as the current that connected listeners and producers. No public radio show did this better than *This American Life*.

---

[75]Shirley Liu, "How atoner's experimentation shaped the early podcasting landscape starting in 2005," *NPR*, August 12, 2022, https://www.npr.org/2022/08/12/1116938798/how-alt-nprs-experimentation-shaped-the-early-podcasting-landscape-starting-in-2.

CHAPTER THREE

Feeling Playful: *This American Life* and Narrative Enchantment

Introduction

In 1995, WBEZ, the NPR affiliate in Chicago, debuted a new sound for public radio with an old-fashioned title: *Your Radio Playhouse*. The very first episode struck an oddly elegiac tone. Host Ira Glass opened with a meditation on the relentlessness of a particular kind of nostalgia, that powerful "force ... so basic to who we are as people," to declare that things were better back when they were new, "before they were popular." Within the first two minutes of the broadcast, in a whimsical demonstration of the reflexive power of this force, Glass imagines some among his audience already looking back to the first thirty seconds of the program with nostalgia for its heady creative era. The accelerated cycle of nostalgia for the no-longer-new had become a familiar cliché of the 1990s hipster stereotype. The episode is, at least implicitly, a meditation on the ironic relationship of the present to the past in an age of mass-mediated culture. On the one hand, a bravura declaration that a new form of radio storytelling had arrived; on the other, a reckoning with the long shadow cast by radio's history of showmanship and intimacy.

Both impulses are on full display in the next segment, in which Glass interviews broadcast legend Joe Franklin, self-proclaimed inventor of the talk show format.[1] It's an Oedipal encounter, part ritual slaying of the father, part homage to the medium and a previous master. Glass records a phone call interview with Franklin, ostensibly asking him for advice about career longevity. Glass, a meticulous editor, leaves in embarrassing interruptions in which Franklin barks instructions to an assistant, including an obvious fib ("Tell them [I'm] with a camera crew,") and uses terms like "emcee" to describe Glass. Through a careful juxtaposition of styles, Glass makes clear the tonal difference of *Radio Playhouse* compared to the talk show of Franklin's era. Franklin's old-fashioned advice to Glass, "get the plug in fast," repeats throughout the episode as a reminder to the audience of the ironic distance separating these radio talk formats represented by Franklin and Glass. What matters most is the current of ambivalence rippling through the segment: now fending off the schmaltzy sentimentality of the talk show format, now embracing it. It is this play of affects, this emotional sleight of hand, that made *Your Radio Playhouse* so instantly distinctive and compelling. In an era in which many local affiliates had shed their locally-produced and locally-oriented programming in favor of "higher[-]quality" programming produced by NPR, PRI, and American Public Media, programs like *Your Radio Playhouse* were met with great enthusiasm by station managers. By the end of the show's first year, it had a new name and a national syndication deal. By 1999, it was airing on 322 stations to an estimated audience of 800,000.[2]

One of the contentions of this book is that *This American Life* (*TAL*) in many ways anticipated much of what became a dominant form of nonfiction narrative podcasting. But it also drew from the radio traditions that preceded it, in part thanks to Glass's long apprenticeship in reporting and editing at NPR. The original title, *Your Radio Playhouse*, nodded to sound-rich public radio

---

[1]Broadcast talk shows, of course, are almost as old as broadcasting itself, and have many claimants to the title of the first. See Sadie Couture, "Forging a format: Advertising, attention and intimacy on the *Mary Margaret McBride Program*, 1941–54," *Radio Journal: International Studies in Broadcast & Audio Media* 21, no. 2 (2023): 155–70.
[2]Marc Fisher, "It's a Wonderful Life," *American Journalism Review*, July/August 1999, https://ajrarchive.org/article.asp?id=326&id=326.

innovations like *NPR Playhouse*, which launched in 1981 and featured experimental storytelling and inventive audio adaptations, like the 1981 version of *Star Wars*. It also seemed to nod to public radio's formally subversive and aesthetically challenging *Earplay*, which debuted in 1972, and featured "the bizarre play of sound and words" as Jeff Porter put it. Sound montages, intersecting soundscapes, and narrative strands of *Playhouse* productions by Joe Frank (not to be confused with Joe Franklin, mentioned above), were like "a fist coming out of the radio," according to actor, writer, and radio host (*Le Show*) Harry Shearer.[3] *Your Radio Playhouse* also has disputed origins in *The Wild Room* (1990–5), an earlier WBEZ program co-hosted by Glass.[4]

Glass likened his first paid job in public radio, as Joe Frank's production assistant, to watching someone perform magic. Glass says he was "hugely influenced" by Frank, admiring the way listeners are "thrown in the middle of the action [of a story] and it's not all going to get resolved." In Porter's terms, Frank's work rode the tension between sound and sense, veering into the "hyperrealistic," according to writer and longtime *This American Life* contributor David Sedaris. "I've never figured out how to imitate that," Glass says of Frank's willingness to leave listeners in the narrative lurch. Instead, Glass mastered his own form of magic, at once playful and didactic. Every anecdote, every reflection, every note in the music bed that punctuated an utterance, was in the service of a single dominant theme. The unity of sound and sense, and the thematic consistency from Act to Act in each episode revealed Glass as a very different sort of virtuoso than Frank, and a much more successful one.

Glass's *Playhouse* would be more accessible than these precursors, but as the ritual slaying of Joe Franklin in the debut episode suggests, it would also mark itself as distinct from the commercial and promotional traditions of mainstream radio,

---

[3]Jeff Porter, *Lost Sound: The Forgotten Art of Radio Storytelling* (Chapel Hill: University of North Carolina Press, 2016), 202–3.
[4]Michael Miner, "What Becomes of the Brokenhearted?," *Chicago Reader*, November 19, 1998, https://chicagoreader.com/news-politics/what-becomes-of-the-brokenhearted/.

and ironically so. Another episode from the debut season entitled "Quitting" (1995), featured an interview with an articulate young woman named Evan Harris who made an art form out of giving up on jobs, relationships, and places to live. The profile nicely captured the program's quirky spirit and ironic approach to the social and financial commitments that defined Gen X entry into adulthood and the habitus of the professional middle class. Harris's preoccupation with perfecting the art of "the quit," spoke to a popular rendition of the zeitgeist of the time, what Chuck Klosterman called an "adversarial relationship to the unseemliness of trying too hard."[5]

In a 1998 profile, Glass admits he hides his own meticulous approach to work behind a posture of nonchalance and serendipity; the show should seem to listeners like "a great find from a flea market," rather than something more earnestly curated.[6] A 1999 profile noting this same disconnect described Glass as "agonizing over every stammer, every breath, every millisecond between words in what originally was a 12-minute chat."[7] Such themes echoed across the popular culture of the 1990s in music, television, and risible attempts at "alt" product launches, like the ill-fated campaign for OK Cola, a 1993 attempt by Coca-Cola to bring the shrugging insouciance of the age to a product that had historically inspired overwrought anthems to world peace. Glass doesn't apologize for hiding effort behind a veil of natural conversation, no more than a magician would apologize for laboring over a trick until it looked effortless. Glass says, "The listener's experience should be, 'God, these people are just sitting down telling these interesting stories.' Whereas our experience is like every 16th of a second is planned and pretimed."[8]

---

[5] Chuck Klosterman, *The Nineties: A Book* (New York: Penguin, 2022), 2.
[6] Steve Johnson, "Ira Glass and 'This American Life': Putting The Public Back In Public Radio," *The Chicago Tribune*, October 18, 1998, https://www.chicagotribune.com/news/ct-xpm-1998-10-18-9810180472-story.html.
[7] Marshall Sella, "The Glow at the End of the Dial; Ira Glass Is, Um (Pause, Delete) ... Listening: The Perfectly Edited World of His 'American Life.'" *The New York Times*, April 11, 1999, section 6, 68, https://www.nytimes.com/1999/04/11/magazine/glow-end-dial-ira-glass-um-pause-delete-listening-perfectly-edited-world-his.html.
[8] Johnson, "Ira Glass and 'This American Life'."

Diffidence and Wonder

This chapter explores three dominant themes running through episodes in the early years of *This American Life*: empathy for strangers, gothic family mysteries, and bathos, which is the abrupt narrative shift from the sublime to the ridiculous, as a hedge against empathy. In each, we can see the sideways rippling of emotions across signs, objects, and subjects, producing moments of feeling before scuttling away. For most of its first ten years, *TAL* specialized in a particular kind of storytelling: first-person singular, filled with wonder and empathy, toggling between a journalist's meticulous attention to detail and grand generalizations about what it means to be human. Mixed into this formula, however, especially in the first five or six years, was an archness, a sense of critical distance on the whole thing, as if Glass could, at any point, decide that he too, would up and quit. Mastering this balance of "characteristic elements of impulse, restraint, and tone," to use Raymond Williams's description of elements that constitute a structure of feeling, required both an aesthetic discipline and an ideological one.[9]

In a 2009 episode of *TAL*, Glass articulated this structure of feeling, with a characteristically autobiographical narrative that fit the parameters of the house style: "this happened, then this happened. And then a reflection."

> I had a lot of really strong beliefs about stuff when I was a kid. I had a religious phase and I had a very strong like, atheist phase. And then I had a very political phase. And I was politically correct for years. The kind of politically correct where when I was in my 20s, I went to Nicaragua and I called it *Nicaragua* (Spanish accent). (laughs). And you know what I mean? I was horrible! (laughs) ... I got older and I saw that things seemed more complicated than the way I believed them.[10]

It's a useful parable for the growth to "maturity" of public radio, and the urgent political commitments that drove the journalists

---

[9]Raymond Williams, *Marxism and Literature* (Oxford: Oxford University Press, 1977), 128–35; and *Preface to Film* (London: Film Drama, [1954] 2003), 21–3.
[10]"This I Used to Believe," *TAL*, episode #378, April 17, 2009.

who got their start there in the 1970s. Importantly, it's not merely the content of his former beliefs that Glass repudiated in this episode, but their intensity. The phrase "it's complicated" (or "nuanced" or "surprising") acted to ward off politics, history, any system of ordering experience that isn't, at heart, based on the universality of individual experience, the aesthetics of surprise, and the ambivalent emotional stew of empathy and disjunctive irony. The bloody history of US involvement in Central America in the 1980s provides a striking backdrop for an anecdote essentially justifying the exchange of politics for a gauzy curiosity about human "complexity." Glass's throwaway line, that pronouncing "Nicaragua" with a Nicaraguan accent made him "horrible," speaks to a very specific kind of American commonsense, in which the mere recognition of social difference represents a kind of capitulation to something suspiciously political and therefore inauthentic. Klosterman referred to the 1990s as "the last period in American history when personal and political engagement was still viewed as optional."[11] This sensibility, based in a loss of faith in politics among other things, suffused *This American Life*'s exploration of human experience.

The embrace of complexity as a fending action against commitment also serves as a model for many of the stories the program featured in its first decade or so. Certainties gave way to surprises; committed engagement gave way to bemusement; politics gave way to storytelling. Glass's list of abandoned passions, or quits, also served as the narrative, in miniature, of the collapse of public radio's Great Society liberalism, articulated in NPR's Founding Principles, into the neoliberalism that characterized its form and content at the turn of the century. The conversation was part of an episode called "This I Used to Believe," a play on longtime public radio producer Jay Allison's radio series and subsequent best-selling book, *This I Believe*, a series of short essays on faith authored by the famous and the unknown alike.

Allison's series was a revival of the early 1950s radio series hosted by Edward R. Murrow. It tends to highlight the virtues commonly associated with modern liberalism, like "listening" and "empathy." Many follow the formula of leading with seemingly

---

[11]Klosterman, *The Nineties*, 2.

mundane objects of faith, like "I believe in Barbie," in order to get at something larger, like "the power of imagination." In this way, the series is similar to the *TAL* formula of anecdote followed by reflection. However, Glass was reluctant to commit to believing in anything, and as in the quotation above, seemed to regard the concept of belief to be naïve and retrograde. When pressed by Allison to admit to a belief, any belief, Glass offers with apologetic uptalk, "Well, I believe that listening to the radio in the car is the best place to listen to the radio?"[12]

The notion of abandoning previously held beliefs, like the theme of quitting and the retrospective impulse that Glass argues is so central to human nature, captures a telltale fending off from commitments and their intensities. Together, they represent an ironic distancing from the political and social investments in the "now" that seemed to be beckoning so insistently in the early broadcasts of *All Things Considered* twenty-five years earlier. Indeed, *This American Life* was as much a creature of the mid-1990s as Susan Stamberg's *All Things Considered* was of the early 1970s. In the midst of a booming American economy, at the height of *Seinfeld*'s reign on television, and as Francis Fukuyama declared the "End of History," Ira Glass took to the air with a quirky little radio show that privileged the microscopic, the personal, and the now over the social, the political, and the historical.

Less charitably, it has been said that *TAL* was "never designed to accommodate harsh economic truths, much less to promote any kind of critical art or intelligence."[13] If NPR's founding purpose was dedicated to "celebrating the human experience as infinitely varied," Glass's approach to "reinventing radio" could be summed up thus: "Noticing when you're bored is really really important."[14] Both statements center the affective experience of nonfiction storytelling, but Bill Siemering's crackles with the unselfconscious idealism of his progressive midwestern heritage and the Great Society liberalism of his moment. Glass, on the other hand, acknowledges up front the

---

[12]Vivi Merrick, Dan Gedimena, Jay Allison, eds., *This I Believe: The Personal Philosophies of Remarkable Men and Women* (New York: Henry Holt, 2004); "This I Used to Believe," *TAL*, episode #378, April 17, 2009.

[13]Eugenia Williams, "Oh the Pathos," *The Baffler*, no. 20 (July 2020), https://thebaffler.com/salvos/oh-the-pathos.

[14]Jessica Abel, *Out on the Wire: The Storytelling Secrets of the New Masters of Radio* (New York: Broadway Books, 2015), 55.

near impossibility of saying something new when everything, even the first moments of his new show, almost instantly becomes stale.

The impossibility of the new haunts the early episodes, as in the 1995 episode "Vacations," which Glass begins with a discussion of the overdetermination of Hawaii as a "paradise," which interferes with his ability to enjoy a family vacation there. The 1998 episode "Road Trip!" begins with Glass's lament over the impossible weight of literary and cinematic references, which he dutifully recites, that burden our own road trips with impossible expectations. In the first act, he commiserates with a young man named Jamie whose inner monologue on his first road trip, "am I feeling something? Am I experiencing something?" tapped into this fear. "I'm supposed to be having a revelation!" Glass added, finishing Jamie's thought. Mainstream media culture, with its endless promise of spectacle, failed to deliver. "Here in 1996," Glass complained, "no one in American life seems bigger than life. Not the president, not Michael Jordan, not Courtney or Madonna." Even the Beatles, the subject of a recent documentary, Glass argues, with a mixture of vindication and ruefulness, are nothing more "than a bunch of aging dullards without any particular magic to them at all."[15]

Weariness in the face of the conventional is part of a longer tradition that Bourdieu has identified as critical to reproducing class distinctions.[16] When asked why she liked *TAL*, one early podcast listener replied, "I listen to it because I'd rather read Vonnegut than *The Da Vinci Code*," a bit of praise which reflects on the cultural capital of the show and the listener at the same time. Another early podcast listener shared with a mix of admiration and hipster rue that *TAL* "seems like something that wouldn't be widely available by conventional means, yet it is."[17] Another said "if it were a place, it'd be a hip coffee shop, where there's no pretense. If it were a place, it'd be the anti-Wal-Mart."[18] "Taste classifies," Bourdieu reminds us, "and it classifies the classifier."[19]

---

[15] "New Year," *TAL* #8, January 3, 1996.
[16] Pierre Bourdieu, *Distinction: A Social Critique of the Judgement of Taste* (Cambridge, MA: Harvard University Press, 1984), 6.
[17] Johnson, *Imagine This*, 68.
[18] Johnson, *Imagine This*, 71.
[19] Bourdieu, *Distinction*, 6.

In the next chapter, I'll explore in more detail the class dimensions of *This American Life* and the Bourdieuan notion of taste. But more than this, *This American Life* seemed a response to a media landscape in which expectations of "magic" had been disappointed. In early interviews, Glass frequently framed his approach to radio as a reaction to the boring nature of mainstream media, public and otherwise.[20] The sentiment persisted into later episodes, along with the affective posture of "the quit," as when *TAL* producer Tobin Low argues, "If you're with the right person, it should just feel like magic. If you don't feel magic, then it's time to bail."[21] In its earlier years, Glass would introduce stories as parts of "life in these United States," an allusion to the corny humor column in the *Reader's Digest*, which was, in the mid-1990s, still the most popular magazine in the US, and an artifact of an era of mass media culture that *TAL* felt compelled to ironize.

The program was an alchemical transformation of public radio's approach to both structure and feeling; through a rigid adherence to a format for storytelling, editing, and themes, Glass modeled intimacy and diffidence, surprise and irony, accessibility and taste. Like Siemering before him, he believed that "listeners should feel," and was ambitious and even didactic in his approach to making them do so. "The mission of public broadcasting," Glass opined in his travelling stage show throughout the late 1990s, is "to tell us stories that help us empathize."[22]

Unlike Siemering, Glass seemed less concerned with the civic value of an affective education than in affect as succor for the loneliness and disaffection of contemporary life. He is careful to ironize anything sounding too much like a permanent solution. "Every episode of *This American Life* has a point, a moment of reflection: a larger meaning of some sort," observed audio documentarian John Biewen, "but rarely is the meaning tied directly to politics or

---

[20]Miner, "What Becomes of the Brokenhearted?"
[21]"Math or Magic," *TAL* #791, February 10, 2023.
[22]Ira Glass, "Mo' Better Radio," 1998. Transcript available: https://current.org/1998/05/mo-better-radio/.

public policy."[23] Glass focused on producing unexpected feelings of empathy, by documenting "these real moments that surprise me and that amuse me, and that just *gesture* at some bigger truth" (emphasis added). Bottomley has recognized as well that "surprise" is recognized as one of the core strategies for creating narrative pleasure" in the audio narrative tradition that podcasting took from radio.[24]

This play between feeling and irony is central to the narrative sleight of hand that made *This American Life* distinctive. As the shows' producers put in their frequent self-nominations to the Peabody Awards, "What makes the program different is partly a matter of tone. There's a friendly intimacy to the show."[25] Such an approach requires playfulness and subtlety, the ability to "gesture at" ideas without being bound too tightly to them. "Friendly" is an odd modifier of intimacy, implying that there are other forms of intimacy, perhaps more intensely felt ones. Friendly here seems to imply "just friends," another fending action from commitment. Empathy is often produced in these early broadcasts through narrative enchantment, a conjurer's trick, in which listeners are invited halfway into the game, the better to appreciate their own manipulation. What made *TAL*'s producers unique, argued Biewen, was their ability to "make their listeners feel something."[26]

Ira Glass's mode of address, by turns vulnerable and arch, sets the tone for the early stories; however, the oscillating current of identification with and alienation from the audience can be traced back to earlier NPR program and market research strategies and insights, covered in the previous chapter. It also set a new standard for audio auteur, a new kind of celebrity voice: authentic and mannered, emotionally present and deftly manipulative, "intensely literary and surprisingly irreverent," per a *Mother Jones* 1998

---

[23]John Biewen, *Reality Radio: Telling True Stories in Sound* (Chapel Hill, NC: University of North Carolina Press, 2010), 10.
[24]Andrew Bottomley, *Sound Streams: A Cultural History of Radio-Internet Convergence* (Ann Arbor: University of Michigan Press, 2020), 219.
[25]WBEZ, 2005. Application materials submitted to the 2005 Peabody Awards.
[26]Biewen, *Reality Radio*, 8.

profile of Glass, that became ubiquitous in the "American style" of podcasting.[27]

In Ahmed's notion of an affective economy, feelings develop and accrue power through their movement among signs, rather than from their residence within any particular subjects. Stories that "help us empathize" did so through the traffic in feelings, rippling across an imagined community of listeners, first on the radio, and eventually on both podcast feeds and radio broadcasts. "In such affective economies," Ahmed argues, "emotions *do things*, and they align individuals with communities ... through the very intensity of their attachments," helping us temporarily, at least, feel "less crazy and less separate," as Glass put it.[28]

For Ahmed, it is precisely the "nonresidence of emotions" that makes them "binding." At the same time, the traffic in feelings must remain brisk in order to produce and sustain powerful affective responses. Perhaps the community drawn together by *TAL* required more than episodic moments of feeling less crazy and less separate; perhaps they required permanent solutions. Stories about permanent solutions, however, proved less compelling in *TAL*'s first decade than did stories featuring moments of shared feeling, artfully told, magical and transient.

Radio Magic

In 1999, Glass's then-wife thought it important to tell Mary Wiltenberg, an intern new to *TAL*, that in order to understand Glass and the show, she needed to know that Glass had only ever had two jobs in his life: a birthday party magician, starting at around twelve years old; and a radio producer, starting at around nineteen years old.[29] This anecdote sheds light on Wiltenberg's early struggle at *TAL*, documented in not one, but two illustrated books by Jessica Abel, *Radio: An Illustrated Guide* and *Out on the Wire: The Storytelling Secrets of the New Masters of Audio*.

---

[27] Ana Marie Cox and Joanna Dionis, "Ira Glass: Radio Turn-On," *Mother Jones*, https://www.motherjones.com/media/1998/09/ira-glassradio-turn/; Abel, *Out on the Wire*.
[28] Glass, "Mo' Better Radio."
[29] Mary Wiltenburg, personal communication, August 12, 2015.

As part of her application process for the internship, Wiltenberg pitched a story about a successful labor action by Black and white sharecroppers in Southeastern Missouri in 1939. Executive producer Julie Snyder told her that the story was "great," but "not what we do." Wiltenberg came back with a different story, which was eventually featured in the episode "Do-Gooders" (#126, 1999), about an affluent older couple who try and fail spectacularly to revitalize a run-down working-class small town, Canalou, Missouri. Their well-intentioned efforts to work towards the civic good backfire, ending in gunplay and hurt feelings: a fiasco. The two clear differences between the stories, it seems, are that the first is plainly "political" in the way that pronouncing "Nicaragua" as it's pronounced in Nicaragua is political; the second is that it lacks a sense of "surprise." "We are really careful to build surprises into *This American Life*," Glass averred in 1998, the year before this episode. "We kill so many stories ... that we just do not find surprising enough."

Fiascos represent a very specific form of surprise that *TAL* producers are especially fond of. In fact, in 1997, they dedicated an entire episode to stories on the theme, entitled simply "Fiasco" (#61), a theme so compelling they remixed it several times over the next fifteen years.[30] In Canalou, the fiasco represents a bemused take on the idea of urban renewal, class mobility, and liberal interventionism, a better fit for the show's "apolitical" ethos than an inspiring story of class unity across racial lines.

The no-good-deed moral is dramatically underscored in the episode's next act, which examines the disastrous results of international humanitarian aid in Rwanda, when international do-gooders supported the Hutus, who were in the process of slaughtering the Tutsi by the hundreds of thousands. It's a horrible story, on a scale that strains against its thematic inclusion with the Canalou Fiasco. A brief reference to Paul Rusesabagina, the hotel manager who saved hundreds of Tutsis, and was the hero of the movie *Hotel Rwanda*, adds a much needed but flimsy counterweight to the program's main thrust that attempts to help others are doomed to failure. Importantly, this exception to the rule acts alone. Glass compares Rusesabagina admiringly to Humphrey Bogart in *Casablanca*, citing his pragmatism and lack of idealism.

---

[30] Versions of "Fiasco" aired in 1997, 2013, and 2020.

The story of Glass's two jobs, first as a magician and then as a radio producer, helps to frame this chapter's analysis of *This American Life*, which centers the role of narrative enchantment, alchemy, and affective play in the first decade or so of the show. Because of Glass's well-documented didactic and formulaic approach to storytelling, these themes come to us largely pre-captioned. Because he spent years "cutting tape" as an editorial assistant at NPR prior to *Radio Playhouse*, the presentations are immaculately edited and structured. Because he spent his youth doing card and rope tricks at birthday parties, he cannot resist the lure of the flourish, the "ta-da" that communicates "delight," "amusement," and "surprise" as counterpoints to empathy.

The play of these opposing affects frames the show's early years and sets a tonal precedent for the American style of podcasting, while simultaneously hinting at its emotional and political limits and contradictions. These early themes represent the warp and woof of the show's production of "liberal feeling" or, as an early listener to the podcast version of the radio show put it, the magic of "staying in tune with the world in the tiniest way possible."[31]

> It's utterly pleasurable [*sic*]... it feels like you're doing something good, staying in tune with the world, in the tiniest way possible and yet without being frivolous about it. It's unlike anything else out there.[32]

In the sections below, we'll examine the narrative elements that helped to produce this pleasure across stories about strangers and families, and in moments of affective retreat into bathos. "Staying in touch with the world in the tiniest way possible" required not just discipline, but an internalized ambivalence that had been wrought through the alchemy of taste into narrative formula. Riding the line between frivolity and ponderousness was like a magic trick, an exercise in dexterity, deception, and affective economy. It required an audience willing to suspend belief, hungry for homeopathic moments of "good," the better perhaps to forestall larger commitments. Perhaps nowhere in American life at the end of the

---

[31] Kristine Johnson, *Imagine This: Radio revisited through podcasting*, MA thesis (Fort Worth, TX: Texas Christian University, 2007), 37.
[32] Johnson, *Imagine This*.

twentieth century was this formula more compelling than in stories about strangers, a category of people for whom empathy could be measured out in moments of surprise, delight, and amusement.

The words "magic" and "magical" occurred over eight hundred times across the show's transcripts and "magic" is mentioned at least once in 219 episodes, or about 27 percent of the entire *TAL* oeuvre.[33] The word magic evokes a sense of unguarded wonder and refusal of critical distance that is very much in tension with the fending, world-weary archness mentioned above. It is in this tension that *TAL* manages to have it both ways, a kind of magic trick of its own. A 2017 episode entitled "Magic Show" (#619) makes implicit, then explicit, the point of the anecdote about Glass's two jobs, magician and radio producer. The repetitive structure of Glass's storytelling—anecdote-observation; anecdote-observation—is mirrored in his recollection of magic tricks as a matter of disciplined formal repetition: "I did my act so many times it got kind of carved into me."

A family friend drives home the point in an interview: "You think you're doing something different [now]?" she asks him. "Wait, wait, wait. You're saying when you hear me on the radio, it reminds you of my magic act?" Glass responds, seemingly taken aback. It's the same showmanship she observes, the same "spiel." Glass performatively resists the idea in the interview but proceeds to liken magic and storytelling, particularly the appeal of the "psychology" of the well-turned surprise. Elsewhere, Glass has admitted that "there was something about [putting on] shows [as a child] that got me into media, and that was what got me to radio. Every trick had a principle behind it," Glass recalls about his magic show, "and it was cool to think about the principles."[34]

Glass developed the analogy between magic and storytelling across many episodes, paying particular attention to the tension between expectations and surprise and to the moments of "delight" that occur when the two collide. Unpeeling expectations to find layers of surprise, Glass is at his best when most transparent, laying

---

[33]This calculation is based on the first 800 episodes of *TAL*, taking us to mid-year 2023.
[34]Claudia Dreifus, "To Get Things More Real: An Interview with Ira Glass," *The New York Review of Books*, August 8, 2019, https://www.nybooks.com/online/2019/08/08/to-get-things-more-real-an-interview-with-ira-glass/.

bare the tricks behind the flourishes. "The Magical Mystifier," as he called himself at age twelve, Glass reflects on the many ways performing magic is itself an occasion for surprises, reversals, and moments of insight. "I thought I was the one who was in charge of the situation during the magic show," Glass admits. But the joke was on him, he understands, in the episode's first big epiphany: perhaps his adult audience had been indulging him a bit years ago. "I thought I ... was controlling everybody's minds with my mind and my magic," he shares with sheepish wonder. But his ability to enchant as a storyteller quickly became part of the media narrative in the early years of TAL. A *New York Times* interviewer gushed that "there are two people in America who so deliberately mesmerize: Ira Glass and Philip Glass. And they're related (first cousins once removed)."[35]

Magic served as a controlling metaphor in other stories on a theme, like romantic love ("Math vs. Magic," 2023); the power of names ("Name Change," 1997); the power of language ("Magic Words," 2014); the failure of language ("Say Anything," 2003); celebrity ("New Year," 1996); libraries ("The Room of Requirement," 2018); and elsewhere. But it proved most useful as a way to evoke a theory of feelings: the appearance, as if by magic, of a rippling through and among strangers of a surprising affective state, a moment of shared feeling. Perhaps nothing better captures the notion of the public radio structure of feeling than the idea of a magic moment in which a story about strangers pulls a listener out of themselves for a temporary spell of empathy. Stories that evoked such moments for listeners often featured storytellers having magical moments of their own. In another episode ("To Be Real," 2017), he realizes that for professional magicians like David Blaine, the goal is not "creating a fake world," but instead to get to "real, raw emotion," in himself and in his audience, an ambition he plainly shares as a storyteller.

In Act II of "To Be Real," Glass talks to a magician named Derek DelGaudio whose act concludes with a bit in which "he walks up to people [in the audience] and stares in their eyes and tells them

---

[35] Dreifus, "To Get Things More Real."

something about themselves." It's a moment, Glass says, in which "the magic is all in service to this very human thing that's happening." DelGaudio calls one audience member "a good Christian"; another, "a ninja"; then "a ray of sunshine"; "a wallflower"; and so on. "Watching him do this," Glass marvels, changes the experience of being in a room of strangers. "It makes you look at them differently … it stops feeling like a room of anonymous strangers." The last woman he encounters he calls "a failure," which sends a ripple of "awwws" through the room and makes the woman cry. DelGaudio "choked up" as well: "I called them a failure in front of a bunch of strangers," he says, as if surprised by his own trick and its affective impact. Glass seems impressed by this new kind of magic, which dispenses with pretense in order "to get to something utterly real, unfaked, and emotional."

> "I'm trying to make perfect moments," he says. And those generate meaning. If you go deep enough in how to make a moment, very quickly you come to how narrative works – to what we are as a species, how we've come up with telling stories in scenes and images.[36]

This remarkable scene, and its unacknowledged cruelty, helps to contextualize the role of "magic" as a way of thinking about feelings in *TAL*, their temporality and their circulation through social bodies. The standard unit of measure for feelings on the show is the moment. They are produced in and by stories in the moment of telling. In that way, the magic trick is an apt metaphor, as they are produced, serially, in moments, typically before a room full of strangers. In the case of DelGaudio's act, a room of strangers transformed by the simple act of naming ("a good Christian, a ninja, a ray of sunshine"). As in the Driveway Moment, such moments are both moving and arresting. They stop the narrative to make way for a narrator's extra-diegetical insight, which is designed as a caption for listeners' own emotional response. In their affective power, they stop time, or at least stop us in time, the better to feel moved.

---

[36] Sella, "The Glow at the End of the Dial."

Such moments are often represented as moments of human connection, of empathy or fellow feeling. Here, a room of strangers comes together in the shared moment of recognition that one of their number is "a failure," which seems like a violation of some basic agreement about how strangers behave to one another. But for Glass, it's an epiphany, an occasion for an ironic kind of empathy: "It's the sort of moment, watching it, all you can think about is her and her life and what that must be about."

It's an unusually stark example of the at-times ruthless formula that produces stories on a theme. As we'll see below, the strangers theme is a bit of a procrustean bed, now stretching this story to fit the criteria, now lopping off a bit of that story that doesn't quite fit. This stretching and trimming can be likened to sleight-of-hand, making things appear not quite as they are, or to editing (i.e., the cutting and splicing necessary to produce a desired effect). Nowhere in the piece on DelGaudio is there any evidence to support that the magician was correct in his designation of each audience member ("a ninja, a good Christian, a failure"). His power came not in accuracy but in putting feelings into social circulation, producing affect out of thin air, and joining strangers, temporarily, into intimates. In the following section, we explore further the many ways that the stranger proves an exceptionally flexible trope, evoking the power and limits of empathy and speaking to the loneliness of modern social life and the hollowness of mediated sociability.

Strangers Like Us

In order for the focus on strangers to resonate, listeners had to have a sense of who *we* are. And here again, *This American Life* exemplified public radio's genius for creating a sense of intimate connection between radio voices and a certain kind of listener. In 1999, Richard Ohmann identified NPR as the quintessential cultural home for the professional managerial class.[37] As discussed

---

[37] Richard Ohmann, "Public Radio: A Cultural Medium for the Professional-Managerial Class," *Chronicle of Higher Education*, November 14, 1999.

in the previous chapter, producers, hosts and reporters in public radio have been consistent in their claims that the people who make the programs and the people who listen to the programs are essentially the same kind of people. In a pledge-drive pitch played all over the country, Ira Glass hailed listeners as like him, "middle class, with responsibilities."

Perhaps it's because of this sense of shared identity around the universality of being "normal and middle class" that Glass has structured so many of his interviews and live performances, even a comic book, around the idea of teaching his audience to tell stories just like he does. It has shaped a generation of podcasters in their approach to storytelling. For example, Chana Joffe-Walt, a producer for *TAL* and then *Planet Money* and *Serial*, is fond of framing a story with this formulation, "this is the *thing* that we all do ... that's a smaller way of saying 'this is universal.'"[38] Glass has been downright didactic about his formulaic approach to storytelling; demonstrating the simple procedures for inducing a feeling of suspense in audiences was a stock part of his talking tour for more than a decade.

> This is the structure of the stories on our show: There's an anecdote—a sequence of events. This happened, and then this happened, and then this happened... Then, there's the part of the story where I make some really big statement like "there's something about the kindness of strangers."

This specific example, of a "really big statement," is instructive. "The Kindness of Strangers" is in fact the title of a 1998 episode that limns this territory from several perspectives. Across a prologue and four acts, strangers are depicted as powerful talismans of human connection, though their status as unknowable ciphers persists. The prologue tells the story of a stranger on a subway platform in New York loudly yelling "you're out" or "you can stay" to the other commuters, inspiring Glass to the following reflection:

---

[38] Abel, *Out on the Wire*, 69.

There is something about the judgment of strangers. When the clerk in the record store seems unimpressed by your choice of CDs. When the one cute person on the bus gives you a look like, out of my way. It's as if, by their status as strangers, they have some special instantaneous insight into who we are.[39]

This episode was also featured in Jessica Abel's illustrated book, *Out on the Wire*, because of how neatly it captures the show's approach to storytelling structure and because of its thematic preoccupation with epiphanous encounters with strangers.[40] In Glass's hands, this anecdote recalls the magical moment DelGaudio's act, when he confronts the strangers in his audience with peremptory judgments ("a good Christian, a ninja, a ray of sunshine"). The "special" quality of strangers, as vectors for insights about who we are, runs through the episode and beyond, shaping the contours of the social imaginary of the show and the "American Life" that it purports to represent. The words "stranger" or "strangers" appear 832 times in the show's transcripts across 312 episodes, 39 percent of the total oeuvre. In the above anecdotes and in the following acts, however, strangers are only occasionally kind. They are often mysterious or querulous, ciphers for great big truths about ourselves and our world.

Act I of "The Kindness of Strangers," narrated by a soft-spoken white man in the flat affect peculiar to many early-period *TAL* storytellers, could easily qualify as a "fiasco." A late-night attempt at urban chivalry goes bad, a car window is broken, and a flirtatious encounter between strangers on a New York street ends abruptly, punctuated by the eerie yowls of another stranger in an empty parking structure across the street, imitating Tarzan. The coda, the narrator yelling "Shut the hell up!" at Tarzan, suggests that the magical spell of strangers in the night has snapped shut and the old rules of city life have resumed.

Act II features Dr. Jack Geiger, recalling in an interview with Glass his experience of running away from his white suburban home in the 1940s and being taken in by Canada Lee, the Black

---

[39]Ira Glass, "The Kindness of Strangers," *TAL* #75, September 12, 1997.
[40]Abel, *Out on the Wire*, 201.

actor, in his Harlem apartment, where he lived for a year. Geiger had met Lee several times by sneaking backstage at his Broadway shows prior to landing on his doorstep, and so this act feels as if it's been stretched a bit to fit the episode's theme. At the end, Glass summarizes for Geiger, "Well you're saying in a way that that Black culture at that time was more conducive to extending a kindness to strangers than white culture." Glass often provided this kind of gift-wrapped reflection that lets listeners know a story is over. This moral, though, gives a bit more away, making explicit some of the race and class distinctions baked into the idea of "strangers" and "empathy" that power so many stories across the *TAL* oeuvre. Strangers, in this case, Black Americans, are different (more generous) from *us* (White Americans). Encounters between us and them can, if told right, be magical, conferring a sprinkling of Geiger's adventurousness and Lee's generosity, on listeners.

The magic of strangers is their capacity to produce these moments of surprise, delight, and connection. Their strangeness lies less in the suddenness or unexpectedness of their appearance in our lives than in the social distance they occupy from the default "normal, middle class" listener and producer of public radio. Many more tap into the idea that momentary breaches of boundaries, "feeling less separate," across social difference produce such moments. Respondents to Johnson's 2007 survey confirm the pleasure that comes from stories that enable them to "identify with complete, random strangers."[41] Still another confides that listening to the show "makes you feel connected and not so isolated in an often-lonely cold world." *This American Life* "helps re-affirm my faith in the universality of the human experience," offers another.[42] "I feel like I'm doing something good," adds another. That same year, in his introduction to an edited collection of nonfiction stories, Glass averred that upon losing himself in stories with a similar "empathy mission" to those featured on his show, "I come out of them feeling like a better person."[43]

Stories narrated by writers feeling alone and crazy were in high rotation in the early years of the program. Frequent contributor

---

[41] Johnson, *Imagine This*, 66, 67.
[42] Johnson, *Imagine This*, 36–7.
[43] Ira Glass, ed., *The New Kings of Nonfiction* (New York: Riverhead Books, 2007), 14.

Scott Carrier performed the house style as well as anyone. At once affectless and overwrought, Carrier's narrative delivery evoked a world so saturated by artifice that human interactions, especially mediated ones, rendered us all strangers to one another. In his stories, Carrier moves through and among a veil of shadowy forms, as in "The Friendly Man" (1995), an early story on his work for a popular syndicated radio personality producing feel-good features. The radio host is only ever referred to as "The Friendly Man," which in Carrier's flat delivery comes across as terrifying at first, some gothic horror with the face of a clown, but is meant to evoke the banality and artifice at the heart of "the media."

Nobody aside from Carrier merits a name in his story. One anecdote concerns "some people in Tucson"; in another, he's menaced by his "executive producer" then aided by "an audio engineer," and so on. Each one is a cog in the merciless grinding of the Friendly Man's story machine. His alienation from the people he works for deepens as he comes to understand the nature of the work: the friendly man never meets the subjects of the stories he reports each week; even Carrier's preliminary interviews are conducted by phone to save on travel costs.

The relentlessly upbeat theme of the stories he helps produce require that nobody actually know one another, he seems to imply. From a twenty-first century perspective, his critique can come off as petulant, a Holden Caufield–esque tantrum about "phonies." When he discovers that one of the stories isn't quite as uplifting as he'd hoped, he dismisses the entire assignment as a "total sham." As alienated media labor, however, he knows he must "do just as I was told, because the audience, the twelve million listeners had something that they wanted to be told: That America was a good place, never mind the screaming coming from the basement."[44]

"The Friendly Man" confronts with suspicion a crucial trope in feel-good liberal journalism that became a staple in the golden era of broadcast journalism. Here a story about "young Black people improving their lives through basketball"; there a story about "old Black people" helping one another out; across small town America, he's hunting for "friendly ladies ... neighbors helping neighbors." Carrier's refusal to name names lays bare the politics of

---

[44]Scott Carrier, in Ira Glass, "The Friendly Man," *TAL*, episode #5, April 6, 2001.

the feel-good genre. People remain strangers because of their two-dimensional representation in the Friendly Man's formulaic stories. The deadpan repetition of the phrase "Black people," "old people," and "these people" make clear that, in the hands of corporate radio, these are production elements rather than knowable humans. He yearns to "go on the road, drive around, and collect interviews, actually meet the people." The story ends when he recognizes that this is impossible; in order to work for the Friendly Man, Carrier must remain a stranger among strangers.

Strangers works as a theme in the early years of the program to index the personal alienation of the storytellers and to call attention to social divisions that seem to require journalistic attention. Carrier's reference to "the screams coming from the basement" restores the initial impression of the Friendly Man as a horror trope, even as he fends off any affective charge, with the performatively blank affect of his delivery. The screaming is an aside almost too obvious to mention, in part as an index of Carrier's own alienation from political discourse, which is why he's an ideal chronicler for early *TAL*. The vocal aesthetics of emotive restraint running through many of the stranger stories rhyme with the appeal of "staying in touch in the tiniest way possible," as if an excess of connection runs the risk of corruption by expectations, especially those that have been built into corporate media formats.

For its one-hundredth episode, "Radio" (1998), *TAL* embraced radio's "inherent" and opposite properties of intimacy and strangeness across four acts. Each act represents a love letter to the medium, capturing the magic of listening as a series of "moments," of human connection across space and time. Stripped down to its basics, "a radio signal whose source is impossible to figure out and the intimacy of one voice," Glass enthuses, is a recipe for the magic of human connection across space and time. Each act also demonstrates, in various ways, the one-way flow in the show's traffic in feelings, in which some of us are objects and other subjects, of empathy.

In the prologue, Glass simulates the sonic experience, common to all his listeners in 1998, of spinning up and down the radio dial, pulling in a bit of advertisement, the sting of a pop song, the unctuous voice of a radio preacher, each bit punctuated by a blast of static. The last bit he plays, from a tape of found radio recordings, is of an Inuit man speaking over, and apparently translating, the Rolling

Stones' "This Heart of Stone." "[Imagine] you know, just stumbling on this?" marvels Glass, who of course, didn't stumble upon it, but actively sought it out and curated it along with the other examples. But the point lands. "It's so ephemeral, this moment just happening and passing and about to actually evaporate into nothing forever. And that's part of what makes radio different from other media, I think."[45]

Radio's real-time serendipity and its unknowability exert a strange pull on the imagination and the senses, even for Glass, who knows how meticulously each moment of his own show has been researched, recorded, edited, and produced. Even so, the voice of the Inuit man coming out of nowhere in improbable dialogue with a classic rock standard is precisely the kind of radio romance that attracted poets, scholars, and listeners for whom the ether exerted a kind of magic.

Susan Douglas chronicled the romance of "DXing," tuning in long-distance signals, as one of the earliest forms of radio listening preceding even broadcasting. Ever since, she has noted, listening along at night in a car, we "engage with a phantom, whose voice and presence [we] welcomed, needed."[46] Stranger stories partook in this romance of phantom voices, an appeal both atavistic and modern. The voice of the Inuit man, its random and exotic distance, reinforces the phantom quality of the other, an object, a curio, a spur to the imagination. "Something like this," Glass marvels, "What could be more personal?"

The radio-dial sound collage is one of the oldest tricks in the sound art book, whether it's authentic or a bespoke composition stitched together in the editing room with ersatz static interspersed after the fact. Either way, it's an elegant way to communicate so much about radio's exceptional role in everyday life. First of all, it remediates and sublimates a quotidian chore—finding a good station—into pop art. Secondly, it's a widely understood signifier of contemporary media culture—its follies, vulgarities, and corniness—stacked on top of one another, an easy shorthand for anthropological insights. It's amusing enough and almost infinitely

---

[45]Ira Glass, "Radio," *TAL*, episode #100, April 24, 1998.
[46]Susan Douglas, *Listening In: Radio and the American Imagination* (New York: Times Books, 1999), 40.

variable in its component parts that it can make exactly opposite ideological points, depending upon the context. The crass nature of commercialism; the numbing sameness of pop music; the lively heteroglossia of the metropolis; the ubiquity across musical genres and their respective subcultures of loneliness or love or both. The spinning of the radio dial speaks also to the specificity of the car radio, of listening while driving, of moving and being moved.

Act I of the radio episode, titled "Brigadoon," is narrated by Iggy Scam, a white zine-writer who opts for a soft-spoken deadpan delivery similar to Scott Carrier and Jack Hit. It's a reminiscence of his time living and driving around in northside Miami, tuned into the mysterious pirate radio signal he refers to as "Black Liberation Radio." The name is a sardonic and condescending response to an offhand comment made by one of the Black DJs about "the man." The station, whose call letters are actually WEED, is low-fi and playful, spinning a mixture of local and national rap artists and crowded with chaotic DJ patter and listener call-ins from Little Haiti and other northside neighborhoods. Iggy becomes preoccupied with finding the origin of the signal, which means zipping up and down the radio dial as he criss-crosses northside Miami.

In a formula that is now familiar to podcast listeners, we get to know the narrator through his obsessive search for something hidden or implicit or furtive, in this case, the intermittent signal of an unlicensed radio station. We learn next to nothing about the pirate station, its DJs, or the precarious nature of its existence; it's a mysterious, elusive signal, like the mythical town of "Brigadoon," of 1940s Broadway fame or the anonymous Inuit DJ spinning classic rock standards. He exists as a quaint foil for someone else's journey of self-discovery. In the 1990s, there were nonlicensed stations calling themselves Black Liberation Radio, including one in Springfield, Illinois, not far from WBEZ where *This American Life* was produced.[47] The decision not to profile community stations that were less mysterious, more knowable, and more able to speak for themselves, suggests that keeping radio strange was more important than the strangers themselves.

---

[47] Black Liberation Radio, also known as KTRA was a 1-watt radio station run by Mbanna Kantako in Springfield Illinois. See Christina Dunbar-Hester, *Low Power to the People: Pirates, Protest, and Politics in FM Radio Activism* (Cambridge, Mass: MIT Press, 2014).

TAL stories on the theme of strangers were often tales of estrangement from other people and from the mass mediation of human interaction. Deadpan narration, paired with a lonely journey by car or bus or train, helped to convey this anomie. For narrators like Carrier, close-mic'd flat affect delivery conveyed the irreducible singularity of perspective, aloneness. For Sarah Vowell and David Sedaris, the vocal performance of hipster-nerd insouciance conveyed an ironic confidence that listeners would cover the ground to meet them more than halfway.[48]

Finding shared humanity in encounters between strangers and snatching moments of authenticity from the maw of the media machine were explicit themes throughout the early years of *TAL*. The stories on this theme are often tightly crafted gems evoking allegories or parables in the service of emotional succor rather than moral or civic instruction. It is in these stories that *TAL* serves most indelibly as a relay point from radio's vox populi past to the podcast structure of feeling that it was helping to invent. The dramatic pull of stories on this theme implicitly acknowledges the critique of empathy as an ethos and a method. As critics have noted, empathy is a limited resource: We don't typically extend empathy to strangers as much as to those closest to us.[49] Radio is an empathy machine to the extent that it helps to produce exceptions to this rule or at least to the extent that it produces the illusion of an exception.

Stories on the stranger theme tend, as I've said, towards the picaresque: storytellers in cars, in buses, on their own, and on the move. Many stranger-themed episodes are featured on the show's website under the "For Your Road Trip" category under the "Recommended" tab, suggesting that, per Glass, the best way to listen to a road trip story is when on the road oneself. In Act III of the "Radio" episode, Glass interviews an itinerant DJ whose heart has been broken by corporate radio's formulaic playlists, whose search for radio's lost magic leads her to replicate the dial spinning quest for authenticity on the new medium of internet radio where she ranges far and wide looking for a real signal. "The most any of

---

[48] Jason Loviglio, "Sound Effects: Gender, Voice and the Cultural Work of NPR," *Radio Journal: International Studies in Broadcast and Audio Media* 5, no. 2 (2008): 77.
[49] Paul Bloom, *Against Empathy: The Case for Radical Compassion* (New York: Ecco, 2016).

us can hope for in this environment," Glass narrates sympathetically, "is pockets of individuality." For Glass, these central contradictions (e.g., empathy for strangers, mediated authenticity) have been productive ones; their unwinding also helps to explain the series of shocks and scandals that rocked the audio storytelling industry in the late 2010s, which I explore in Chapter 7.

Glass has admitted that it takes a significant amount of narrative enchantment to produce empathy for strangers and that it comes, mostly, in "moments." From his earliest speaking tour, Glass has boasted of the deftness with which he deploys this trickery in the service of empathy between his listeners and the Other. In a lengthy analysis of a story on the mundane travails of a Mexican American teenager named Sylvia, navigating parental expectations and peer pressure, Glass makes clear the sleights of hand necessary to produce empathy. "I don't say that she's Mexican until a ways in … because … people conjure images and they think that 'that's not me,' and it just pushes you away." "We tried to structure the story to make these [poor, Mexican, immigrant] kids seem like just your kids, if you live in the suburbs, and to try to create empathy, to say that this person is just like you." Lest his audience misunderstand, Glass doubles down on the assumed class and racial distance he expects them to travel:

> And then [Sylvia] goes on to, you know, tell the story about what's going on with her and her Mom and how they disagree on her future, and the whole thing is designed to make her sound exactly like you and me.[50]

The shifting pronouns Glass uses for his assumed audience, from "they" ("they think that's not me") to "you" ("pushes you away") to "you and me" at the end, aptly ranges across public radio's understanding of its audience, covered in the previous chapter. Listeners are people like us who need a nudge. The *public* of public radio refers to these people and to that nudge, the obligation to extend empathy to that other, broader public: the people who aren't, "like us," white and middle class.

---

[50] Glass, "Mo' Better Radio."

The working assumption of a public radio audience composed of people-like-us, provides a framework for pursuing moments of personal surprise and amusement that will likely land similarly for listeners. Glass encourages would-be radio storytellers to "follow a kind of pleasure and feeling and instinct and build around that."[51] The passage is instructive in its embrace of affect as a moving thing, a thing that connects people, moves them, and moves between them. Following feeling depends upon "instinct," and those instincts, Glass insists, are often rooted in habitus, the intersection of class and taste. Referring to another early story, Glass described his exultation when a young unnamed Black boy living in a poverty-stricken housing project in Chicago tells him that he yearns to live in a house with a basement: "Thank you. Thank you. Thank you. Thank you. Thank you. Thank you." The reason? "Because his dream is so normal and middle class, and anybody can understand it. And empathize."[52]

Recreating such moments of joyful surprise requires narrative trickery: "we consciously manipulate the facts to allow you entrance," Glass confides. This movement from instinctive pursuit of pleasure and conscious manipulation is in the service of "making you relate to characters you normally would *not* relate to," acknowledges the difficulty of extending empathy outside of tight networks of intimacy. It also concedes the stickiness of association between non-white ethnicities and anti-social stereotypes, as Ahmed does in her analysis of how metonymy in political speech can "stick words like terrorist and Islam together," even when the connection is being explicitly rejected. Only here, Glass is trying to unstick the association between Mexican and "not-like-me." "The sliding between signs," Ahmed notes, "also involves sticking signs to bodies." This stickiness works "even when arguments are made that seem to unmake those links," like President Bush's insistence that "this is not a war against Islam" in Ahmed's example.[53]

Glass's hope that keeping the word "Mexican" at bay for as long as possible means "you would *become* them more," both acknowledges the stickiness while failing perhaps to account for

---

[51] Glass, "Mo' Better Radio."
[52] Glass, "Mo' Better Radio."
[53] Sara Ahmed, "Affective Economies," *Social Text* 22, no. 2 (2014): 132.

the ways in which the entire construction of strangers depends upon and strengthens the racist and classist metonymies at play. For instance, the feature on the Black boy in the housing project, turns on the notion that the projects "operate just like a small town," a conceit that assumes the obligation of unsticking urban poverty from Otherness and adhering it instead to a much rarer and nostalgic conception of community, which is often mobilized in the American imaginary as White.

Such efforts, in the name of empathy, may work to valorize one kind of American life to the detriment of others. Lisa Nakamura sees a similar dynamic in the "toxic empathy" at play in journalistic and documentary use of virtual reality (VR) putatively in service of "anti-sexist, anti-racist," ends. The kind of place-taking made possible by technologies of feeling are, she argues, "founded on the concept of toxic re-embodiment: occupying the body of an other who might not even own their own body."[54]

The production of empathy for strangers required technical prowess; "on our show there's an edit every eight seconds," Glass boasted in an early interview. It often required fairly elaborate stunts like Dishwasher Pete's cross-country errand into the wilds of Greyhound America.[55] Other stunts were more narrative in nature. In "Didn't Ask to Be Born" (2002), the shocking fact that two runaway teenaged girls who narrate as adults, many years later, a harrowing story of homelessness and addiction, are so young (one is only fourteen years old), is held back until well into the story. This produces an effect Glass later admits is purely for the aesthetics of surprise, rather than in an attempt at producing empathy for exotic protagonists.[56] If anything, the revelation produces an almost prurient exoticism of the narrators whom we've already come to know as articulate and insightful adults, looking back on the events of their childhood.

Other episodes stretched and trimmed stories to fit the stranger theme in procrustean fashion, that were largely successful, to the

---

[54]Lisa Nakamura, "Feeling good about feeling bad: virtuous virtual reality and the automation of racial empathy," *Journal of Visual Culture* 19, no. 2 (2020): 47–64.
[55]https://www.chicagotribune.com/news/ct-xpm-1998-10-18-9810180472-story.html.
[56]This story wasn't produced by *TAL* but instead by Sandy Tolan of Homeland Productions. However, its selection as for this episode speaks to its structural and stylistic fit with the themes and strategies mentioned here.

extent that they sounded powerful, if paradoxical themes about the shared loneliness that connects people, nestled in a musical bed evoking a certain "yearniness."[57] "Some of us spend a lot of time thinking about people we don't know," notes Lily Sullivan in an episode entitled "The Lives of Others." "And some of us are the thought about."[58] Creating "these moments where there can be this portal of emotion from your life to someone else's," as Sullivan puts it, requires art, sticking and unsticking words and associations to feelings. For Sullivan, as for Ahmed, emotions move across lives, through portals, rather than residing within us.

The theme of strangers made way for stories about loss, estrangement, and the singularity of human consciousness, exploring and evoking powerful feelings. The programs' formal emphasis on personal narratives and interiority may well function both as antidote to and a model of the anomie of contemporary neoliberal life. Strangers evoke "the sense of longing, and the distances there always are between people" in ways that provide temporary respite and an acknowledgment of the distance.[59]

And as the gauzy indifference to politics of the early years of *TAL* gave way to more politically engaged stories, strangers was a capacious enough theme to embrace topics like the US immigration crisis and a range of social justice "reckonings" that rippled across the political media landscape during the first Trump presidency. Estrangement and strangeness were no longer merely features of modern existence, but instead, tied to specific contemporary problems. Trading universal truths for the currency of the topical was part of a wholesale shift in *TAL*'s approach to storytelling whose origins and implications I'll explore in later chapters.

A 2018 episode on libraries provides an instructive example of the power of strangers to resonate with contemporary political controversies. Producer Zoe Chace extols libraries as charmed spaces that, like the "Room of Requirement" in the *Harry Potter* novels, "magically change to suit the needs of its users." Perhaps at any other time in American life, a program dedicated to appreciating public libraries would sound like pabulum, a slow-

---

[57] Abel, *Out on the Wire*, 152.
[58] "The Lives of Others," *This American Life* #799, May 12, 2023.
[59] "The Lives of Others," *TAL* #799, May 12, 2023.

news-day feature, or an eye-rolling bit of Cold War propaganda for American democratic institutions.

But by the second decade of the twenty-first century, the episode represented an intervention into a simmering culture war. On one side, the long neoliberal struggle to replace public institutions with private ones seemed to have joined forces with the atavistic conservatism that energized many Trump supporters who approved his attacks on liberal tolerance of sexual and racial minorities. By 2018, the public library, like public education, the paying of taxes, and the federal government itself, had become endangered species, newly prized for their increasingly rare status as "resting places of decency."[60]

The first story leans into the notion of library as political refuge with a story about a library perched between the United States and Canada. Its border-straddling location—the entrance in the United States; the reading room in Canada—enables Muslim international students banned from re-entering the United States if they leave to visit family, to visit them in a quirkily trans-national space. "Like imagine these Iranian students in America, thinking, I need a magical place that is somehow in America but also outside it," Chace marvels, in a vocal style that owes a bit to Glass's ingenious delivery. "It's like a very particular, only-at-this-time-in-America would this exact requirement exist." The intimate reunions Chace describes between adult children and their parents are moving in part because they are played out in public, among strangers, and therefore more restrained and cinematic. Reporter Yeganeh Torbati, who chats with Chace about the interactions, is struck by this tension as well.[61] Families reunited in this magical room are "expressing all their emotions, but also trying to fit them into the space."

---

[60] Radio historian David Hendy uses this elegant phrase when describing the BBC, an institution he's has chronicled, often critically, but which he has also had to defend, in recent years, against similar neoliberal and conservative attacks on its right to exist. Hendy is riffing on Timothy Snyder: *Tyranny: Twenty Lessons from The Twentieth Century* (New York: Crown, 2017). David Hendy, *Routledge Companion for Radio and Podcast Studies*, online book launch: European Communication Research and Education Association, online, November 3.

[61] Ira Glass, "Room of Requirement," *TAL*, episode #664, December 28, 2018.

Media Strangers

As I've mentioned above, one of radio's earliest and more durable genres, the audience participation program, demonstrates the enduring appeal of briefly extending empathy to strangers. From syndicated advice columns to *Queen for A Day*–style giveaway shows, to "vox pops" or "man-in-the-street" clips, this genre of entertainment has been successful and compelling across media platforms. In the 1960s, Tony Schwartz's *Adventures in Sound*, an occasional series on WNYC in New York, included interviews with strangers and intimates alike, capturing their (and his) strangeness and humanity. Talk radio, which became a dominant form in the 1970s and 1980s in the United States, also represented the implicit democratic argument of audience participation genres.

Joe Richman's *Radio Diaries*, another public radio innovation which launched the same year as *TAL* and now can be heard as a podcast on the Radiotopia network, represents an important instance of this formula. Richman's method of gathering tape was starkly different. He'd hand over a recorder to a subject to make regular audio entries and collect and edit the tape afterwards. But he shared the ambition of bringing listeners into intimate contact with strangers, for "lucky moments" in which they become "characters."[62]

StoryCorps, an audio nonprofit project started by public radio veteran David Isay in 2002, and frequently aired on NPR's *Morning Edition*, brought together people already connected through bonds of intimacy for recorded conversations or interviews, which were then, like Richman's *Radio Diaries* and like interviews on *This American Life*, heavily edited. The tightly formatted and emotionally compelling StoryCorps segments have been singled out for being especially productive of driveway moments and for bringing drivers to tears on their Friday morning commutes.[63]

---

[62] Abel, *Out on the Wire*, 80.
[63] "StoryCorp's New Season Also Features A New Host: Kamilah Kashanie," *Inside Radio*, December 15, 2020, https://www.insideradio.com/podcastnewsdaily/storycorp-s-new-season-also-features-a-new-host-kamilah-kashanie/article_541a907a-3efb-11eb-a7b7-d3b580ad7bb5.html.

Even so, *TAL*'s specific formula of spinning stories in which encounters with strangers work a kind of magic spell for listeners and for the people inside the story has had a deep and widespread impact on the American style of podcasting. In part, this was because of Glass's hand-holding narrative style and his give-it-away ethos of sharing (and promoting) the show and encouraging cross-promotions with kindred shows, like *The Moth*, a live theatrical amateur storytelling series that pre-dated *TAL* and which spun off *The Moth Radio Hour*.

A *Moth* spinoff of sorts, Lea Thau's award-winning podcast *Strangers* (PRX, 2012–19) doubled down on the idea that audio storytelling was precisely for making us all "strangers no more," a motto that suggests a more ambitious movement beyond momentary intimacy. *Next Door Stranger* (NPR, 2018) was "a podcast about finding connection in a time of division," in part a response to the violent racism in the 2017 Charlottesville Unite the Right rally. It was another explicit attempt at using audio to find a more enduring common ground for political as well as emotional reasons. That same year, *Storycorps* launched *One Small Step*, which paired two strangers on opposite sides of the political divide. "This is what public radio is for," Isay said by way of introduction. "As we know from StoryCorps," he continued, "it's hard to hate up close."[64] The goal of the project seemed to be the kind of intimate alchemy championed by Glass in stories about strangers with whom you are forced to identify: "you *become* them more" (emphasis in the original).[65]

Invisiblia[66] represents a good example of a show that took up and extended elements of *TAL*'s formal and tonal storytelling elements, along with the commitment to audio narrative as a mechanism for creating moments of empathy for strange people. Like *TAL*, it began as a public radio show before shifting to podcasting, first as a catch-up mechanism, then as its main format. Like *TAL*, it organized episodes around a single theme; like *TAL*, it zeroed in on character, often dedicating an entire episode to a

---

[64]Lisa Rayam, "StoryCorps Founder Dave Isay's New Project to Bring Americans Together," *Wabe*, March 25, 2021, https://www.wabe.org/storycorps-founder-dave-isays-new-project-to-bring-americans-together/.
[65]Glass, "Mo' Better Radio."
[66]NPR, 2015–23.

single person's story. In a move that marked its extension of the public radio storytelling method into a more properly podcasting format, it developed overarching themes for entire seasons; while each episode could stand alone, thematic unity rewarded multi-episode bingeing sessions.

Invisibilia was developed by *This American Life* veteran Alix Spiegel, and Lulu Miller, a former *Radiolab* producer. In 2016, Hanna Rosin, co-founder of the *DoubleX* website and podcast from *Slate* joined, taking over for Miller. Unlike *TAL*, which emphasized the universal in every human story, often holding back elements of particularity (ethnic, racial, sexuality) until the last moment, *Invisibilia* led with the strangeness. Finding extreme cases to make a point about shared humanity and challenging listeners to cover the distance. I will explore *Invisibilia*'s extension of the public radio structure of feeling, particularly in regard to gender and storytelling, in later chapters. Other radio-to-podcast programs (*Planet Money, Radiolab, Snap Judgment*) created lushly produced nonfiction stories with sympathetically drawn protagonists and explored similar terrain. Across the emerging podcasting universe that grew out of United States public radio in the 2000s, strangers played a key role in introducing listeners to aspects of humanity, the strange and the familiar, in the service of moments of recognition and empathy.

One of the controlling ideas of *TAL*'s early years was that Americans felt "crazy and separate," and that part of the problem was the distance that contemporary media forms and institutions put between us. Another perhaps even stronger theme from this era was that the seat of American estrangement lay in the modern nuclear family. It was the very intimacy of this family home that was responsible for the secrets and lies that made us feel alone. Uncovering those secrets, revealing those lies powered some of the most haunting and cited episodes of the early era.

These episodes and stories, taken together, provide a powerful exploration of the idea that the movement of affects, rather than the forces of history and politics, were the most important factors in understanding contemporary American life. This message, delivered again and again, shaped an understanding of human interiority as both inscrutable and lonely. Intimate revelations, family secrets unspooled, were presented as a balm for broken hearts and invitations for moments of connection, even if only at a remove.

Gothic Family Mysteries

The obverse of finding elements of familiarity in the lives of strangers is encountering strangeness within one's own domestic scene. Gothic family dramas have been another durable and generative theme in the *This American Life* oeuvre, even if sometimes they require what Glass calls "a strange finesse" to cohere, stretching backwards from literary origins, and forward to the memoiristic turn in audio storytelling. *TAL*'s early preoccupation with the family scene remediated the novel and theatre,[67] providing psychic material for modern emotional interiority. It also provided literary cover for popular genre preoccupations like stolen babies, uncertain paternity, true crime, and sexual awakenings (i.e., the stuff of daytime talk shows, soap operas, and romance novels), topics that have spread across many podcasts produced by *TAL* and NPR alumni.

The durability of this theme was tested by the mostly white male perspective of these stories, a seemingly arbitrary limitation, but one that proved, on closer inspection, to be constitutive of the genre. (Chapter 6 explores in greater depth the show's preoccupation with masculine performance and its consequences for audio storytelling.) This next section examines the theme of families, their gothic mysteries, and the alchemy needed to make them into emotionally compelling long-form audio narratives.

TAL's transcripts of its 850+ episodes reveal the centrality of family as a theme. The word "family" appears almost 5,000 times across the transcripts and in 726 episodes, over 90 percent of all episodes. A look at the episodes under the "Recommended" tab on the program's website also bears out the idea that family stories have been privileged topics. More than half of the featured episodes touch on one or more of the following themes: family, teens, childhood, children, mothers, marriage, fathers, divorce, siblings. Audience reception research on the show's podcast listeners circa 2008 turned up a list of favorites that heavily overlaps with those featured on the program website.[68]

---

[67]Lynne Spiegel, *Make Room For TV: Television and The Family Ideal in Postwar America* (Chicago: University of Chicago Press, 1992).
[68]Johnson, *Imagine This*, 33.

Often narrated from the perspective of an adult remembering childhood, the tone of *TAL*'s Gothic Family Mysteries mixed the wide-eyed innocence of youth with the arch distance of the adult narrator. Many such narrators spoke like preternaturally articulate children: perceptive, sensitive, ironically recalling the contextlessness of youth. In this way, formally at least, the stories bear some resemblance to *The Wonder Years*, a popular television show of the era (ABC, 1988–93), which was swathed in Cold War nostalgia and voiced in the universalizing perspective of a white heterosexual middle-class boy and his older, sardonic retrospective self. Such stories were in tune with larger cultural narratives about family dysfunction, changing ideas about gender and sexuality, but almost always expressed in ways that centered feelings that could be described, per Raymond Williams, as "private, idiosyncratic, and even isolating."[69]

Unlike the biting, subversive satire that Henry James used to lay bare adult secrets through the innocent child's perspective in *What Maisie Knew* (1897), *TAL*'s children narrators paired their adult insights with the passivity of remembered youth and innocence. Like David Sedaris's family stories, the adult child assumed a passive, if petulant, childlike posture while revealing contemporary family secrets.[70] Dave Eggers's 1999 *TAL* story of supporting his Republican brother's campaign for local office conveyed the power of family loyalty over politics, while still communicating the essential haplessness and passivity of remembered youth.[71]

At times, the thematic cramming of multiple acts in a single episode also constrains the emotional intensity of each story, except for those occasions when a grand family mystery takes up the entire hour-long episode and becomes its own theme, as in *The Ghost of Bobby Dunbar, The House on Loon Lake,* and *Switched at Birth*. Central to these and many other episodes and individual stories in the gothic family style, is a narrative of loss set in the mysterious crucible of the nuclear family. The most common emotional themes

---

[69] Williams, *Marxism and Literature*, 128–35; and *Preface to Film*, 21–3.
[70] Jon Ronson, "The Trajectory and Force of Bodies in Orbit," *TAL*, episode #214, May 31, 2002.
[71] Dave Eggers, "Family Photo Op," *TAL*, episode #140, September 24, 1999.

include nostalgia for lost certainties, lost or confused identities, and mysterious family ties. The resolution of these crises of loss sometimes requires the intervention of science in the form of paternity tests, or DNA testing, or hushed confessions, as in "The Ghost of Bobby Dunbar" (2008), "Switched at Birth" (2008) and "Go Ask Your Father" (2005). The narrative preoccupation with paternity, as I've said, has been shared with soap operas, *The Jerry Springer Show*, and melodramas from time immemorial.[72]

Questions of identity and parentage animate "The Ghost of Bobby Dunbar," "Switched at Birth," and "The House on Loon Lake" (2001).[73] Absent fathers, distant fathers, mysterious, unknowable fathers and substitute fathers (and mothers) loom in "Go Ask Your Father" "Make Him Say Uncle" (2005), "Babysitting" (2001), "Sissies," "Origin Story" (2008), "Social Engineering" (2008), and "Valentine's Day" (2008).[74] Fathers sought redemption in "Who Do You Think You Are" (2008), "The Wrong Side of History" (2009), and "Long Shot" (2010); fathers leaving home lead to the disintegration of the rest of the family in "Didn't Ask to be Born" (2002); and others.[75] In "Baltimore, Circa 1956" (1996), Barry Glass, Ira's father, is represented as a distant "workaholic" father.[76]

Stories of switched babies, lost fathers, invented families, and other family mysteries sound the twin themes of the search for lost identity and the relief of re-discovering them in DNA, in the nuclear family, and in narrative closure. Even when multiple perspectives on a family mystery are explored, as in "Searching for Bobby Dunbar," "Switched at Birth," and "Babysitting," family members

---

[72]Ira Glass, "The Ghost of Bobby Dunbar," *TAL*, episode #352, March 14, 2008; Ira Glass, "Go Ask Your Father," *TAL*, episode #289, May 13, 2005; Ira Glass, "Switched at Birth," *TAL*, episode #360, July 25, 2008.

[73]Ira Glass, "The House on Loon Lake," *TAL*, episode #199, November 16, 2001.

[74]Ira Glass, "Make Him Say Uncle," *TAL*, episode #289, May 13, 2005; "Babysitting," *TAL*, episode #175, January 5, 2001; "Sissies," *TAL*, episode #46, December 13, 1996; "Origin Story," *TAL*, epsisode #383, June 19, 2009; "Social Engineering," *TAL*, episode #358, June 27, 2008; "Valentines Day," *TAL*, episode #349, February 15, 2008.

[75]Ira Glass, "Who Do You Think You Are," *TAL*, episode #368, November 7, 2008; "The Wrong Side of History," *TAL*, episode #376, March 13, 2009; "Long Shot," *TAL*, episode #398, January 8, 2010; "Didn't Ask To Be Born," *TAL*, episode #209, March 29, 2002.

[76]Ira Glass, "Baltimore, Circa 1956," *TAL*, episode #14, February 21, 1996.

are interviewed one at a time, never together in conversation, which gives each perspective its own kind of lonely authority, closing it off from the social, collaborative, and partial way that family memories are often processed and shared. Such collaborative storytelling is messier and less useful for abstracting big ideas and big feelings. A program so focused on family secrets and dysfunction might have occasionally explored the political or social stakes of the family as an inherently opaque social institution with a history of structural and affective flux. But that would have required conforming to the already written narratives of the political, which, Glass had let us know again and again, in interviews and in episode introductions, are boring.

Math and Magic in Narrative Audio

To hear Glass or any of the other longtime producers tell it, the show's success (i.e., what Jessica Abel calls "making these stories solid gold") depends precisely on the formula of word-centered narrative guidance.[77] Articulating a very specific "big idea" takeaway is crucial for radio stories for "taking it out of the province of bar story and into the world of literature," Glass tells Abel, "which is where you know, where you want to be ... at this end of the radio dial!"[78] Much of this "finesse" comes in post-production, in the editing phase, a skill that Glass credits to his years as a tape cutter for *All Things Considered* host Noah Adams. It's noteworthy that as late as 2015, at the start of the "proleptic" hype around podcasting, both Abel and Glass still referred to long-form narrative audio as "radio" rather than podcasting.[79]

Word-centered storytelling, while often didactic, could also pack a powerful emotional punch. With *TAL* and its imitators, the sticking and unsticking of words and feelings to objects and social bodies was accomplished largely through editing, which was its own kind of alchemy. Abel keys in on the idea that the real magic

---

[77] Abel, *Out on the Wire*, 3.
[78] Abel, *Out on the Wire*, 28.
[79] On podcasting's "proleptic imaginary," see Neil Verma, *Narrative Podcasting in an Age of Obsession* (Ann Arbor: University of Michigan Press, 2024), 2.

happens in the editing booth—in the creation of a simulated reality. Pat Mesiti-Miller, a sound editor for *Snap Judgment*, likens his work to that of his counterparts in Hollywood. "This is the same kind of magic that the Foley artists work on the movies we love. It sounds more real than real."[80] Joe Richman's process, of breaking down hours of radio diary tape into its "atoms" and then rebuilding it into a story, is according to Abel, "magic!"

In breaking down the process by which stories become "solid gold," Glass and other podcast producers featured in Abel's book and in other sources oscillate between math and magic. Glass frames it as "planning versus chance," admitting that one must rely on both; "the magic of being out in the field," he says, is the element of chance. John Biewen keyed in on the metaphor of radio as alchemy, "turning base materials into gold" all in the service of "making … listeners feel something."[81] It's not enough for them to convey facts. They gather words and sounds and music, and assemble them, painstakingly, into an *experience*, or as Soren Wheeler, one of *Radiolab*'s producers, puts it, "turn it into a feeling."[82] The metaphor of alchemy speaks to the mixture of science and magic in a mobilizing feeling.

Rob Rosenstein, of the Transom Story Workshop, puts his stories ideas into a formula: "Somebody does something because _____ but _____."[83] Wheeler uses another: "This happened_____, then this_____, then this_____ and then you wouldn't #$%&*! believe it but _____. And the reason that is interesting to every person walking the face of the earth is _____."[84] *TAL* veteran and Gimlet founder Alex Blumberg is similarly attached to understanding storytelling's magic in terms of mathematical formulas. "For example, I'm doing a story about a homeless guy who lives on the streets for ten years and what's interesting is, he didn't get off the streets until he got into a treatment program. Wrong track. Solve for a different Y." He runs through several other possible scenarios that are, he believes, better tracks. "Y= … and what's interesting is there's a small part of him that misses being

---

[80] Abel, *Out on the Wire*, 161.
[81] Biewen, *Reality Radio*, 8.
[82] Abel, *Out on the Wire*, 187
[83] Abel, *Out on the Wire*, 52.
[84] Abel, *Out on the Wire*, 60.

homeless." Or "Y = and what's interesting is he fell in love while homeless and is haunted by that love still."

The importance of a predictable formula with a reliable "twist," says Wheeler, is key to moving stories. In Blumberg's example, the key is to shift from policy and politics and towards something more universally relatable. The wrong Y, in these examples, are those which make the cardinal sin of being predictable, sociological, boring. They're not surprising. A homeless person in love, is for some reason, fascinating, perhaps because it makes the valuable point that they are "just like us." Like successful labor actions, stories about treatment programs are "not what we do."

Some of this occurs in pre-production, as well. Glass relates the instruction he got as a young reporter at NPR from Mike Shuster, then head of the New York bureau, to think about what his first bit of tape would be *before* he went out to do the interviews. After initially balking at this seemingly unethical approach to objective truth-finding, Glass admits that this was a conversion point, a moment when he realized that the storytelling trumps the gathering of information. "I've had times where I will make up a character and then send people looking for them," Glass admits.[85]

There is artistic and ethical danger in approaching the subjects of human-interest journalism as base materials, which could be acknowledged and forgiven sometimes by the larger mission statement of building empathy. It is in this tension between humanity and art, journalism and alchemy, where the close-up magic trick of empathy begins to unravel. The imperative to "move people," through math and magic can be persuasive. It can also appear manipulative. When Soren Wheeler, producer for *Radiolab* tells Abel, "I want impact on bodies ... to literally move people ... their faces, their bodies," it's not clear if there's a larger principle being served.[86] In the concluding section, I explore some of the limits of this impulse to move, to mobilize feelings and meanings.

---

[85] Abel, *Out on the Wire*, 64.
[86] Abel, *Out on the Wire*, 185.

Bathos: Who Laughs Last (Whose Laughs Last?)

Bathos, the literary gimmick of moving abruptly from the sublime to the ridiculous, is the term that best captures the tonal sleight-of-hand that suffused *TAL*'s ambivalent playfulness, especially in its early years. This is most apparent in the show's closing gag, in which a bit of tape from the just-ended program is re-purposed to poke fun at Torey Malatia, WBEZ's program director for much of the show's run. The move, a bit of fun between Glass and his boss, places the latter in the place of ventriloquist dummy, giving voice to a snatch of dialogue or narration, with a new and gently emasculating context provided by Glass, the ventriloquist-editor. One of the earliest and most consistent features of *TAL*, the gag first appeared in 1996 in episode #9. Often these bits of audio tape employed a woman's voice or perhaps that of a child, or the set up positioned Malatia a feminized addressee. The goal, each time, is to produce a moment's comic relief through the absurdity of a new juxtaposition, although rather than an arbitrary juxtaposition, the humor comes from innuendo.

> Glass: WBEZ management oversight by Torey Malatia, who turned to me after our first time on the air and said—
> David Sedaris: Relax, sugar, you're a woman now.

The teasing manages to be laddish in its not-so-subtle implication that it's funny to ventriloquize a man in a woman's voice or to repurpose audio to refer to a man as a woman. The potential hostility of such a move is successfully mitigated by several factors: the Mad-Libs crudeness of the gag conveys silliness rather than malice. In later seasons, Glass introduces Malatia as "our boss," his voice reliably cracking on this deferential phrase into a boyish falsetto evocative of the Magical Mystifier he was at twelve years old; the honorific, along with the designation of "*Mr.* Torey Malatia," cues both the punching-up nature of the gag while also clarifying Malatia's masculinity just as it is punctured for comic effect. Often the set-ups imply a sexual kink or a campy kind of performatively queer sexuality, which again lands as friendly in the idiom of laddishness, a jocular teasing in which the derogation of one's manliness as gay or effeminate serves as a shared ritual of reconstituting it (in ways that radio's earliest comedians employed to similar effect, by "going swish," as Matthew McAllister has demonstrated).[87] Take episode #99, for example,

> Glass: WBEZ management oversight from Mr. Torey Malatia who says that from now on he wants to be called—
> Rebecca: Ghetto hoochie mama.
> Glass: Or perhaps you would prefer—
> Rebecca: Booty house girl.

In this case, the borrowed voice comes from a girl at a racially diverse urban high school, so it's not just gender and sexuality that are in play here. In addition, Glass and Malatia, two middle-aged professional white men, signify their relationship through the comic travesties of race, sexuality, class, and age. Once again, the humor is centered in a sudden juxtaposition of identities across social difference. It qualifies as bathos because the rest of the program is dedicated to precisely the opposite movement, using audio narrative, first person singular, the human voice, to bridge these social differences and to emphasize common humanity, rather than the categories that make us feel "so separate," as Glass puts it. The self-deprecation of identifying himself and/or Malatia with socially subordinate identities functions as an alibi against mean-spiritedness; yet it is precisely the association with the feminine, queer, and nonwhite that constitutes the deprecation, an implicit hostility. In a characteristic fending maneuver, Glass has defended the bit by distancing himself from it, "It's totally sophomoric and indefensible, but hard to let go of."[88]

The repositioning of audio testimony into a gag also suggests a certain disregard for the original emotional context of the interviews, narratives, stories. Take for instance, the credits sequence from a 2006 episode, which used the borrowed audio of Jake Royko, a teenager, describing the horror of a particular tantrum thrown by his autistic twin brother, Ben, that helped convince his family to institutionalize him. It's an unusually heavy story emotionally,

---

[87] Matthew Murray, "The Tendency to Deprave and Corrupt Morals: Regulation and Irregular Sexuality in Golden Age Radio Comedy," in *Radio Reader: Essays in the Cultural History of Radio*, eds. Michele Hilmes and Jason Loviglio (New York and London: Routledge 2002), 135–56.
[88] Ira Glass, "Ask Me Anything," Reddit, October 10, 2012, https://www.reddit.com/r/IAmA/comments/13gox8/iam_ira_glass_back_for_another_ama_with_something/.

mostly told by Dave Royko, the boys' father, with no moments of levity.

> Glass: Thanks, as always, to our program's co-founder, Mr. Torey Malatia, who I tried—I tried talking him into starting this radio show for years, until that one night—
> Jake Royko: That one night with the banging, and the hitting, and the screaming, and the sobbing, and the more sobbing, and the scratching, and the banging, and the pounding through doors that—you remember it.[89]

The repurposed bit asks listeners to move from trauma to joking very quickly. Some listeners have defended the gags as a lighthearted way to conclude episodes which often feature heavy themes, difficult stories, and references to emotional or physical violence.[90] Others have pointed out that the handful of episodes eschew the gimmick feature stories about which Glass himself is personally emotionally invested. Most notoriously, "Retraction" (2012), in which an audibly distressed Glass confronts disgraced documentarian Mike Daisey for fabricating facts in his report, aired the previous week on *TAL*, on working conditions at the iPhone manufacturing plant in China ("Mr. Daisy and the Apple Factory").

> Glass: WBEZ Management oversight by our boss, Mr. Torey Malatia, and I think this is a week I am just not in the mood for an extra quote here from Torey.[91]

Glass later admitted that he feared the scandal could mean the end of the show.[92] The first episode to air after the September 11 attacks was another one for which the particular comic relief of this kind of bathetic repurposing of voices seemed inapt. Such exceptions

---

[89] Ira Glass, "Unconditional Love," *TAL*, episode #317, September 15, 2006.
[90] Glass, "Ask Me Anything."
[91] Ira Glass, "Retraction," *TAL*, episode #460, March 16, 2021.
[92] Delia Cabe, "Truthiness: *This American Life* and the Monologist," *Case Consortium at Columbia University* (Columbia University, 2012), https://ccnmtl.columbia.edu/projects/caseconsortium/casestudies/108/casestudy/www/layout/case_id_108_id_739.html.

demonstrate a keen sense of the disrespect and potential for hurt feelings inherent in such revoicing, as well as a selective sense of when to worry about it. When, in 2013, Malatia resigned (under pressure) from WBEZ, Glass changed the sign off, acknowledging him as "our co-founder, Mr. Torey Malatia," a not-so-subtle way for Glass to protest the change and reinforce Malatia's importance to the program.

The appeal of the bathetic, at least in the context of the Torey Malatia gag, is in the hypermobility it displays across affective registers: now heartfelt, now cynical; now feminist, now post-feminist. The play of fending off any fixed emotional position requires a light touch and Glass's mastery represents his version of an audio high-wire act.[93] It serves as a reminder of the constructedness and temporary nature of the show's conjuration of feeling, the "grandeur" that Glass says he aspires to in the production process.[94] Moving bits of tape around, organizing stories around anecdotes and reflections, the audio storyteller is a wizard, a magical mystifier, whose spells can be done and undone.

This mobility across affective registers demonstrates the sideways movement of affect, the sticking and unsticking of feelings and signs, forming social bodies and dispersing them. The figure of empathy in the story, a child living through a chaotic childhood, becomes in the episode's coda, a figure of fun, source code for an insiders' gag. His words, his voice, his experience can be attached and re-attached at will, demonstrating the editor's power to "turn it into a feeling." The feeling is necessarily a transient one, and even perhaps an arbitrary one.

Such a game requires narrative enchantment, the constant production and repurposing of moments of heightened feeling. It demonstrates the way that affect forms and reforms through the circulation of signs among social bodies. The same audience brought together affectively in shared empathy at one moment, can be united in laughter the next moment by the very same audio, undermining Glass's statement "I feel very protective of the interviewees."[95] It is

---

[93] Abel, *Out on the Wire*, 81; Neil Verma, "Radio's Oblong Blur: On the Corwinesque in the Critical Ear," in *Anatomy of Sound: Norman Corwin and Media Authorship*, eds. Jacob Smith and Neil Verma (Oakland, CA: University of California Press, 2016), 42.
[94] Abel, *Out on the Wire*, 28.
[95] Abel, *Out on the Wire*, 27.

also part of the besetting contradiction at the heart of the public radio structure of feeling: empathy is fleeting, a pleasurable ripple of shared feeling, momentary. It's an essential part of feeling liberal, even while fending off liberalism in favor of something a bit more detached and a bit meaner.

In some ways, the central aesthetic insight of the popular culture version of postmodernism was this brand of disjunctive irony. Simply moving a signifier from one context to another for the purposes of satire or merely for the passing *frisson* of an arbitrary collision of meanings was popular because it was easy to accomplish with modern editing tools. It also appealed to the sensibility of exhaustion at the surfeit of media signifiers, their constant circulation, and the seemingly endless mix-and-match aesthetic of forms and genres. In such an environment, where moments of genuine feeling felt impossible, bathos represented a small respite, an oasis of feeling. It was also a way to perform the kind of hyperliteracy that postmodern forms of popular culture rewarded.[96]

The fending off from unseemly displays of emotional excess is, in some ways, tied to the earliest impulse of the program—a sardonic exhaustion with the saturation of experience by media hype. Carving out bits of joy and wonder in such a world required a keen sense of what Glass would call, in another context, "taste," but it also meant paying attention to the changing mores, changing understandings of who and what can be made fun of. Glass punctuated serious moments in his live 1998 presentation "Mo' Better Radio" with facetious asides like, "Sylvia Poggioli: a man *and a Canadian*," a gender identity gag that would not be attempted today. In a 1998 interview, Glass admits to bathetic pleasure in the "offhand casualness" of Raymond Chandler's account of a suicidal girlfriend. "If you read it the right way, out loud, you can actually get a laugh."[97]

Several podcasts that followed *TAL* developed their own versions of this closing gag. *Invisibilia* repurposed bits of audio

---

[96]John Seabrook, *Nobrow: The Culture of Marketing, the Marketing of Culture* (New York: Vintage, 2001).

[97]Ana Marie Coz and Joanna Dionis, "Ira Glass Live and Uncut," *Mother Jones*, August 11, 1998, https://www.motherjones.com/politics/1998/08/ira-glass-live-and-uncut/.

from the show, or more often, outtakes or audio bloopers during the credits preceded by: "Here it is, your moment of non-Zen." The "moment of Zen" references Jon Stewart's *Daily Show* (Comedy Central, 1999–2015) closing bit of de-contextualized video, which often featured powerful political figures in an embarrassing light, a punching-up scenario quite different from this one. *Reply All*, the hit show from Gimlet, did a version of the Malatia gag as well, signing off with a twee bit of praise for the show's co-founder, Matt Lieber. For example, "Matt Lieber is the smell of my grandma's kitchen" and "Matt Lieber is candy you've never heard of from some other country."

Car Talk, NPR's long-running and iconic call-in automotive advice show, was a likely inspiration for many of the self-deprecating sign-offs. Since the mid-1980s, hosts Tom and Ray Magliozzi ended the program, in variations on the same formula, by pointing out with performative humility how out of place the show was on the otherwise highbrow network. For instance, "And although George Foster Peabody would roll over in his grave if he heard us say it, this is NPR" (#2296) and "Even though Uncle Sam starts beating his radio with his hat whenever he hears us say it, this is NPR" (#1527).

Such moments of self-deprecating levity are of course a staple of modern electronic entertainment going back decades. One thinks, for example, of Mary Tyler Moore's production company logo, featuring a kitten meowing sweetly in its center, a sly wink at the MGM's iconic roaring lion. *TAL*'s re-use of an interviewee's voice merits attention because of the way that it directly undermines the program's mission of deeply researched, richly contextualized stories that center the human voice as guarantor of authenticity and as vector of empathy and understanding. "By dragging a serious quote out of context and turning it into a joke," one listener complained on a 2019 sub-Reddit, "it's as if the producers are so cocooned in their radio show bubble that they paradoxically lose sight of the real, human, emotional situations their stories are about."[98]

Executive Producer Julie Snyder has suggested that the terrorist attacks of September 11, 2001, marked a turning point in *TAL*'s

---

[98] TJ Fox, "Does anyone else find some of the 'Torey Malatia quotes' grating?" Reddit, March 11, 2019, https://www.reddit.com/r/ThisAmericanLife/comments/b01vok/does_anyone_else_find_some_of_the_torey_malatia/.

storytelling.[99] While the Torey Malatia gag survived this shift, there is evidence of stories that reflect a more sober encounter with the conflicts of the world starting at the turn of the century. But it's a very gradual process, taking nearly a decade; and this process provides a remarkably sensitive gauge for listening to the changing soundscape of American liberal feeling as it is buffeted by the emerging common-sense of neoliberal policies, disastrous wars, economic crises, and a growing chasm of political division and violence. In the next several chapters, we'll explore how increasingly fraught categories of class, national identity, gender, and race shaped *TAL*'s storytelling and its approach to empathy and to feeling liberal in illiberal times.

---

[99]Julie Snyder, "It's a Long Story," Sydney Opera House, Podcast, April 17, 2017, https://soundcloud.com/sydneyoperahouse/its-a-long-story-julie-snyder.

CHAPTER FOUR

Feeling American

In the final act of "Before and After" (2001), WBEZ reporter Shirley Jahad describes for Ira Glass a scene in which two Chicagoans shout at each other while holding opposite ends of a massive American flag. It's the week of September 17, 2001, only days since the September 11 attacks. One man is Arab-American; the other man is described, variously, as "white" and "Anglo-American." The flag, which had been carried by two white protesters, has come quite literally between them during a street demonstration in response to the attacks of September 11:

> in the crush of the crowd, somehow the other end of the flag gets transferred to another young man... an Arab-American. Now here's the white guy and the Arab-American and they're both holding the same American flag.

The two men belong to different groups that have been coming out every evening since the attacks to demonstrate their very different understandings of patriotism in a Chicago neighborhood near a large mosque, "one of the biggest Arab-American communities in the region." Neither man can let down his side of the flag even as they debate its meaning and its relevance to the nature of their debate.

According to Jahad, "the white guy is completely at a loss as to what to do, because he's put in this perplexing situation. He's holding Old Glory, and he can't let go, but he's holding it with an Arab-American." The two sides are both yelling "USA USA," but Jahad reports with some prompting from Glass, "they mean completely different things." "The white guys... they're basically saying, this is my country, and we want you out of here. You get out." Jahad then plays some actualities from the demonstration featuring the white man holding the flag: "We're sick of the Arabians (*sic*) being here and we want them out of our country." An Arab American man counters, "Did you know my father served in the Korean War and I'm a Palestinian? My grandfather served in World War II? Can you believe that?... this should not be happening."

Another white man responds to questions of citizenship with "Well, I don't give a [BLEEP] whether they were born here or not. Their ancestors weren't born here." Amid the conflict and anger, the two men, representing opposed sides, are stuck with each other, holding up their end of an enormous flag, a problem especially for the white man, according to Jahad. "He can't drop it. He can't let go of it. Because it's the flag. So he's got to hold it. So he holds it, and after a while, they walked together." Eventually, Jahad tells an incredulous Glass that the two men, still holding the flag, "walked together and they actually started a conversation."[1]

It's a stunning scene, a tableau vivant of the country at an inflection point. It's also an early example of *This American Life*'s ability to bring its journalistic and storytelling prowess to bear on political conflict with as much impact as it had on stories of families and personal journeys. This scene, and all the stories in the first full post-9/11 episode, demonstrated Glass's literary approach to understanding human communication and conflict. The episode represents a remarkably apposite declaration of intent for the show's evolving orientation towards engagement with politics, conflict, and dialogue and away from "things that were so small and personal that other journalists wouldn't have touched them."[2]

Executive producer Julie Snyder has said that September 11 was a turning point for the show, inverting the formula of using the tools of journalism to tell everyday stories to using storytelling tools to do journalism.[3] It was a pivot away from the smaller, more intimate stories and towards the urgency of the political. "I got sick of

---

[1] Ira Glass, "Before and After," *TAL*, episode #194, September 21, 2001.
[2] Claudia Dreifus, "To Get Things More Real: An Interview with Ira Glass," *The New York Review of Books*, August 8, 2019, https://www.nybooks.com/online/2019/08/08/to-get-things-more-real-an-interview-with-ira-glass/.
[3] Julie Snyder, "It's a Long Story," Sydney Opera House, Podcast, April 17, 2017, https://soundcloud.com/sydneyoperahouse/its-a-long-story-julie-snyder.

everyone's bothers," Snyder explains, justifying her shift in interest as just another shrugging off of something whose popularity had expired, a recursive gesture, a quit of the quit.[4] The shift proved durable, however, and *TAL* became interested in a wider world, moving from a "microscopic perspective" to a "telescopic" one.[5] This media-savvy exhaustion with its own successful formula anticipated *The Onion*'s send-up of the show only a few years later. "There is not a single existential crisis or self-congratulatory epiphany that has been or could be experienced by a left-leaning agnostic that we have not exhaustively documented and grouped by theme," it satirically attributed to Snyder.[6]

Two Americans on either side of tribal hostilities, holding up and laying claim to either side of a giant flag, forced into cooperation and eventually dialogue and forward progress down the street, is an almost too on-the-nose anecdote to be credible. It captures, almost perfectly, the explicit mission of public radio and podcast storytellers in the years following the attacks: dialogue will save democracy. And narratives about "the other" will produce the necessary empathy to enable such dialogue. Posed against the dominant worldview available at the time, a world-historical clash of civilizations, this perspective had much to recommend it. The mission borrowed a great deal from public radio's earlier commitments to affective education and pluralism, of course. But in its new incarnation, it signalled a self-conscious attention to the very idea of democracy and in particular, the precarious hold that liberal institutions had on maintaining a functioning and inclusive civic sphere.

It also signalled a new kind of national context for stories with particular attention to the implicit inclusions and exclusions built into notions of American identity. This shift in the show's approach

---

[4]IG: At the beginning, we mostly apply the tools of journalism to things that were so small and personal that other journalists wouldn't have touched them: things going on in people's families, a weird personal quest. But now, so much of what we do is things that are in the news.
[5]Susan Douglas, "The Turn Within: The Irony of Technology in a Globalized World," *American Quarterly* 58, no. 3, (2006): 619–38.
[6]"This American Life Completes Documentation of Liberal Upper Middle Class Existence," *The Onion*, April 20, 2007, https://www.theonion.com/this-american-life-completes-documentation-of-liberal-1819569071.

to storytelling acknowledged, in the very framing of the stories, that the liberal impulse to hear all sides was increasingly difficult and increasingly necessary at the same time. Braiding this new understanding of the historical moment with the show's trademark whimsy required work, but it was work that was becoming increasingly compelling across media platforms.

The September 11 attacks inspired a new age of nonfiction storytelling. Journalists, documentarians, educators, and prestigious award-granting organizations responded accordingly. *The New York Times* won a Pulitzer Prize for its "Portraits of Grief," a series of obituaries for the hundreds of victims of the attacks, which read like quirky little stories, a slightly irreverent innovation on the traditional biography formula common to newspaper obits. The miniature narrative arcs in these short stories, most between one hundred and two hundred words long, were presumably in the service of "a compassionate style."[7] They brought a human scale to an atrocity known for its spectacular televised images and for its unprecedented body count. "Portraits of Grief" was part of its special coverage called "A Nation Challenged," whose title provides a clue as to how mainstream prestige journalism followed political elites in understanding the attacks in terms of national identity and storytelling as a process of national recovery.

Public radio pioneers, the Kitchen Sisters, produced "The Sonic Memorial: Remembering 9/11," part of its *Lost Sounds* series. The Peabody-winning production featured a collection of interviews, voicemails, and "small shards of sounds," gathered from collaborating producers "across the nation," as a way to remember the humanity of the dead through recordings of their voices.[8] StoryCorps, the venerable public radio outfit, launched its Peabody-winning "September 11th Initiative" in 2005 with the ambition "to record at least one story to honor each of the lives lost" during the attacks on the World Trade Center.[9]

Portraits, vignettes, and recollections of the lives lost in the attacks seemed to thrive most in radio. NPR's Robert Siegel's essay

---

[7] Roy J. Harris Jr., "'Portraits of Grief' 10 years later: Lessons from the original New York Times 9/11 coverage", *Poynter*, August 31, 2011, https://www.poynter.org/reporting-editing/2011/portraits-of-grief-10-years-later-lessons-from-the-original-new-york-times-911-coverage/.
[8] https://kitchensisters.org/present/sonic-memorial/.
[9] "September 11th Initiative," *StoryCorps*, https://storycorps.org/discover/september-11th/.

on September 12, in which he picked up and read aloud from scraps of paper found amid the rubble and dust in lower Manhattan, is an often-recalled example of the power of public radio to move and inform in ways that visual and print media could not.[10] Indeed, listeners and industry veterans alike have openly credited the attacks as a boon to NPR's fortunes, precisely because of its combination of intimate storytelling and national and nationalizing address.

In her short story, "Senseless," Allison Elliot Dark describes her character Clara White's reluctant conversion to NPR listener in the days after September 11; her desperate loneliness and anxiety only fueled by horrific televised coverage of the disaster leads her to overcome her resistance to NPR's "smug... East Coast liberal" sound. Listening to public radio in the car provides Clara with "the first real relief she'd had in days. She wanted to be spoken to in that quiet reassuring voice." In just a few hundred words, the story recapitulates the role of "inoculation" that Roland Barthes describes as one of the key techniques of modern myths: "admitting the accidental evil of a class-bound institution in order to conceal its principal evil."[11] "She'd accede to NPR," writes Dark, a necessary surrender. But the relief is temporary and soon, even NPR's "calm burring voice," droning and relentless, overwhelms her.

The story ends with her weeping in the parked car, moved but not moving, seeking but not finding peace in the "voice that knows everything." The ending recalls John Cheever's classic short story "The Enormous Radio," which also ends with an overwhelmed New York woman breaking down as she searches vainly for succor in the radio announcer's "suave and noncommittal" voice as he relays the headlines and the weather.[12] It also recalls the notion, explored in the previous chapters, of the car radio as a means of emotional transportation; only in the new context of the post-9/11 world, public radio, "that voice," would need to sound less knowing and more feeling if it was going to move Clara where she needed to go.

In an echo of the radio/crisis synergy explored in the first chapter, many people saw public radio as newly vital in the wake of the

---

[10]Tim Eby, "Three Things: The Impact of September 11 on Public Radio," *Three Things*, September 11, 2023, https://timjeby.substack.com/p/three-things-the-impact-of-september.
[11]Roland Barthes, *Mythologies*, 150.
[12]John Cheever, "The Enormous Radio," *The New Yorker*, May 10, 1947, https://www.newyorker.com/magazine/1947/05/17/the-enormous-radio.

terror attacks and the war years that followed. Tim Eby, a veteran public radio executive, recently opined that

> the value of NPR as a national institution was... firmly established during those months in the fall of 2001. I don't believe it was a mere coincidence that Joan Kroc's transformative estate gift to NPR occurred two years later.[13]

Joan Kroc's bequest of $200 million in 2003 symbolically and materially joined McDonald's and NPR, two iconic institutions representing opposite ends of the American cultural hierarchy. While McDonald's had, by the early twenty-first century, become a fully global empire, moments such as the 9/11 attacks had the power to temporarily renationalize such iconic brands. Books like Benjamin Barber's 1995 *Jihad vs. McWorld*, anticipated the Manichean political worldview of the post-9/11 world and McDonald's symbolic centrality to one side of it.

NPR, always a mostly nation-bound brand, renationalized itself in its sympathetic coverage of American military activities in the years that followed. Its centrality to a new kind of vigilant monitorial citizenship, as dramatized in Dark's short story, reinforced this elision of NPR and the nation for which it was named.[14] Likewise, *TAL* embraced a more literal interpretation of its own title in the months and years that followed. From embedding reporters on a US aircraft carrier in "Somewhere in the Arabian Sea" (2002), to exploring the tenuous bonds of shared identity that held neighborhoods and the national community together ("Enemy Camp," 2003), the show's traditional format (several stories on a theme) gave way, for a time, to the overarching theme of Americanness itself.

Even so, a residual commitment to entertainment, to moments of surprise and amusement, persisted. Storytelling, quirky and irreverent, prevented topics from the ponderous and agonistic tone of documentary or investigation, two journalistic genres responsible for uncovering some of the nation's greatest scandals, meting out bold reckonings, and pushing through significant reforms. On *TAL*,

---

[13]Eby, "Three Things."
[14]Michael Schudson, *The Good Citizen: A History of American Civic Life* (New York: Free Press, 1998), 310–11.

traces of the original fending off from politics and history remained. In the place of politics *tout court*, the show doubled down on form. It increased its emphasis on reaching listeners where they were and transporting them somewhere new emotionally, even if only temporarily.

Jason DeRose, a longtime NPR editor, recently emphasized to me the necessity of "building on-ramps" for listeners who happen upon a live broadcast, "while driving to work," so that they will be moved to care about a story.[15] Perhaps he was thinking of Clara White, who found NPR news delivery too fast, too overwhelming, too much at once. "*Stop,* she thought. *Please slow down. Please repeat that, go back to the beginning… please start over.*" DeRose understood that "you have to bring them along," echoing Glass's 1998 exhortation to "make them empathize" and Nikole Hannah-Jones and Chana Joffe-Walt's 2016 Third Coast presentation, "make them care," revealing the extent to which this approach had become an industry standard and a formal mandate by the end of podcasting's first decade.

Reduced to a formal exercise and divorced from any politics in particular, storytelling quickly became a way to make people care about lots of things. By the end of the 2000s, "storytelling" had become unavoidable across industries. Universities began offering courses in "digital storytelling" and marketing and PR professionals adopted the term with the reverence of Walter Benjamin acolytes to connote wholesome truth-telling rather than mercenary mendacity.[16] At a time when educators and media critics focused on the short attention span wrought by quick edits in film/video, sound-bite journalism, and new digital media trends, the first decade of the century was also a time in which long-form storytelling, on blogs, public radio, and eventually in podcasting, became wildly influential for corporate communications and for the sprawling

---

[15]Jason DeRose, personal communication, July 13, 2023; A fuller version of his comments, and how they related to NPR's flagship *All Things Considered*, can be found at Jason Loviglio, "From China to 'Our Land,' how Melissa Block set a standard for NPR's reporting," *Current*, August 18, 2023.

[16]Laurence Dessart, "Do ads that tell a story always perform better? The role of character identification and character type in storytelling ads," *International Journal of Research in Marketing* 35, no. 2 (2018): 289–304.

world-building cinematic, video game, and fantasy fiction universes that rewarded Talmudic attention to massive oeuvres.

Glass focused on the formal elements involved in getting listeners to care, building on-ramps for empathy, curiosity, and narrative engagement. By 2001, this was already a hallmark of *TAL* stories, something Glass says he learned from NPR and for which he was now getting credit for reminding his former colleagues at the network. A semiotics major at university, Glass has recounted the impact of the work of Roland Barthes on his understanding of language and culture and how to tell stories that engage and entrance. What most captured his attention, he has told multiple interviewers, was the way Barthes identified the structural logic of creating a compelling narrative in his essay "S/Z," a painstaking and joyful structuralist dissection of Balzac's novella *Sarrasine*.

One big idea he credits to Barthes is to bring listeners along by offering a mystery or puzzle to solve, wrapped in "the *hermeneutic*, or *enigma code*." As we saw in the previous chapter, the enigma code was central to *TAL*'s gothic family mysteries where questions of biological paternity abounded. As we'll explore in Chapter 6, "Fellow Feeling," questions about fathers and the nature of manhood in general have been a preoccupying enigma for much of the show's run. The best stories, according to Glass's formula, also step out of the narrative and address what he calls "big ideas," concepts typically rooted in the culture's most salient learned debates, or "*the cultural code*" according to Barthes's schema.[17] These interludes of abstract theoretical consideration engage a macro level of awareness, emerging out of storytelling dense with data and delivered through self-reflexive, highly transparent humanistic narrative.[18]

When it came to Barthes's structuralist and Marxist notions of the essentially conservative, even coercive, power of language in hands of dominant institutions, Glass adopted his trademark diffidence. With the characteristic insouciance of one who quit politics as a young man, he says he simply "ignored" the semiotic implication that "narrative is part of the general conspiracy of language to

---

[17] For more on this, see Martin Spinelli and Lance Dann, *Podcasting: The New Audio Media Revolution* (London: Bloomsbury, 2019), 180.
[18] David O. Dowling, *Podcast Journalism: The Promise and Perils of Audio Reporting* (New York: Columbia University Press, 2024), 29.

imprison us in our place in society."[19] It was, he said, an "incredibly pretentious" argument. Glass's avoidance of the political was an extension of his earlier posture, explored in the previous chapter, a fending off from an adolescent version of himself that he had quit.

In talks and interviews, Glass has told multiple versions of his own origin story as a storyteller, magician, semiotician. In each, however, he emphasizes the importance of structure, repetition, and payoff. "A sequence... a sequence is like a sermon," he told the *New York Times* in 1999. There must be a big idea, something to take away. In the previous chapter, I explored *TAL*'s early years and the show's preoccupation with producing magical moments of connection between people in an often-explicit challenge to the hyper-mediated nature of late-twentieth-century experience. Conjuring moments of shared connection, common humanity, and unlikely empathy, Glass and company generally eschewed any political or historical frameworks for these narrative gems, preferring a vaguely universalizing human perspective.

In this chapter, the focus is on the dozen or so years following the September 11 attacks, a time of heightened national and economic anxiety in the United States. Stories in this era explore questions of identity and division within inescapably historical and political contexts of terrorism, war, dissent, and recession. Questions of ethnicity, race, and class identity shaped questions of American identity and American belonging in a moment when the idea of "this American life" felt less pat and more precarious than it perhaps had in 1996.

TAL was uniquely situated in 2001 to provide weekly meditations on such questions, and the decade following proved to be a period of impressive growth and critical accolades. While legacy print and television media leaned into the debates between the emerging neoconservative commonsense embodied in the invasions of Afghanistan and Iraq and a more diffuse anti-war opposition, *TAL* steered towards stories about irreconcilably opposed forces in moments and zones of uncertainty. The September 11 attacks and the ensuing wars helped create a new context for the telling

---

[19]Dreifus, "To Get Things More Real."

of small, personal stories in which historical and political forces buffeted regular people.

In a time of starkly divided opposites, with a president committed to ridding the world of "evil-doers," and a largely docile fourth estate, little dramas situated at the crossroads of antithetical absolutes became a vehicle for *TAL* to use the tools of storytelling to do journalism. In building themes around the idea of conflict, the show also addressed questions of ethnicity, religion, and class as elements of identity beyond the universalizing of earlier episodes. As the scene with two American men arguing over the flag demonstrated, these stories would center conflicting ideas about national identity and belonging.

In addition to "Before and After," the months following September 11, 2001, featured episodes organized around a series of identity-based antitheses: "us and them" ("Them"); "Faith" and the loss of it; "Hearts and Minds," in which feeling and thinking are artfully posed as in conflict.[20] "Rashomon," explores the multiplicity of irreconcilable narratives about the same event, Plan A vs "Plan B"; and "Suckers" vs Non-Suckers.[21] With very few exceptions, the stories gathered under these themes were not explicitly political in the sense of taking clear sides in the debates swirling around say, the Patriot Act (2001) or the invasions of Afghanistan (2001) and Iraq (2003).

Instead, they seemed to be interventions into the *idea* of conflict and meditations on the narrative and real-world possibilities of mediation. The political and historical tumult of the era provided a stage on which the show explored irreducible contradictions at the center of human experience and national identity. In such a world, it was not magicians who were needed, but translators with fewer joyful epiphanies and more courage to acknowledge impasses.

With its focus on "the pairing of antitheses," French theorist Roland Barthes's essay "S/Z" proved a handy guide for such ambitions. Glass has said that he was drawn to the essay because of the way it broke a story down, line by line, in ways that helped

---

[20]Ira Glass, "Them," *TAL*, episode #201, December 7, 2001; "Faith," *TAL*, episode #202, December 21, 2001; "Hearts and Minds," *TAL*, episode #200, November 30, 2001.
[21]Ira Glass, "Rashomon," *TAL*, episode #196, October 5, 2001; "Plan B," *TAL*, episode #205, February 1, 2002; "Suckers," *TAL*, episode #222, September 27, 2002.

explain how to build tension through the play of several codes, which together form a network or "topos." Or rather, Barthes clarifies, it is by *passing through* this topos, that these various codes cohere into a text. Barthes's essay provided for Glass "an enormously useful" map for building on-ramps into narrative: "How does this story pull you in, engage, and give you pleasure?"

Barthes identified the "symbolic code" as a web of signifiers "through which the meanings of the story unfold." In *Sarrasine*, the first signifiers that fall under the symbolic code concern paired antitheses like inside/outside, warm/cold, life/death, which Barthes argues, helps to set up a story of the castrato's secret life at the forbidden intersection of the male/female boundary. It also sets up the position of the narrator as an observer of antithetical states of being. The narrator, whose name we never learn, observes the goings on of a Paris soiree from an alcove from which he can observe and feel the light/dark and warmth/cold of the party and of the world outside the window. This position grants him a kind of neutrality and thus a kind of authority. Barthes sees the narrator's mediation as the "site of a transgression effected by the narrative." It is the mediating narrator who physically embodies this position between antitheses, who sets the entire story into motion, a necessary mediation through which "something can be told and the narrative can begin."²²

It is precisely this narrative mediation that lies at the heart of *TAL*'s exploration of paired opposites in the post-9/11 era. Glass and reporter Jahad together fulfill this function in the final act of the "Before and After" scene where two men struggle to keep a flag aloft and their opposite understandings of its meaning intact. The flag too operates as a kind of symbolic mediator, its physical embodiment now doing double symbolic duty. Its first obligation is to the nation for which it stands and secondly it serves as the site of a transgression, a meeting of opposites, what Barthes would call

---

²²Roland Barthes, *S/Z: An Essay* (New York: Hill and Wang, 1975), 28. The first hermeneutic code in Balzac's story, according to Barthes, was the hermeneutic or "enigma code," that provided the signifiers necessary to establish a mystery, for the unspooling of a secret: "who or what is 'Sarrasine?'" The story's title turns out to be the first riddle—readers learn that Sarrasine is the name of a man, who falls in love with an opera diva who is a castrato in disguise, who eventually becomes a curiously spectral patriarchal figure at the Paris soiree.

"a battle of plenitudes set ritually face to face like two fully armed warriors."[23]

The journalists telling this story are also navigating a strange and necessary contradiction, keeping aloft a thing that is both heavy and unwieldy, a thing that seems to want to fall down. The possibility of a liberal pluralistic America itself seems suspended precariously between two edges of the flag. Telling this story of an America held together by contradicting forces over and over again, *TAL* turned storytelling into a kind of ritual, as if mediating opposites could keep them suspended in tension.

In a slightly different context, John Durham Peters has analyzed this phenomenon in "Courting the Abyss," his 2005 exploration of the contradictions within modern liberalism. Peters examines the liberal impulse to "fraternize with the enemy," by which he means the liberal tolerance for the intolerant to the point that it seems, to him, to become a kind of perverse celebration of its own opposite. "[Liberalism] always needs to be losing itself in the other, imbibing the other's potions, scouting the regions of sin, passing thru the bower of bliss."[24] Peters was most concerned with the intellectual and moral knots that liberalism ties itself in when say, civil libertarians defend the First Amendment Rights of the KKK. His analysis is helpful for considering the challenges of a liberal media committed to finding the middle ground, the narrative alcove, in any face-off between combatants.

This chapter focuses on the post-9/11 War on Terror era which forced questions of American identity to the fore. In the process, *TAL* evolved its approach to storytelling, mediating struggles between irreconcilable differences that threatened national identity. The transformation was galvanizing and perilous, more of a Corwinian "tightrope act" than the close-up magic of Glass's earlier work. Peters's idea of "courting the abyss," the liberal "attitude of warming oneself in the fires of hell" is an apt one for thinking about the challenges of narrative journalism in the post-9/11 era and the unspooling horrors that followed.[25]

---

[23]Barthes, *S/Z*, 27.
[24]John Durham Peters, *Courting the Abyss: Free Speech and the Liberal Tradition* (Chicago: The University of Chicago Press, 2005), 136.
[25]Peters, *Courting the Abyss*, 6.

The use of high-concept thematic organization for each episode, mitigated the risk considerably by asking listeners to attend not to the unrelenting drone of news, which drove Clara White mad in Dark's short story, nor to documentaries of the dire consequences of war and poverty, but instead to stories about people caught betwixt and between. In these stories, the journalist, above all, emerges as the master mediator. No longer a magician working on intimate surprises, the journalist is an "abyss artist," to use Peters' phrase, straddling impossible prerogatives in the name of liberal democracy. Listeners are asked increasingly not to empathize with the exotic dispossessed "other," but instead with "those with advanced training in self-doubt and stoic self-control" (i.e., the journalist, the liberal caught betwixt and between).[26]

There is some textual support for aligning Peters' "abyss artists" with journalism. Liberalism, he complains, "never met a secret it could keep." It is "fine with almost everything being spoken from the rooftops."[27] Publicity, of course, is journalism's stock in trade. And the conceit of *TAL* in the post-9/11 era especially is that we are all capable of "speaking into the air," to borrow another line from Peters.[28] Spinelli and Dann explore the ethical limits of "humane and empathetic podcast values" through close analysis of the technical and narrative tricks used in a 2015 episode of *Love+Radio*. The producers of *Love+Radio*, *The Heart*, and other intimacy-based podcasts, are, they argue, "empathy artists," exploiting and questioning the power of narratives of liberal feeling. While they are more sanguine and less sardonic about this kind of artistry than Peters seems to be, they acknowledge that perhaps empathy is less, "an unerring human predisposition and more... a malleable artistic material."[29]

The journalist as quintessential middleman becomes an important way to drive home two points, one formal and one political, that ripple across the show's stories in the post-9/11 era.

---

[26] Peters, *Courting the Abyss*, 86–97.
[27] Peters, *Courting the Abyss*, 130.
[28] John Durham Peters, *Speaking into the Air: a History of the Idea of Communication* (Chicago: University of Chicago Press, 1999).
[29] Spinelli and Dann, *Podcasting*, 98.

The formal message is that *stories work best when the telling is the tale*, a meta-cognitive approach to narrative in tune with the residual postmodern mood of early twenty-first-century American culture. The second concept was *we're all liberals now*—stuck in the middle, holding up something we can't drop and can't fully defend.

"Middlemen" (2002) provides a good example of how *TAL* navigated this new era in narrative journalism.[30] Like the narrator of *Sarrasine*, the episode's protagonists find themselves in narrative alcoves, perched between things in order to mediate and so that "something can be told and the narrative can begin."[31] Act II of the episode focuses on Sal Princiatta, a firefighter from Engine House 33, Ladder Company 9, in lower Manhattan. His group lost ten of their forty-man crew on September 11 and now, wherever they go, they're treated as heroes, which makes Princiatta uneasy.

> It's hard to conceive. You're getting gifts while your friends are dead…. How does that equate? What's the purpose?

The story features lots of natural sound of Princiatta and his coworkers navigating award presentations and cliché-ridden small talk with admirers, punctuated by Princiatta's attempt, in voiceover, to understand it and his own uneasiness, the weight it adds to his own burden. A visit to Memphis, Tennessee, proves awkward as he and his colleagues encounter overwrought introductions and celebrations that sound patronizing to Princiatta's ears. "That's pity out there. That's pity treatment." At other times, the hero-worship feels "embarrassing and unnerving" because it's too much to live up to.

> I've seen a cartoon in the newspaper of Spider-man, Batman, and Superman asking for a firefighter's autograph. And that's just too much to be put on any average Joe's back. Because that's really what we are.

The piece is cut together deftly so that Princiatta provides the gloss, after the fact, on a series of awkward encounters, almost as if he's

---

[30] Ira Glass, "Middlemen," *TAL*, episode #224, October 25, 2002.
[31] Glass, "Middlemen."

authoring the piece, instead of responding to interview questions. He becomes the journalist, the mediator, of his own story. His attitude is ambivalent. He knows that they mean well but can't help feeling alienated from the hero-worship, gifts, and thank-yous. The rubbing together of New York and Tennessee accents in these encounters provides an additional sonic layer to the sense of distance and tenuous connection. The irreducible distance between these figures, in sentiment, experience, and language also anticipates the Red State/Blue State divide over domestic and foreign policy in the wake of the 9/11 attacks, and the deepening ideological divides that follow in the 2010s.

When the Fire Department of Memphis, Tennessee, invites the New York firefighters to play softball against them, the gesture backfires, reminding Sal of the talented outfielders from his firehouse lost to the attacks, a subsequent suicide, and lingering health issues after months of working in the dust and smoke of Ground Zero. It isn't until late in the story that Princiatta shares his realization that he and his surviving colleagues are middlemen between 9/11 and everyone else. The Memphis man who can't stop saying "we're proud of you" over and over, he realizes, isn't talking to him so much as he's trying to come to understand the incomprehensible. "We're his connection to what went down. And that's it. We connect him to what went down. And we give him a vehicle to feel it."

It's a generous insight and it recasts the awkward and false-sounding notes of the previous interactions in a new light. The piece ends with Princiatta balancing this newfound realization of how far most people are from 9/11 with the heavy pall cast by his own proximity to it. The recent suicide of colleague, Gary, "a regular Joe [who] didn't show any signs of hurting any more than any one of us," means more than just an undermanned outfield. He's become the middleman for his own story, occupying a no-man's land between Ground Zero and Middle America. The irreconcilability of the two perspectives, the awkward praise on one side and the abyss on the other, is what persists.

Coming only a year or so after 9/11, the story is still raw, as is Princiatta. The nine-minute piece is a *meditation on mediation*, attesting to the impossibility of representing 9/11 in journalism or in a human life. Princiatta leans on terms like "average Joe" and "regular Joe" as a way to resist the hero treatment, but also perhaps as a way to recast himself and his fellows as the center of their own story, rather than as figures in someone else's. The struggle to reposition oneself from object to subject became an increasingly powerful theme in the years that followed. As part of this struggle, the very idea of mediation, of the value of middlemen, came under new scrutiny.

The other acts in "Middlemen" adopt a lighter tone. Glass interviews a man named "Chris" who recalled working as an interpreter between hearing and deaf people on phone calls. He read what was typed and typed what was spoken so that the two parties could communicate. Glass and Chris play this mostly for laughs, focusing on times when the mediation itself became the topic of conversation, as when two African American callers, one hearing and one deaf, made Chris, who is white, repeat Black vernacular speech, including the "n-word" over and over, for their own amusement. Glass and Chris find their own amusement in the retelling of this transgression with a built-in alibi. Sometimes, two people arguing would demand Chris intervene on one side or the other. Chris's job required a common form of mediation, one that depended upon its own invisibility to function properly. His awkward encounters echo Richard Howard's preface to the 1975 English translation of S/Z, to wit: "all telling modifies what is being told." This sort of self-conscious mediation plays with broadcasting's oldest conceit: that we can listen in on other people's lives without changing the conversation. Glass's interview with Chris echoes one of the oldest themes in human-interest journalism: the "interesting jobs" interview, which juxtaposes the voice of an affable, slightly condescending host with a typically plain-spoken elevator operator or mortician's assistant as a proxy for the lurid curiosity of the home audience. These interviews were often played for laughs in radio's Golden Age too. The hearing-impaired people who rely on Chris's phone service lurk in the shadows as hilarious anecdotes, rather than subjects. As with the phone service, they cannot speak for themselves.

The next story features another "interesting jobs" type interview with a man recalling his days as a ticket scalper based in Chicago during the heyday of Michael Jordan's dominance in the NBA to *TAL* producer Alex Blumberg. Identified only as "our ticket scalper" (and "Man" in the transcript), his buying and selling of tickets for basketball games and other events represents an iconic form of middleman, occupying the space between producers and consumers as well as the demi-monde between formal and informal economies. Outside the stadium, but inside the perimeter of its marketplace, the scalper navigates the impossibility of visibility and invisibility necessary to do the job. He also represents the way mediation of antitheses "constitutes a transgression," in Barthes' words. "Our ticket scalper" takes over Blumberg's role halfway through the story, introducing his own interviewees, miking up his friend Leo to record him scalp outside a University of Michigan football game in Ann Arbor. The sound of Leo working the crowd comes up in the mix. He buys and sells. Blumberg and our ticket scalper then discuss Leo's strategy, explaining crucial terms and maneuvers, an additional layer of mediation, as the theme of mediation moves from Glass to Blumberg to our scalper to Leo, all circulating between production and consumption, inside and outside, narrative and metanarrative.

The stories on this theme demonstrate mediation as the *sine qua non* for a narrative to begin. Because of the layering of mediators, mediation becomes the story. Each telling, each layer necessarily produces something left over or left out, unfinished business, a remainder. This theme drew attention to journalism and the narrative arts required to report on big stories. What is it like to be in betwixt and between, always a powerful narrative bid, became in this period a way to explore the political crisis, ideological polarization, and the role of the storyteller as central to the action.[32]

Mediation stories immediately post-9/11 made it clear that translators, interpreters, journalists, and other go-betweens were actually the most important figures in a story, precisely because their jobs were nearly impossible. Like the massive US flag, go-betweens were doomed to stand in the place where two opposites met, unreconciled, and implacable. In the prologue to "Why We Fight" in late 2002, at the moment when the debate about invading Iraq had seemed to shift to a certainty, Glass hazarded a rare moment of ideological insight. Members of Congress, along with other members of the press and public had made a compelling case that invading Iraq would be both unnecessary and a fiasco. Glass is bemused by the impotence of these dissents and unsettled about

---

[32]Richard Howard, "Preface," in Barthes, S/Z, ix–xii.

how to mediate something hardening so quickly into a dubious consensus. "After a couple of days, (anti-war argument) just kind of dropped off the radar."

Readers of Chomsky and Hermann's *Manufacturing Consent* may not have been as caught off guard, but such a position would require a political, rather than a journalistic approach, which Glass acknowledges, albeit from a perspective that longs for a middle ground:

> it's hard to bring it up without seeming like you're just out there on the fringe, like you're some kind of extremist, pinko, unpatriotic, like you're against the President, like you are picking a fight.[33]

It's as close as *TAL* comes to a genuine politically left position in the run-up to the wars. For much of the early years of the wars, the show hews carefully to an America-first version of neutrality, assuming patriotism and support for the troops as a commonsense starting point. But it is important to note that Glass's beef here is less with the policy and the probability of war, than it is with the disappearance of a middle ground in which to neutrally mediate competing sides. To "bring something up," which seemed utterly reasonable a couple of weeks ago, and which is now "on the fringe," means it *feels* impossible to speak it without risking criticism of partisan bias, "like you're against the president."

At the same time, these episodes seem to argue, stories that are hard to tell were the most interesting and worthwhile. "Before It Had a Name" (2001) featured stories set in the temporal and affective gap between an event and its explanation and narrativization or diagnosis.[34] How to understand the incomprehensible (e.g., the Holocaust, a dire medical diagnosis, and even love) before the media and historical framing solidifies around them? Like "Before and After," mentioned above, which came out only the month before, this story is clearly shaped by the sense in the immediate aftermath of 9/11, that America was experiencing something that hadn't been fully mediated yet. In this way, Glass and company were

---

[33]Ira Glass, "Why We Fight," *TAL*, episode #227, December 20, 2002.
[34]Ira Glass, "Before It Had a Name," *TAL*, episode #197, October 26, 2001.

describing the precise dilemma Raymond Williams wrestled with in his account of the structure of feeling. It required paying attention to "specific feelings, specific rhythms" before they hardened into a commonsense narrative.[35]

"It's hard to be in the middle," Basim, an Iraqi interpreter for the US military forces in Iraq, reports to Glass in 2007. The episode "By Proxy," like other shows in the dozen or so years after 9/11, like "Lost in Translation" (2003), "Shouting Across the Divide" (2006), "The Bridge" (2010), "Surrogates" (2013), "Stuck in the Middle" (2014) (among others), features stories of "people substituting themselves for other people, and the difficulties that can create."[36] Translating for the US military and the Iraqi officials and civilians was crucial but impossible and sometimes, deadly. Basim describes for Glass a 2004 incident where he had to lie to achieve the best outcome in a tense standoff between US soldiers, Iraqi police, and an angry crowd of citizens. Glass, ingenuous, marvels that "it's not just that you're just interpreting what's going on, but in a sense, you're taking over for whoever's in charge." Translators moving outside their impossible role in order to move beyond an impasse represents a powerful recurring theme across many of the stories.

After the images of prisoner abuse at Abu Ghraib were leaked, the danger for Basim and other translators increased. "If you have the chance to kill a translator or American," Basim explains the new logic on the street, "kill the translator first." While this raised the stakes for Basim on what translating might cost him and his family, it also served to underscore for Glass an underlying dilemma facing journalists and others who "step into the breach," as he says of his team's efforts to understand the debate over invading Iraq in "Why We Fight"(2002).[37]

---

[35]Raymond Williams, *Marxism and Literature* (Oxford: Oxford University Press, 1977), 128–35; *Preface to Film* (London: Film Drama, [1954] 2003), 21–3.
[36]Ira Glass, "By Proxy," *TAL*, episode #327, March 9, 2007; "Lost in Translation," *TAL*, episode #238, May 30, 2003; "Surrogates" *TAL*, episode #485, January 25, 2006; "Shouting Across the Divide," *TAL*, episode #322, December 16, 2006; "The Bridge," *TAL*, episode #407, May 10, 2010; "Surrogates," *TAL*, episode #485, January 25, 2013; "Stuck in the Middle," *TAL*, episode #516, January 14, 2014.
[37]Glass, "Why We Fight."

Translating, serving as a proxy, finding oneself betwixt and between is both universal and very specific to the communicators working in a war zone. While the stakes are wildly asymmetrical, the show emphasizes the feeling of uncertainty and ambivalence animating all proxy work. Adam Davidson, reporting from Amman Jordan in December 2002, finds himself at pains to explain the logic of the looming US invasion of Iraq to a group of Iraqis, Jordanians, and Egyptians, who are confused about and suspicious of American motives.

> And I don't know what to say to him. So I stumble around like an idiot, trying to explain American foreign policy, which I honestly don't understand myself, to people whose families might be bombed– another conversation you find yourself in with disturbing frequency around here.[38]

He finds that he can't explain the rationale for war, which is part of the unusually critical approach of the "Why We Fight" episode. But he's also making clear that though his reporting is difficult and involves feeling uncertain most of the time, an important tonal shift develops as the wars drag on and the contradictions multiply and refract into more than just two-sided fights. Davidson is skeptical of the Bush plan to invade Iraq but horrified by the first-person accounts of Saddam Hussein's brutality. "I find it very confusing myself," he blathers, when trying to explain US policy to Iraqi expats in Jordan. "I think there are few... George Bush... it's very hard to fight."

Including such unflattering tape of a reporter's voice is a calculated part of the show's embrace of ambivalence and uncertainty; it's a harbinger of a particular kind of "recessive epistemology," to use Neil Verma's term, that will become dominant in the true crime podcasting of the following decade. While Verma sees the aesthetic of uncertainty as a break from, and perhaps an antidote to, the "hand holding" or "realist regime" associated with *TAL*, the show's approach to the wars, as they slip into quagmires, provides some early examples of how this voice developed a new kind of cultural authority in narrative audio.[39]

---

[38] Glass, "By Proxy."
[39] Neil Verma, *Narrative Podcasting in an Age of Obsession* (Ann Arbor: The University of Michigan Press, 2024), 193

Davidson foregrounds his haplessness in the face of the logistical and moral obstacles that dog his research trip. He dithers about whether or not to use the names and images of his Iraqi expat sources, despite their giving him permission. He frets about how to verify their credibility when they claim Saddam does have weapons of mass destruction. He enumerates the hours he's spent trying in vain to get a visa to enter Iraq. He centers his own difficulties, emotional and practical, in getting the story, perhaps as a way to hedge his bets on the question of Saddam's weapons, which Americans were debating hotly at the time as well. But tonally, it's quite similar to the highly personal narration that made podcasts like *Serial* so distinctive. "I've probably spent 30 hours at the embassy waiting to get a visa," Davidson complains.

Of course, all procedural narratives, fictional or otherwise, take pains to convey the due diligence and admirable commitment of the investigator. The hazards and discomforts of war correspondents already represented a distinctive genre long before these recent conflicts. Even so, there is an element here in which Davidson's journey becomes a vehicle for the show's ambivalence about the coming war and about America's war on terror, in general. It also becomes a way to gesture toward the limits of clarity as the quarry of narrative journalism.

More than a kind of fog-of-war-type confusion, Davidson's trip and the narrative framing provided by Glass, who claims to be "constantly surprised" by his reporting on Iraq, seem to be making the case that Americans' understanding the reasons for war was an impossibility. "I came to the Middle East thinking this war that's coming is a terrible thing. I still don't really understand why we're doing it," Davidson shares towards the end of the piece. "But I feel sick, I really, actually have had a stomachache this whole week talking to these Iraqis, hearing what Saddam Hussein has done to his people." Rather than presenting rational arguments for competing positions on the wisdom of invading Iraq, Davidson's reporting has only intensified his confusion and its emotional impact. America is going to war and to be American is to not know if it's a good idea or not, Davidson seems to conclude. In this way, Davidson becomes our proxy.

In "Lost in Translation" (2003), *TAL* turns again to the metaphor of translation to plumb the mysteries of in-betweenness. The longest story, "The Star of Bethlehem," features a Palestinian TV journalist who translates the Israeli TV newscast into Arabic in real time every evening, in service of peace, understanding, and "compassionate listening," a job that seems both necessary and doomed. Nasser Laham's broadcast provides insights into a world that is both very close and impossibly far. This story is told with an appreciation for the sheer idealism of its subject. The other acts are lighter in tone, trading on the wry, self-deprecating humor of the failed translators of genre, as in the story of a comic trying out Borscht Belt-era stand-up comedy at a hip NY karaoke bar.

A 2003 episode titled "Enemy Camp" reprises, in an odd sort of way, the trope of the US flag as a locus for uncomfortable mediation. In this case, two middle-age women in a small town find themselves on opposite sides of the US invasion of Iraq. Kathy, a Presbyterian minister, demonstrates every week on the town green against the war. Joy, a homemaker, counter-demonstrates in favor of it. At one point, Joy lends Kathy her American flag for her protest, helping to spawn an unlikely friendship along with a model for the rest of us. Joy's patriotism is expansive, unlike that of the white mob demonstrating in Chicago against Arab Muslims in the days following 9/11. Sharing the flag with an opponent was "fine with me. It's the flag, why not share it? It's a symbol of everything that we hold dear. And don't we want to share that with everybody?"

Kathy, for her part, saw in their friendship a way forward for the nation: "I think that we're doing in this small town between us what I'd like to see our political leaders do more of." The friendship flounders quickly, first in a spat over the state of the grass being trampled on the town green, then in hurt feelings about Joy's husband's pro-war sign, which Kathy feels is mocking her. Like the flag and the trampled town green, which had been re-seeded and fenced off, the friendship was a symbol of a meeting space in between extremes. Its bitter end is an occasion for reflection less for the former friends than for reporter Blue Chevigny, whose bemusement at the schism moves to center stage.[40]

One of the most unusual formal approaches to the theme of mediation that the show attempted in the post-9/11 era was the long dispatches recorded by Hyder Akbar, an American teen who travels to Afghanistan in 2003 to join his father who, after years

---

[40] Ira Glass, "Enemy Camp," *TAL*, episode #242, July 18, 2003.

in the United States, has secured a position in the Hamid Karzai's new post-Taliban government. Drawing on the approach of *Radio Diaries,* producer Susan Burton outfits Akbar with a tape recorder for his three-month visit. The excerpts from the tapes featured on the program are long, lightly edited, and unlike most of the show's signature hand-holding style of storytelling, unstructured, taking on the shapelessness of time spent alone far from home.

The child of immigrants, Akbar felt Afghan in the United States and feels keenly his Americanness in Afghanistan. His account of the violence, poverty, and wartime vigilance are filtered through his perspective as a southern Californian teen, coming of age. The following year, in "Teenaged Embed" (2003), the show features Akbar again, now serving as an informal emissary for his father, who is now governor of Kunar Province. Akbar serves as a translator for the US army. He has become more adept at recording his narration and his tapes have been edited so as to layer his observations over the conversations, mostly in Dari, that he is translating. Akbar translates his translations, layering mediations on top of each other, providing that real-time illusion of listening in mastered by broadcast journalists, especially NPR. Kunar is a borderland, nestled in mountains near Pakistan, far from Kabul and the shaky new order his father represents and so Akbar finds himself uncomfortably in between even as he becomes more fully part of the country. In a particularly harrowing encounter between the US forces and an Afghan man suspected of vague wrongdoings, Akbar tells us how he reassured the man, who later died in US custody.[41]

What is most striking about these episodes, which won several top journalism prizes, is how lightly touched they are by *TAL*'s house style—the handholding editorial interruptions in which events are punctuated by grand insights or "takeaways" provided by the host. But they very much fit the model of post-9/11 stories in which the protagonists are the middlemen, translating, interpreting, and mediating opposite sides. The very first lines are loaded with layers of contrasts: the Kabul of a year before with burkas everywhere to the Kabul of today with women in jeans; Akbar's American desire to hug his father upon his return versus the expected Afghan

---

[41] Ira Glass, "Come Back to Afghanistan," *TAL*, episode #230, January 31, 2003; "Teenage Embed, Part 2," *TAL*, episode #254, December 12, 2003.

protocol of kissing his hand; and the beauty of the land in contrast to its violent history.

These observations serve the purpose of placing Akbar in a narrative alcove from which he can see the play of opposites, and set up his account, which quickly follows the introduction of his encounter with the deadly legacy of Soviet-era collaborators on the contemporary political landscape, a layer of division laying on top of the problems facing his father in navigating Taliban, Al Qaeda, and US forces. With surprising economy, Akbar's trip to a Soviet-era mass grave and the rippling resentments around it, offer both backstory and prophecy for the current conflict. Akbar's sotto voce narration in the middle of the night confers a diarylike intimacy to some sections of the story. His youthful enthusiasms and worries remind us that he is also navigating the borderlands of late adolescence. These charming elements obscure his implication in US military violence and by extension, that of *TAL*, which has, in some ways, underwritten the legitimacy of his errand into the wilderness. Seen in this light, Peters' concerns about liberal abyss artistry come into tighter focus.

If 9/11 served as a prod to bring *TAL* out of its inward gaze, the wars that followed sharpened its journalistic chops and ambitions. It also provided a new set of challenges for reporting on and narrating difficult subjects, while maintaining a tonal lightness, "a friendly intimacy," that distinguished it from documentary. The struggle to wrest clarity from the fog of war and the tug of deadly opposites required an interpreter whose authority and uncertainty were so intertwined as to assume a kind of heroic persona. Nobody managed this tone better than Sarah Koenig, whose stories from this era manage to balance self-awareness with taut procedural plotting and journalistic gravitas. Her 2014 interview with "Adam," a US veteran, stands out as exemplary for its ability to point to terrible things from a position not of neutrality but from the narrative alcove of ambivalence. "Adam" had approached *TAL* years before, when still a US soldier in Afghanistan, and sent audio dispatches, like Hyder Akbar, another embed. Koenig's carefully selected excerpts of those tapes and subsequent interview with him in "The Deepest, Darkest Open Secret" reveals "something way more personal and honest than anyone expected." The story comes swathed in caveats and disclaimers from Glass, including "Adam doesn't speak for the military in general" and it's "probably not something children should be hearing."

Koenig's introduction is long and careful, as well, but delivered in an even more informal idiom than is typical for the program. "So yeah, he said he'd give us an outsider's perspective." Like the caveats, the casual way of talking is how Koenig manages to distance herself from recordings that she said she found "jarring." In the first excerpt, Adam declares with equanimity his theory on how the volunteer armed forces work: "Men want to kill other young men." Koenig clarifies Adam's point: "these guys didn't want to just kill the enemy. They wanted the opportunity to kill another human being period." Koenig plays another bit of tape from Adam which concludes with "I'm pretty sure I just want the opportunity to kill someone too."

The tried-and-true formula for *TAL* is to deliver a sequence of events, typically involving someone else's story or testimony, followed by a grand reflection from Glass, or another host or producer, universalizing a feeling contained therein. In this case, for obvious reasons, Koenig makes a different choice, and in the process, develops the persona she used later that year to great effect as host and reporter for the first season of *Serial*. She moves the focus off Adam, and the shocking thing he has just confessed on tape, to address her own shortcomings as the teller of this story.

> Take a moment to compile all your stereotypes of a public radio producer. And now apply them to me. And you're right. I'm exactly who you think I am. I'm a peaceful sort. I don't come from a military family. I've not been in a lot of bar fights. I haven't lived in the South, playing violent video games.

These facts about her, she explains, distance her from Adam's background and, she suggests, means she's "coming to this a little more wide-eyed than a lot of other people might." The problem, in other words, is going to be one of translation. "What are we supposed to make of that?" she asks, inviting us into her quandary and implying, perhaps that we are more like her—wide-eyed and ingenuous—but also Northern, liberal, and, implicitly, well educated. This little aside implicates the listener several times in ways that are both canny and manipulative.

By immediately following Adam's shocking admission with an invitation to poke fun at her stereotypical public radio habitus, Koenig inserts herself protectively between listeners and her interview subject, even if it requires a touch of bathos, a tote bag joke in response to a homicidal confession. At the same time, she sets up a binary in which Southern, military families might somehow feel closer to Adam's sociopathic sentiments. That Koenig, a veteran reporter, is somehow less worldly than say, Southerners who play violent video games, is another odd rhetorical move. These seem like unintentional, though telling, distinctions. She's far more interested in setting herself up as a stranger in a strange land. It is precisely her confession that sets up her authority to tell this story in a particular way, a head scratching, self-conscious way, in which her own social distance from Adam represents an invitation of identification with her listeners. "What are *we* supposed to make of that?"

The rest of the interview consists of Koenig asking Adam to help her get her mind around what he's saying in the original audio tapes, to help put it into some kind of universal context for us. She starts with a warning that there would be no easy answers. "This is a story where I just ask that you hear him out." But really, she's asking us to hear both of them out, to become like her, an abyss artist. The interplay between Koenig and Adam in real time, listening together to his audio dispatches from 2009 in Afghanistan, gives them a common text to decipher. This is the same device used in the stories featuring the firefighter Sal Princiatta and "teenage embed" Hyder Akbar. The distance between the recordings and the interview also sets up a distinction between the soldier in country and the veteran reflecting upon his former self now that he's stateside.

But Adam refuses this distinction and the "fog of war" alibi. Koenig plays an excerpt in which he says, about a couple of Al Qaeda prisoners, "It's a hey, I want to shoot you, because I want to know what it's like, what it feels like to shoot you." Adam refuses to recant and describes his account as "accurate." What follows is a tense back and forth in which Adam systematically shoots down every easy assumption or interpretation on offer. No, it wasn't feelings of revenge, nor was it a sense of justice. No, it wasn't to protect society from demonstrably bad actors, like Al Qaeda. And no, it's not just him. "This is not an abnormal mindset," he says with equanimity. Koenig tries again to redirect: "And you're saying that's because these are the enemy, and they're going to kill us, and so...." but he cuts her off. "It's not like that." His desire to kill was neither strategic nor temporary. Stateside and out of the service, his desire to kill another person persists.

After confirming again that Adam is talking about killing for its own sake, Koenig dutifully rounds out the story with the *de rigeur* check-in with psychological experts providing context. Yes, the military does break down the taboo against killing; no that doesn't mean all soldiers yearn to kill for killing's sake. "Of course, the US military doesn't want sociopaths or would-be murderers in its ranks," Koenig offers, which at this point, feels obligatory and it's notable that she doesn't run this by Adam, only us. She talks to other soldiers in his unit who essentially confirm Adam's account that such feelings are in fact common, if perhaps a bit exaggerated in his case. The entire piece feels like a cautious, thrice-checked journalistic swathing around the assertion at the center of Adam's testimony: "This is not me. I am merely a representative of the community that I was working in." Poised between the inescapable testimony of war's terrible impact on those who fight it and the post-9/11 taboo against wholesale criticism of US militarism, Koenig and company's decision to hold off on airing this story for five years comes as no surprise. It may also help to explain Koenig's decision to center her own unreliability as a way into the story.

The elements of storytelling most prized by Glass (amusement, empathy, and joy) are hard to come by in such episodes. The only moment of emotional reprieve, of amusement, comes from Koenig's self-conscious distancing of herself and her kind, from the darkness of Adam's testimony. It is this distancing move that establishes her as the narrator that such post-9/11 stories called for. At the same time, it provides an alibi for getting closer to difficult truths and dark corners than might otherwise be possible. It is only by getting this close to the abyss that the journalist can tell the story. Finally, Koenig's narrative choices and *TAL*'s editorial decisions provide a kind of primer on what liberal feeling might sound like in a time of war and war's challenge to empathy.

In a matter of months, Koenig would reprise her head-scratching middleman role as she explored the murder of Hae Min Lee and the conviction of Adnan Syed for the crime. As many others have shown, *Serial*'s first season became a phenomenon in part because of how deftly Julie Snyder and Sarah Koenig centered Koenig's obsessive uncertainty, inviting her own ambivalence as an attractive proxy for the millions of listeners who waited for the Thursday morning drop of new episodes throughout the autumn of 2014.

As has been well documented, the producers and reporters and editors who staffed *TAL* and the flagship NPR shows, pioneered nonfiction narrative podcasting, setting the tone, mastering the sound, and selecting the agendas for newsworthy content. Performing the cognitive struggle to maintain an open mind was key to this sound. And it was, at one level, merely an extension of Bill Siemering's decision in 1970 to emphasize curiosity over authority and young enthusiastic voices like Susan Stamberg over stentorian voice-of-God baritones for *All Things Considered*. Glass was eager to generalize the impulse as part of the modern liberal condition: a necessary but risky attempt at empathy:

> But I think that we don't just become proxies for the people that we love. It happens with friends. It happens in business settings. It happens in politics. And when it happens, things can get very confusing. When you really step in for somebody else, substitute yourself for somebody else, it can be hard to tell if you're doing the right thing at all.

In some ways, this is a gloss on the show's central concern before and after 9/11: how to care about other people artfully, how to "do good in the tiniest way possible," per a 2007 listener mentioned in the previous chapter, the same year that Glass waxed universal on proxies. For John Durham Peters, in 2005, this is the central problematic of liberalism, which "always needs to be losing itself in the other."[42] Looked at from the perspective of the object of another's regard, Lauren Berlant, writing in 2011, frames it more direly: "recognition is the misrecognition you can bear."[43] From every angle, it seems, the challenge of taking up for the other reached an impasse in the years between 9/11 and the start of the golden age of the podcast.

The emergent pessimism about what can be known and reported also contained a pointed critique of institutions and lived experiences

---

[42]Peters, *Courting the Abyss*, 136.
[43]Lauren Berlant, *Cruel Optimism* (Durham: Duke University Press, 2011), 26.

that were becoming more difficult to report on neutrally. In his prologue to "Why We Fight," Glass shares his confusion about the juggernaut momentum of pro-war media and political discourse in the run-up to the invasion of Iraq in a way that feigns humility but gestures towards ideological critique:

> I've sometimes thought about [recent reports casting doubt on WMD in Iraq] as I've seen other news stories. And it always has this feeling of, did that really happen? Did that really happen? It's like this event which vanished off the face of the earth, never to be spoken of again.

Taken together, the shows about the war on terror demonstrate a stubborn faith in the sense-making responsibility of narrative journalism, even when confronting stories like Adam's that elude easy answers. The performance of self-doubt, as seen in the above quotation from Glass, conveys information and feeling about the inexorable march to war. It also provides a gloss on how dominant media narratives take on a momentum of their own. The episode's title, "Why We Fight," is taken from a series of propaganda films that aired before feature films during World War II. Directed by Frank Capra, the lushly produced documentary-style films were designed to refute the still strong isolationist sentiments that remained as late as 1942. The use of the title here can be called ironic, but it's an uneasy irony.

It's telling that the episode features a collection of traditional broadcast news-type features, a departure for the show. These include a quick round-up of "vox pops," person-in-the-street sound bites of Americans sharing their uneasiness with the coming war on Iraq; expert talking heads providing arguments for and against the war; and the above-mentioned piece in which Adam Davidson interviews Iraqi expats in Jordan. Doing intimate nuanced stories about war and peace made Davidson sick to his stomach. It makes sense that the rest of the episode leans into tried-and-true journalistic formats, which provide some distance and cover.

Confronting the abyss of war proves challenging to Koenig, Glass, and Davidson, encouraging each of them towards greater self-disclosure. Using the tools of storytelling to do journalism provides a set of narrative techniques to navigate high stakes stories by bringing them closer to home. Because of its commitment to feelings over politics and narrative enchantment over public affairs, *TAL*'s approach to war coverage is necessarily impressionistic. But the episodes dedicated to wars and their abysses tend to be heavier, resolving into impasses more than insights, as in the disturbing picture we're left with at the end of Koenig's interview with Adam, an American veteran on the Homefront hungry to make a kill.

The narrative alcoves in which *TAL* reporters find themselves reporting from are more than a journalistic conceit, designed to preserve neutrality and from which to wrest literary themes from real-life stories. As they accumulate across the decade after 9/11, stories on the theme of America come to resemble what Berlant has called "cruel optimism," the desire for "something that is an obstacle to your flourishing." The term may be a useful one for helping us to understand the sense of impasse that so many of these stories end on. It may also help us to re-think the controlling metaphor of this chapter—an enormous American flag held at each end by people whose understandings of its meaning are irreconcilable. Reading that scene not as a symbol of pluralism's hard-won promise but instead as an impasse opens up another approach to what "Feeling American" meant in the show's stories of this decade.

"Shouting Across the Divide" (2006), is one of the episodes from this era that most explicitly examined the impasse at the heart of American identity in the wake of the attacks: "stories of Muslims and non-Muslims shouting across a divide, trying to communicate with each other and not always getting their point across." Each of the episode's three stories end without resolution or satisfactory recognition among the people involved. The longest and most emotionally compelling story, "Which of These is Not Like the Others?," explores a family drama that draws heavily on the consequences of misrecognitions and "attachment to a significantly problematic object." In this case, the object is American identity.

Alix Spiegel, one of *TAL*'s founding producers, tells the story as "a sad story that began as a happy one." A young couple fall in love, "set in a location not usually associated with love stories, the West Bank." A young American woman named Serry falls in love with a Palestinian man, they marry and start a family. Serry insists on raising her children in the US and insists that being American is "her primary identifier." The September 11, 2001, attack interrupts the family's happy suburban life of upward mobility and intentional assimilation. Spiegel's framing sets up a dichotomy between the West

Bank, an unlikely place for a love story, and an American suburban pastoral, but it's clear almost immediately that this opposition was presented in order to be undermined by the rest of the story.

After 9/11, the family begins to experience hostility and prejudice from their largely white, non-Muslim community. Chloe, the eldest daughter, endures the worst of it, from her classmates and her teacher, who teaches the students to fear Muslims, using district-approved materials. A once popular girl, Chloe recedes into herself and eventually drops out of the school. Her father, once outgoing, becomes depressed; his initial concerns about raising a family in the United States have been realized and he pleads with Serry to move the family to the West Bank, where oppression can be more easily borne because it is shared. Serry balks, doubling down on America, insisting that this was a "fluke" rather than business as usual. "I have to believe this is not what America is about." Then Chloe's younger sister Samia is brutalized by the boys in her class. There is a Justice Department investigation resulting in some perfunctory diversity training but little else. By then it's too late for this American Dream. The husband is gone and Serry has relocated to a cramped apartment in a nearby town with her three children. To make ends meet, Serry works two jobs and rarely sees her children. She loves her husband, loves her family, and loves her country but finds herself unable to live properly with any of them.

Berlant describes cruel optimism as "a relation of attachment to compromised conditions of possibility whose realization is discovered either to be impossible, sheer fantasy, or too possible, and toxic." This seems an apt summary of Serry's relationship to her American identity. Berlant explores this concept through close readings of several works of contemporary literature that, she argues, "deliberately remediate singularities into cases of nonuniversal but general abstraction."[44] More simply put, she chooses stories of people at an impasse, "maintaining an attachment to a significantly problematic object," that speak to a broader human experience. Spiegel's story likewise centers a cruel relation, that of Serry's American Dream. Limning an emotional and material terrain between ideology and the unconscious, Berlant's notion of cruel optimism eschews easy answers to the stuckness

---

[44]Berlant, *Cruel Optimism*, 44.

that such attachments produce. They are cruel precisely because people "might not well endure" their loss, even as such attachments present obstacles to thriving. It is the optimism, however dire, that enables "the continuity of the subject's sense of what it means to keep on living and to look forward to being in the world."

Spiegel is at pains to introduce Serry at the outset of the story as someone for whom assimilation in a diverse American culture is crucial. She is devout but to outward appearances, she prefers to present as secular, save for her hijab. She considers the prospect of living only among fellow Muslims "boring." We never learn about her racial or ethnic identity, only that her husband is Palestinian and that they met in the West Bank. The omission points to an early *TAL* strategy of withholding "othering" information about race to inculcate "empathy" for a putatively white audience. In other words, the story is set up to make clear that for Serry, the impasse is not incidental but profound. The story ends with Serry waiting and hoping for her husband's return. She has set up a mailbox in the apartment for her children to share daily letters about their lives, as she is not home when they're awake.

In this way, Serry's life is taken over by what Berlant calls "practices of self-interruption, self-suspension, and self-abeyance."[45] Unlike the subjects of many of the post-9/11 stories examined in this chapter with their focus on the journalist as abyss artist, Serry emerges as the protagonist of her own story. The other stories in this episode feature subjects who also star in their own impasses: the communications director of the Council on American-Islamic Relations, a job in an organization whose very name suggests a condition of cruel optimism, especially in this particular historical moment, and an advertising copywriter tasked with making America palatable to Muslims abroad. But these stories are told with a lighter touch, perhaps because the attachments are less cruelly intense.

For most of the stories, it is the *TAL* storytellers who create attachments to significantly problematic objects, the better to explore impasses in contemporary American life. Their subject matter is serious, but their attachments are less so, which make the

---

[45]Berlant, *Cruel Optimism*, 27.

stories easier to bear perhaps because they have staked out formal positions that do not place their "sovereignty" at risk. An exception is Akbar Hyder, the "teenage embed" who follows his father to the borderlands of post-invasion Afghanistan to literally and figurative translate worlds, whose audio diaries suggest that his optimism has yet to become palpably cruel. Berlant describes such conditions as ones in which one can be "happy in an ordinary, often lovely, way, because the weight of being in the world is being distributed into space, time, noise, and other beings." Such a condition maintains so long as one can manage it and circumstances allow, a temporal precarity that Hyder's audio recordings preserve. His voice in early recordings conveys "the energy of feeling relational, general, reciprocal, and accumulative." We do not know how he bears his proximity to bloodshed nor how he reconciles his nation-building work with the tragic history of the ensuing years.[46]

Nasser Laham, the Jerusalem-based TV news translator from "Lost in Translation" is another subject whose optimism seems, at the time of its telling, to be prior to a reckoning that, with the benefit of hindsight, seems inevitable and cruel. Basim, the translator who found asylum in Norway after his work for US forces in Iraq, is another person whose attachment to the Americanization of his country is undermined, this time by the Abu Ghraib photographs. Like Serry, he struggles to reconcile the two things: "It was the total opposite of everything" he'd believed about the American military presence.

Stories like Serry's and Basim's share a common focus on subjects who are "worn out by the activity of life-building," a condition, Berlant points out, that afflicts "especially the poor and the nonnormative." In the 1990s, *TAL*'s mission was to "create empathy" by making the poor and nonnormative seem "just like us, middle class, with responsibilities." Listeners were thought to be isolated and alienated, but hungry for the intensities that moments of connection could generate.

---

[46] A follow-up interview with Akbar in 2011, on the occasion of the tenth anniversary, does provide him with the chance to reflect on the disappointments and betrayals of the intervening years. Ira Glass, "Ten Years In" Act I, *This American Life*, September 9, 2001, https://www.thisamericanlife.org/445/ten-years-in.

In the following era, the goal seemed to have shifted as *TAL* producers and reporters sought narrative perspectives that reproduced, at least formally, the cruel optimism of the no-win scenarios confronting the most vulnerable subjects of the war on terror. Listeners were called upon to empathize with these journalist proxies, to share "the energy of feeling relational, general, reciprocal, and accumulative." This distribution of feeling among producers and listeners recalls Sara Ahmed's notion of an affective economy in which emotions "align individuals with communities." For Ahmed, feelings gain intensity through circulation while for Berlant, distribution enables a kind of "coasting sentience." *TAL*'s post-9/11 stories assumed intensity and explored instead our capacity for what Berlant calls "ongoingness… our visceral intuition about how to manage living" and feeling liberal in illiberal times.[47]

While plenty of episodes in the 2000s continued to hew to the formulas, stories, and affective strategies of the earlier era, the shift in focus away from the "small personal stories" to "the world" and its problems was palpable enough that Glass and other producers have spoken of it as a major turning point. If 9/11 breathed life into NPR and the longform audio narratives that *TAL* specialized in, it also presented challenges. In fact, by 2011, stories about 9/11 and its aftermath became "an insane narrative puzzle" for Glass because "We're all bored of it," he complains, as if dismissing a trendy Park Slope eatery. "We don't want to dive back into sadness and also, we've heard everything that you could possibly hear about 9/11." In a characteristic pivot, *TAL* began to view its own preoccupation as part of the larger media saturation that first animated *Your Radio Playhouse*. Glass acknowledged the "tactical choices" necessary to keep listeners tuned in: enigmas, enchantments, and all. As the wars in Iraq and Afghanistan lurched deeper into quagmire and the questions of post-9/11 American identity hardened into the kinds of sad impasse confronting Serry's family, journalists turned to another crisis with personal and national stakes, which we'll explore in the next chapter.

---

[47]Berlant, *Cruel Optimism*, 52.

CHAPTER FIVE

Feeling Flush: *Planet Money* and the American Dream

The story of public radio—and its transformation into one of the most dominant narrative and sonic styles of podcasting—is, at one level, the story of a cultural institution narrating the felt experience of the era when liberalism sclerosed into neoliberalism, a harder, meaner, and more cynical philosophy of freedom. In this context, it's hard to imagine a more appropriate title than *Planet Money* for a show dedicated to "the financialization of everything," David Harvey's pithy definition of neoliberalism.[1]

Explaining the economy during the implosion of decades of neoliberal policy (deregulation, leveraged speculation, and shrinking public investment) presented a challenge: elucidating rules, laws, and accepted practices alongside the history of exceptions to them. *Planet Money*, a partnership between *This American Life* and NPR, provided lessons in economics via an "affective education," to borrow from NPR's Founding Purposes document, in order to navigate listeners through a maze of financial schemes and through the logic of "exceptions" that governed market logic.

With its deference to market forces as a solution for problems of poverty and stagnation, neoliberal thinking dominated economic policy starting in the 1970s, reaching its apotheosis with the crisis-

---

[1] David Harvey, *A Brief History of Neoliberalism* (London: Oxford University Press, 2007), 33.

era election of Barack Obama, before beginning to falter with the rise of Trumpist populism and authoritarianism, pandemic-era social spending and urgent, if scattered, leftist political formations. *Planet Money* narrated the logic of neoliberalism as a series of stories in which the global catastrophe became a series of human dramas and in the process, humanized the system that produced the catastrophe.

The emphasis on moments of crisis and epiphany that structured so many of the stories on *This American Life* made literary sense; moving from stasis (home) to crisis (adventure) to a new state (return) represents the outlines of an ancient formula. But on *TAL*, this formula had the perhaps unintended effect of assuming a stable origin point for the lives and systems featured in these stories. This tended to obscure structural origins of crisis and evoked a world in which protagonists traveled from stability into transformative moments and then back again. The unmarked time and space out of which stories and their protagonists emerge reinforced homogenous assumptions about race and class. Making poor and non-white people, for whom precarity can be as common as stability, seem "relatable" to presumed white and middle-class listeners was an explicit objective of Glass's storytelling techniques, as we discussed in Chapter 3. As we have seen, this impulse carries with it the potential for eliding critical political and social distinctions.

This strategy proved to be central to the stories on the economic crisis that became *Planet Money*, the cross-platform project between NPR and *TAL* that formed in 2008, just in time to explain the mortgage debt crisis, bank failures, and the ensuing Great Recession. Answering the question "what happened?" seemed to evoke explanations centering novelty rather than predictable cyclical forces and colorful characters rather than unsustainable structures of inequality. Born of crisis, *Planet Money*, like so many public radio innovations before it, staked out new territory for the medium, anticipating the coming long-form narrative journalism of podcasting's decade of meteoric growth.

It also explored the economics of feeling American at a moment when spiraling inequality and the collapse of the American Dream of homeowning made that feeling uniquely fraught and open to critique, as the ensuing populist movements of left (Occupy Wall Street) and right (Tea Party) soon demonstrated. While *TAL* was willing to probe some of the religious, racial, and ethnic contradictions at the heart of feeling American after 9/11, *Planet*

Money and the *TAL* episodes and NPR stories that developed around it, explored the crisis through the lens of surprise rather than politics, a retrenchment of sorts. Together, the stories of *Planet Money* and *TAL* held up the edges of a sagging American Dream and the economic rules that governed it, attachments that show how the power of cruel optimism rippled through the public radio structure of feeling. *Planet Money* extended the concept of empathy to embrace the apparatus of global finance capitalism itself.

Neoliberalism is the economic philosophy in which public radio's fitful growth and periodic crises took root. With its unreliable public funding, creative merchandising, and coy relationship to advertising, the fortunes of NPR as a going concern have always been a cipher for larger questions about the public investment in the public good. One could say that public radio has always been in a relationship of cruel optimism with the idea of the public. As Bruce Robbins said in 1989 of the concept of the public, it is, for all its theoretical shortcomings and historical failures, a concept "we cannot do without."[2] For Robbins, as for public radio and liberalism writ large, the public is a thing whose loss we "might not well endure," to use Berlant's elegant articulation of this Catch-22.[3]

In its response to the subprime mortgage crisis, *Planet Money* extended this cruel relation by doubling down on the American Dream, one in which rational actors, properly informed and obeying economic laws, thrive. As we'll see below, *Planet Money* explained the economic crisis as a series of exceptions to rules—surprises! Explaining the crisis of 2008 required enormous amounts of reporting and equal amounts of storytelling and editing, with the "characteristic elements of impulse, restraint, and tone" that *This American Life* brought to intimate stories and which Raymond Williams suggested were crucial parts of how a structure of feeling manifests itself.[4]

The story was a compelling one: a new concentration of capital in search of growth opportunities found new and exotic financial instruments whose risks nobody quite understood. To convey the uniqueness of the circumstances that led to the crisis, the status

---

[2]Bruce Robbins, *The Phantom Public Sphere* (Minneapolis: The University of Minnesota Press, 1989), xv.
[3]Lauren Berlant, *Cruel Optimism* (Durham: Duke University Press, 2011), 52.
[4]Raymond Williams, *Marxism and Literature* (Oxford: Oxford University Press, 1977), 128–35; and *Preface to Film* (Film Drama, 2003 [1954]), 21–3.

quo ante was, at least implicitly, represented as stable and, for the purposes of the stories, beyond critical examination. The old way of doing things had given way to runaway investments in debt, highly leveraged transactions, and anemic regulatory oversight. Unprecedented, unique, and extreme, the debt crisis was, at heart, also a crisis of faith. Restoring belief in the system and in market logic, *Planet Money* reporters explained, was critical to the recovery of the economy. Like FDR's May 1933 "Fireside Chat" on the banking crisis, *explaining the problem* was understood to be the best mechanism for restoring confidence in the system.[5]

Straddling the democratic impulse to make clear the mystifications of finance capitalism and the mystification inherent in "explainer journalism" with its reliance on talking head experts and compelling but simplistic analogies, *Planet Money* exemplified the contradictions of public radio's liberal mission. At the same time, the show helped to pioneer podcasting's longform journalism-on-a-theme just as public radio began to fully embrace podcasting as a promising frontier for born-digital audio content. It was the perfect show for the moment, in part because it was not just a show. It was a podcast, a blog, and a periodic feature on *TAL*. It could also be edited into shorter bits for insertion into NPR daily news programs. As NPR transformed from radio network to digital platform, flexible and timely content from *Planet Money* proved useful for insertion into flagship news programs like *Morning Edition* and *All Things Considered*. In what follows, I'll explore the origins of "public radio capitalism" that set the stage for *Planet Money* and then examine the show's defense of free market principles through a logic of neoliberal exception, a brand of cruel optimism uniquely suited to the collapse of the American Dream of homeownership.

Public Radio Capitalism

In 2015, Glass declared in the pages of *Ad Age*, that "public radio was ready for capitalism."[6] At that point, it was hardly a novel

---

[5] Jason Loviglio, *Radio's Intimate Public: Network Broadcasting and Mass-Mediated Democracy* (Minneapolis: University of Minnesota Press, 2005), 1–37.
[6] Felicia Greiff, "Ira Glass: 'Public Radio is Ready for Capitalism,'" *Ad Age*, April 30, 2015, https://adage.com/article/special-report-tv-upfront/ira-glass-public-radio-ready-capitalism/298332.

position to take; in fact, the embrace of market forces has been central to *TAL*'s structure and feeling from the beginning. As argued in Chapter 1, public radio's readiness for capitalism has been the founding contradiction for all public broadcasting ever since the passage of the Public Service Broadcasting Act in 1967. "I see a huge middle ground," Glass clarified, "where we keep our mission and our ideals, and bring in more money using the conventional tools of the market economy." By 2015, Glass could point to a string of examples of this embrace of for-profit ventures across a range of media platforms, like movie deals, books, and TV shows. He was also talking about podcasts like *Serial* and *Invisibilia* and the for-profit networks that were increasingly populated by former producers from public radio.

Glass rightly pointed out that US public broadcasting's long history of genteel underwriting messages effectively blurred the line between public media and the market economy. Listeners were now consuming public radio podcasts alongside audio content from Gimlet, the for-profit company whose producers, reporters, sound design, and vocal performances had all been cheerfully and uncontroversially expropriated from public radio the year before. The public radio brain drain had also created rival networks like Radiotopia, Panoply, Midroll, and Pushkin. Perhaps more than any other program, *TAL* had mastered the formula, in which something that *felt like public radio* took advantage of "market forces." *Serial*, podcasting's breakout hit, was *TAL*'s first and most lucrative public/private venture into podcasting, but it had followed film, television, and publishing initiatives.

Market-friendly public radio has always tried very hard to have it both ways, preserving the social markers of public radio's "alternative" and highbrow distinctiveness while embracing the "financialization of everything" that is the hallmark of the neoliberal era from which it sprang. Glass's celebration of this duality, while not shocking, was an uneasy reminder, perhaps, of how unsavory this public and private intermingling could come across: "Public radio today is entrepreneurial and DIY," Glass offered. "I mean that in the punk rock sense, as well as in the business sense," he added in a surprise departure from irony.[7] This assertion recalled his 1999

---

[7] Jessica Abel, *Out on the Wire: The Storytelling Secrets of the New Masters of Radio* (New York: Broadway Books, 2015), 4.

defense of his preoccupation with revenue: "We wanted stations to make more money off our show than it cost. The flakier your mission, the fiercer you have to be on the business side."[8]

Eleanor Patterson notes that Glass mastered this hybrid public/private business model to profit from commercial ventures while cashing in on the public radio's highbrow reputation, "uncompromised by the constraints of the commercial sponsors or advertisers." She also points to the fact that by this time, *TAL* had become a multi-platform franchise with intellectual property in other media industries.[9] In some ways, it was merely the latest public radio show to venture into the commercial possibilities of this kind of hybrid brand. Minnesota Public Radio's (MPR) decision in the 1980s to monetize and merchandise *A Prairie Home Companion* represents an earlier moment in private/public hybridity, driven by the success of one program, the structural deficits inevitable in the public funding model, and the decentralized nature of US public radio.[10] It was also perceived as an upstart challenge to NPR's reliance on government and grant funding. MPR head Bill Kling relished the role of public radio gadfly in the 1980s as much as Glass did in 2015.[11]

Marketplace, the self-described "business show for the rest of us," also gestured towards a hybrid of high-brow and market-minded approach. Begun in 1989, outside of NPR's orbit in the upstart American Public Media network (later Public Radio International (PRI)), *Marketplace* struggled to find financial support until General Electric, at that time the epitome of the nation's "military-industrial complex," swooped to the rescue. The show's theme song raised eyebrows by incorporating the strains of GE's

---

[8]Marshall Sella, "The Glow at the End of the Dial; Ira Glass Is, Um (Pause, Delete) ... Listening: The Perfectly Edited World of His 'American Life,'" *The New York Times*, April 11, 1999, https://www.nytimes.com/1999/04/11/magazine/glow-end-dial-ira-glass-um-pause-delete-listening-perfectly-edited-world-his.html.

[9]Eleanor Patterson, "This American Franchise: This American Life, public radio franchising, and the cultural work of legitimating economic hybridity," *Media, Culture, & Society* 38, no. 30 (2016): 455.

[10]Jack Mitchell, *Listener Supported: The Culture and History of NPR* (Westport, CT: Praeger, 2005), 154–5.

[11]"As NPR sputters, Kling points to problem under the hood," *Current*, March 14, 2011, https://current.org/2011/03/as-npr-sputters-kling-points-to-problem-under-the-hood/#.

jingle "we bring good things to life," already an ironic motto for a manufacturer of nuclear weapons and dumper of toxic chemicals in the Hudson River.

It was another poke in the eye to the staid and respectable NPR, even as it opened up a new tolerance for the "public-private partnerships" that became a neoliberal mantra as public institutions faced more cuts and private equity stepped into the breach across a range of American institutions. *Marketplace* in many ways set the stage for *Planet Money*'s broad-gauge approach to storytelling about the economy. By focusing on stories that "we don't typically think of as business stories" to find the economic lessons, *Marketplace* made an appeal to the cultural tastes of knowledge workers who prided themselves on a polite distance from the vulgar instrumentality of financial news and perhaps the conservative bent of the *Wall Street Journal*.

By showing how everyday life was economic, it backed its way into promoting neoliberalism under the guise of intellectual curiosity. It is no surprise that *Marketplace* continued to thrive, spawning myriad spin-off shows like *Marketplace Money*, and *Marketplace Morning Show*. Like many public radio shows, it transitioned easily into a thriving podcast ecosystem starting in the mid-2010s, with shows like *This Is Uncomfortable*, *How We Survive*, and *Million Bazillion*, *Codebreaker*, and *Corner Office*, and *The Uncertain Hour*. The latter was a long-form investigative show that took on knotty problems of inequality in ways that reflect the tectonic shifts in the kinds of audio narrative sensibilities that took over in the years after the 2016 election of Donald Trump. It has also developed a smart-speaker version of the brand called "Make Me Smart," which delivers bite-sized factoids about the economy.

For most of its run, *Marketplace* specialized in slickly produced short narrative explorations of businesses large and small, reifying the commonsense notion that stock prices, corporate layoffs, and market innovations were, like the weather and sports scores, changing data points requiring daily updates. The sheer number of variations of the show is a testament to the fact that public radio's preoccupation with its own relationship to market economics aligns with its listeners' appetite for business news swathed in public radio respectability, news that prizes emotional connection and the performance of liberal empathy realized through the conventions of public radio storytelling.

As explored in the first chapter, the history of public radio could be told as a series of market-friendly epiphanies: spells of crisis and leaps of private-sector innovations, each one making possible the next. Because there was never the political will nor the economic mandate to establish a robust public broadcasting service, periodic spasms of financial panic were ensured. This ritual unspooling of the knotty contradiction at the heart of the public radio's founding purposes, funding sources, and constituency is an important context in which to consider how economic matters have been covered on programs like *TAL*. It's unusual to have cultural institutions provide a primer on the political economy of its own industry, and yet that's what *TAL* and NPR did when it launched *Planet Money* in September 2008. As I have argued in Chapter 2, several themes recur in public radio's courtship of an audience of affluent, educated listeners for whom liberalism is a dominant feeling, if not a coherent set of political commitments.

Politically, public radio embraced a blithe confidence that coverage of conditions of inequality in formats palatable to elites represented a progressive impulse in journalism. At the same time, periodic financial crises at NPR demonstrated the fact that a public funding model was insufficient. It also inculcated in public media journalism a sense of its own improbable and unrealistic existence.[12] Its persistence was quirky, an exception to a rule that periodically came knocking. Privatization, the wolf at the door, seemed inevitable. Podcasts that picked up where public radio left off, Christopher Cwynar has argued, proved an ideal host for shows emphasizing "entrepreneurial values."[13]

Market Values and Storytelling

With this context in mind, it's easy to identify how the centrality of market economy values is built into the very bones of *TAL* storytelling. As explored in the previous chapter, *TAL* alumnus Alex Blumberg reduced the neoliberal logic of the personal over

---

[12] David Giovanonni, Leslie Peters, and Jay Youngclaus, *Audience 98: Public Service, Public Support*, 26–32.
[13] Christopher Cwynar, "Self-Service Media: Public Radio Personalities, Reality Podcasting, and Entrepreneurial Culture," *Popular Communication* 17, no. 4 (2019), https://doi.org/10.1080/15405702.2019.1634811. 318.

the political to a storytelling catechism. Stories that centered the obvious role of social justice or political movements were on the "wrong track." Journalists interested in homelessness, for example, who want to explore the role of social welfare programs in helping get people off the street, were exhorted to "solve for a different Y." Instead, why not cover homelessness via a story of a formerly homeless guy "who misses being homeless," or something equally "surprising." By the mid-2010s, Blumberg was not only the author of an instructional video on how to tell stories, he was also the CEO of a for-profit podcasting startup, Gimlet Media, and actively wooing venture capital to the concern. Like Glass, he understood the power of evangelizing storytelling as a means of building the brand. Making stories and making a career depended upon the same market insights.

The Giant Pool of Money

All of this to say that by 2008, when the US economy confronted its greatest economic crisis since the Great Depression, *TAL* was ready with an entire philosophical and narrative approach to understanding the relationship between people and the economy. By 2008, the show had developed formidable journalistic resources for taking on big stories that required months of deep reporting. It had also begun its ambitious franchising projects, developing stories for film projects, touring the nation with live stage versions of the show, and even trying their hand at a short-lived television show. The nation's looming economic crisis looked like an opportunity to expand its storytelling journalism into new areas and with new partnerships. *Planet Money* was the show's first podcast spinoff, but it would not be its last. It set the tone for the show's expansion into born-digital audio programming. A partnership with NPR, it represented an acknowledgement of the limits of *TAL*'s journalistic resources and perhaps a concession that NPR had erred when it passed on a distribution deal for *Your Radio Playhouse* back in the 1990s.

Planet Money also exemplified the way *TAL*-developed talent, like producer Alex Blumberg, found bigger projects and higher profiles as podcasting began to command its own audiences and production cultures, distinct from those of radio. It embraced a chatty casual mode of address that invited listeners into the story.

It was introduced variously as a "project," a "collaboration," a "podcast," a "blog," and "a team," in its first weeks of existence. The shifting of nomenclature and chummy informality was a sign that the producers were still figuring out what cross-platform, cross-organizational audio journalism should sound like. Finally, *Planet Money* experimented with the theme of nerdy curiosity, bordering on obsession, which developed into the default persona for many of the podcast hosts in the public radio diaspora that followed, as Neil Verma has recently demonstrated.[14]

Planet Money's origin story begins on *TAL* with an episode called "The Giant Pool of Money," hailed as a landmark collaboration between NPR and *TAL*. Airing in the weeks following the collapse of the big investment bank Bear Stearns, the show featured Alex Blumberg and NPR reporter Adam Davidson, whose public radio career began at WBEZ in sales and underwriting. The episode out to explain, with forensic detail and literary aplomb "what went wrong" in the 2008 subprime mortgage crisis and the cascading global credit crisis that seemed to be unfolding. It was an impressive contribution to explainer journalism, providing a primer in modern financial products like "credit default swaps" and "collateralized debt obligations," which were, essentially, sliced and distributed bits of mortgage debt that enabled more and riskier trades on real and anticipated growth in housing equity.

They also document, with the confident plotting of a police procedural, the steps in the expansion and implosion of the housing and credit bubble that characterized the crisis between 2000 and 2008. The policy details and economic terminology was always carefully leavened with carefully curated moments of human interest. Glass opens one early show about the crisis with a list of surprising signs of the human impact of the deepening recession: dentists reported an uptick in patients with cracked teeth; shark attacks and traffic jams decreased; and porn video rentals trended down as commuting and discretionary spending and travel plummeted.

There are elements in the debut broadcast reminiscent of the liberal "muckraking" tradition that dates back to the Progressive era a century earlier, when the conditions of the poor were

---

[14]Neil Verma, *Narrative Podcasting in an Age of Obsession* (Ann Arbor: The University of Michigan Press, 2024).

juxtaposed to the outrageous fortunes amassed by aristocratic industrialists to great effect. "The Giant Pool of Money" (2008), which *Planet Money* producer Robert Smith has called "The ur text that started this whole thing," and which Glass has described as "the most popular, commented on, awarded piece of journalism we ever did," begins with two very brief vignettes that put economic inequality into sharp contrast.[15]

The first is set at an awards ceremony held at the Ritz-Carlton Hotel for mortgage brokers like Jim Finkel, honoring the creators of that year's best collateralized debt obligations (CDOs). CDOs are the instrument, reporter Alex Blumberg tells us, at the heart of the mortgage crisis that turned into a global credit crisis. The next scene is at a community college in Brooklyn, where a nonprofit conference for people facing foreclosure is being held. There, Blumberg introduces us to Richard, a Marine and Iraq veteran, who breaks down as he describes the financial ruin he faces, the shame of raiding his son's college fund to make ballooned mortgage payments, and so forth.

But the rest of the episode seems dedicated to providing a neoliberal interpretation of this almost Dickensian opening. "Now it's clear that these two groups are connected: Jim at his black-tie dinner and Richard the Marine," Blumberg concedes cautiously. But in order to lay bare the entire process in detail, Blumberg turns to another homeowner in trouble, this one named Clarence Nathan, at risk of losing his home after borrowing half a million dollars, despite his lack of collateral and a patchwork of part-time and unreliable jobs. Nathan, whose desperate need for money is never explained, and who admits he may have had to do "something more drastic and dramatic" if he hadn't gotten a No Interest, No Asset (NINA) bank loan, and who also admits to what Blumberg calls "shadowy criminal contacts," represents what Blumberg called a "more nuanced" portrait of the mortgage crisis.

> Stories like this have been in the news for months. They often feature an innocent homeowner who is duped by a lying greedy mortgage banker. Or if you're a more of a *Wall Street Journal*

---

[15]Ira Glass, quoted in Abel, *Out on the Wire*, 49.

editorial page type, an innocent mortgage banker duped by a lying greedy homeowner. And no doubt, both kinds of people exist. But Clarence's case is more nuanced. And much more common.

Blumberg's promise of nuance represents a neoliberal triangulation between, on the one hand, the *Wall Street Journal* editorial page's notorious arch-conservatism and, on the other hand, the equally dubious muckraking accounts that juxtapose partying mortgage brokers with weeping veterans fighting to save their homes. "The bank made an imprudent loan and I made an imprudent loan," Clarence confesses with equanimity. This microeconomic explanation provides a both-sides frame that sets the tone for a great deal of *Planet Money*'s subsequent coverage of the economy. To explain the problem at a macro level, Glass, Blumberg, and Davidson attempt to answer the pressing question of *why* banks would make such an imprudent loan. They do not, it should be noted, bother to ask why someone like Richard would take on such massive debt on confusing and even usurious terms. The *Planet Money* version of events suggests that the crisis was an aberration from the normal flow of global capitalism. Banks historically lent money prudently,

> but then suddenly, the early 2000s, everything changed. Banking turned on its head. And went out looking for partnerships with *people like Clarence*. What happened?[16]

The *Planet Money* team bring their formidable resources to explaining what caused people to "throw out the old rules of banking," rules that presumably were basically sound. Such questions obscure a well-documented history of predatory and discriminatory lending, blockbusting, and bouts of wildcat speculation by US financial institutions over the course of the last century.[17] Blumberg and Davidson explain that the surging global economy, thanks to the

---

[16] Alex Blumberg, "The Giant Pool of Money," *TAL*, episode #355. [emphasis added].
[17] Gary Richardson, "Categories and causes of bank distress during the great depression, 1929–1933: The illiquidity versus insolvency debate revisited," *Explorations in Economic History* 44, no. 4 (2007): 588–607; Joseph E. Stiglitz, *Globalization and Its Discontents* (New York: W. W. Norton, 2003).

massive growth in China and other developing countries, created the eponymous "giant pool of money" that investors around the world were keen to invest. That, combined with US monetary policy (a low federal funds rate), led investors to feverishly seek profitable returns from the US mortgage industry. A speculative bubble was fueled by a giant pool of money with an almost anthropomorphic drive to reproduce.

The excesses of this frenzy are documented in the person of Glen LaRusso, a New York mortgage company sales manager who made and blew thousands on fancy Champagne and cars before the inevitable crash. A few modern-day Cassandras are interviewed as well, bankers whose intimations of doom went unheeded by peers. There is a quality of Greek tragedy to the story. The overweening greed of mortals, driven mad by an insatiable, immortal giant pool of money, lead to a brief, heady rise followed by a predictable and pitiless fall. Through it all, like mythology, the account never reckons with certain givens; like the whims of the gods, the rules of capitalism are assumed to be eternal and their origins unknowable. An economy run by the money itself, rather than the social interests that money might, in another world, serve, is the *doneé* of the world-building of *Planet Money*'s storytellers.

Little mention is made of the deregulation of the financial industry in the United States over the years since the 1999 Financial Services Modernization Act, which effectively repealed the 1933 Glass-Steagall Act that was created as a remedy for the kinds of malfeasance that led to the Great Depression. A brief section on a later installment briskly dispenses with the idea that the fault lies with the tax-cutting, deregulation-mad Republican Bush administration. No mention is made of the tremendous transfer of wealth from the poor to the rich that took place in the United States since the 1970s, setting the stage for the kind of speculative frenzy that leveraged the debts of poor Americans into massive profits for banks and global investors. And while the rapid rise of capital markets in India and China is acknowledged as part of the backstory, there's no mention of "the overall dynamics of capitalist accumulation and its propensity to periodic crisis," which had been well documented by economic historians.[18] Indeed, the insight for

---

[18] David McNally, *Global Slump: The Economics and Politics of Crisis and Resistance* (Oakland: PM Press, 2010), 217.

the episode, credited to Glass, is that "the character is the *money*." Just as neoliberalism and the Supreme Court's 2010 *Citizens United* ruling would have us believe that corporations are people, *Planet Money* found its protagonist in a pool, or planet, made of money.

In place of where such context might have gone, however, we do get to hear Blumberg and Davidson chatting amiably, stumbling over each other's new job titles, and the pronunciation of the name of their International Monetary Fund source, Ceyla Pazarbasioglu. Over the next several installments, Blumberg, Davidson, and their colleague Chana Joffe-Walt take this breezy conversational style to new heights. Juxtaposed to the arcana of modern financial products, the intention perhaps is to make economics approachable and fun. But it also works as a fending action, the better to distance listeners from a too-close engagement with the experiences of others. Glass even distanced himself from the recession data about broken teeth and fewer shark attacks, framing it as part of a "mini-industry inside journalism right now... finding unusual signs of the recession all over the place."[19] The oscillation between empathy and ironic distance requires a mastery of tone, a virtuosic precision in delivery, an artful artlessness. If these episodes are meant to be didactic as well, then the lesson is clear: the giant pool of money just is and it's naïve to imagine a world in which such concentration of capital would be subject to social constraints.

Part of what made *Planet Money* so successful was its mastery of the affordances of radio and podcasting at once. Like many podcasts that dominated in the public radio structure of feeling, *Planet Money* was born at the intersection of the two formats and played to their respective strength in often surprising ways. For instance, in between the award-winning, hour-long pieces aired on *TAL*, *Planet Money* was running an almost-daily podcast, which dropped in the afternoon around 4:00 p.m. just after markets had closed. The daily reports conveyed a sense of urgency often associated with radio broadcasts, evoking and assuaging states of crisis in a single report; however, listeners were beginning to habituate to RSS feed downloads and notifications in ways that remediated the appointment listening associated with broadcast news.

---

[19] Ira Glass, "Scenes from a Recession," *TAL*, episode #377, March 27, 2009.

The format of the daily podcasts lent an air of broadcast-style immediacy as well. Co-hosts Laura Conaway and Adam Davidson adopted an informal camaraderie with each other and with listeners that suggested a shared precarity and often, a sense of relief that all in all, the market hadn't been *that* bad today. Conaway played the plain-spoken newcomer to economics, our proxy; Davidson, on the other hand, provided the answers, based on his day's reporting and his growing rolodex of industry, regulatory, and Bush and Obama administration experts.

The short nightly podcasts also provided a way to repurpose segments from the longer episodes and to tease upcoming ones. In the podcast from November 7, 2008, Davidson introduced one of the more ambitious podcast-first stories, a sprawling global epic, in partnership with *The New York Times*. In *TAL*-style, it begins in small-town Wisconsin in which local school boards confronted staggering financial losses thanks to a catastrophic investment in risky (and likely fraudulently marketed) credit default swaps, which were sold to them as simple low-yield corporate bonds. Their gradual collapse in value put them into debt because they borrowed millions to buy them. The story shifts to Germany, then Dublin, then New York, and then the Caribbean, following investment banks and tax shelters that drove the production and circulation of these toxic assets.

Like a social realist novel of the nineteenth century, memorable characters flit in and out; reporters move through dark corners and festive halls. Unlike a Dickens novel, however, the protagonist is the money, not the pauper, and the happy ending involves the restoration of trust in the system, rather than a change of heart. The fact that the school boards made these investments in pursuit of health care and retirement plans for teachers is mentioned as an aside, with little attention to the built-in precarity of existing social systems which set such catastrophes in motion. It was not a giant pool of money, but rather the lack of one, a structural social deficit at the center of this small-town story. The episode's title, "A Tale of Intertwined Misery," reflects both the literary ambitions of the narrative and the role of affective language as a deflection from the economic structures that reliably produce underfunded social institutions.

What lent *Planet Money* both a sense of urgency and a sense of narrative coherence was its consistent emphasis on the role of market confidence in the economic crisis and its resolution. Uncertainty about the future of modern finance capitalism drove each day's news, told through a revolving list of "indicators." As inter-bank lending "froze up" in the fall of 2008, *Planet Money* reporters provided a collection of experts, anecdotes, metaphors, and historical comparisons to drive home the irony of a system built on credit, a word derived from the Latin *credo* (to believe), imploding thanks to a widening gyre of mistrust. Over the course of a year, in blog form, short daily podcasts, and the occasional one-hour special on *TAL*, the *Planet Money* team provided a sparkling exposition on the life and death of credit, as a human drama and an economic phenomenon, as it made its way through the capillaries of everyday life.

Because of the literary allure of the play of antitheses: credit and fear, pools of money and sinkholes of debt, the crisis lent itself to stories in which the poor and rich were juxtaposed. Because the daily podcast charted the government's interventions, there were elements of the tightly plotted police procedural with each installment adding layers of detail and complication, building towards resolution. In short, the show's multimedia platform and vast resources in access journalism and storytelling expertise, combined with the house style of insouciant chat-cast rapport between and among hosts, lent *Planet Money* a kind of magnetism. Its popularity and influence ensured the regular appearance of policy wonks and politicians, eager to assuage listener fears and to influence the national conversation.

Planet Money succeeded in its two most explicit goals; the first was providing entertaining and accessible explanations of financial products and processes. The second was making clear the importance of confidence in the system as the only antidote to the crisis. A third implicit goal was to extend the *TAL* franchise into new directions with new partnerships and on new platforms, a nice synergy with NPR's 2006 declaration of its ambition to "build multimedia experiences around one or more topics of strong public interest."[20]

---

[20] Mike Janssen, "NPR rallies system to jointly build 'trusted space'," *Current*, July 17, 2006, https://current.org/2006/07/npr-rallies-system-to-jointly-build-trusted-space/.

The first two goals worked well together, as each anecdote, metaphor, sympathetic character, and relatable talking head served the goal of clarity and made way for an understanding of the rules of banking and the guard rails around the financial system. The government's intervention into the practical and ethical problems of private debt raised fascinating questions about "moral hazard." These debates provided the reporters and hosts the narrative structure for a dichotomy between economic laws and legitimate exceptions to them. In its explanatory capacity, *Planet Money* defined for listeners what counted as a state of exception to "free market" principles. In so doing, it helped to explain the political conditions in which both the Bush and Obama administrations were obliged to massively intervene in the economy, bailing out financial institutions to the tune of trillions of dollars.

These "exceptions" assertively reinforced the rule that market forces, and the giant pools of money who love them, should normally be left alone. These "exceptions to neoliberalism," as Aihwa Ong would call them, and *Planet Money*'s daily report highlighted, demonstrate Ong's insight that neoliberalism can best be understood as "a technology of government," characterized by "the interplay of exceptions," in diverse settings, rather than as a monolithic and static approach to markets, governments, and citizens. While Ong has in mind East Asian contexts in which western neoliberalism hasn't fully taken hold, it's a useful perspective for thinking more broadly about the flexible ways in which neoliberalism operates through a "logic of exceptions" during crises in Western liberal democracies as well.[21]

In the September 24, 2008, podcast, "Moral Hazard, Meet Naked Short Selling," Adam Davidson explores at length the practical and ethical problems involved in bailing out financial institutions like Bear Stearns, which were saved with federal funds, compared to the macroeconomic risks involved in letting banks like Lehman Brothers fail, which set off a wave of panic and a loss of faith in the financial system. Davidson outlines the exceptional danger of bailouts in terms of moral hazard theory; removing consequences from financial actors increases their tolerance for risky investments in the future.

---

[21] Aihwa Ong, *Neoliberalism as Exception: Mutations in Citizenship and Sovereignty*, (Durham, NC: Duke University Press, 2006), 3.

Davidson says he's glad that the government has allowed Lehman Brothers to fail because of the moral hazard involved in bailing out reckless banks and because it represents a return to market forces as arbiter of winners and losers. In another episode that fall, Davidson declares that he's "a big believer in market incentives." The free market works precisely as designed when banks' bad bets cause them to fail, Davidson and numerous talking head interviews insist, because "better banks" scoop up their assets and behavior improves for the entire sector. When pressed to explain why the government's bailout of Freddie Mac and Sallie Mae, the massive home mortgage companies, didn't raise alarm bells, Davidson dismisses the idea in ways that illustrate the logic of neoliberal exception:

> Well, Fannie and Freddie are creatures of moral hazard. They're different from Lehman Brothers, Bear Stearns, all these others... They are a walking, talking, screaming moral hazard that's been in the American financial system for 70 years.

Government created, privately held, Freddie and Sallie were created in the 1930s to make it easier for Americans to buy houses. Because the profits are privatized and the risks are publicly borne, these entities, which were responsible for 70 percent of all US mortgages, give the lie to the notion that the modern finance capital system has ever been a market free from massive government participation. As Ong and others have pointed out, neoliberalism is itself a form of governmentality. Davidson explains these massive organizations in terms of the logic of exception. "They've always been a very odd organization—private, profit seeking companies trying to get money for their shareholders and at the same time they are trying to promote a public mission. There are many people who think this is a bad idea since their creation in 1938." Because they have been central to the entire homeowning economy of the United States since the Great Depression, it is a remarkable position to frame them as an exception to a free market that somehow exists outside and around these massive exceptional institutions.

Again and again, in the podcast, radio feature, and blog versions of the *Planet Money* coverage of the crisis, the phantom of "the free market" is invoked by hosts and guest alike as the default way things work, a heuristic for explaining the exceptions taking place during the housing and credit crisis. This invocation often takes the form of an economist, often from Cato Institute, George Mason University, or another repository of neoliberal economics, hectoring the hosts about how markets work best when left alone in one breath and in the next, acknowledging the necessity of emergency actions of the federal reserve in propping up financial markets.

"Today is another massive expansion of the role of the federal reserve and the central bank of the US," Davidson declares on the October 22 podcast. The federal government had just announced a new program to back transactions of "commercial paper," overnight loans between banks for millions of dollars. As buyer or seller of last resort, the Fed was essentially taking over the market in credit to jumpstart an economy whose loss of trust in itself had essentially frozen these overnight loans, bringing financial activity to a standstill.

Thematically, *Planet Money*'s beat was the restoration of the natural laws of free markets. Fed interventions were occasions for scholarly reminders for the unsoundness of government intervention. Zigzagging back and forth about when the moral hazard rule applies and why, *Planet Money* reporters and their guests describe a system in which exceptions are the rule. In so doing, they simultaneously promote confidence in the credit system and reinforce the rhetoric of market freedoms.

The contradictions in this approach were managed and contained largely through affect. On the Wednesday November 19 podcast, Adam Davidson and Laura Conaway interview a financial analyst, Charles Peabody, about the role of political influence on which banks are receiving government funding in the form of the Fed buying shares of their stock. The conversation begins with the typical critique of the folly of government interference with the equally typical hedging about the exceptional circumstances. "Should I be angry?" Davidson asks Peabody, towards the end of the interview, when it's unclear how to measure the exigencies of the emergency against the corruption of the process. "I feel angry," he concludes.

The December 3 podcast, titled "Layoffs Are Good for Us," cold opens with an audio clip from *Office Space*, the classic Gen X movie about workplace anomie in which an older paranoiac worker rages about his inevitable layoff. The clip dissolves into a screaming guitar riff of "The Warriors Code" by the Dropkick Murphys, the left-wing punk band. These sonic markers caption a lengthy defense of layoffs by economist Don Boudreau, a Friedrich Hayek acolyte. Layoffs act as a tonic to the overall economy, freeing up capital, and ultimately benefitting all, Boudreau contends. The end of the episode features a couple of recently laid-off professionals who laughingly deride this perspective. Conaway and Davidson sheepishly consign the debate to a matter of "head versus heart." In this way, *Planet Money*'s coverage of the crisis rehearsed, on a daily basis, the zigzag logic of exception through tonal shifts and sonic markers of emotional dissonance.

Economists were asked to locate the credit crisis within a larger historical context, which the hosts mediated through shifting affective responses. Because of the scale of the downturn in the economy and the seemingly overnight disappearance of trillions of dollars of value from the global economy, *Planet Money* was at pains to explain the crisis as extraordinary while shoring up belief in the market fundamentals undergirding the financial system. During Thanksgiving week 2008, *Planet Money* borrowed explicitly from *TAL* in taking on a weeklong theme "What is Money?" in an attempt to define money in ways that both explained the crisis while naturalizing the very system that produced it. In an episode titled "Money is a Relationship" (November 24, 2008), British economist Niall Ferguson explains that ever since ancient Mesopotamian documents bore the words "pay to the bearer," money has always been primarily "a relationship between people" built on trust. This sweeping narrative locates the credit crisis in a larger history of money as a set of human relationships.

The episode begins with The Flying Lizards' punk cover of the Motown classic "I Want Money," the singer's flat affect operating as a set of air quotes around the sentiment behind the lyrics. Announcing the day's big headline—Citibank has been bailed out with a $20 billion dollar stock injection by the federal government, signals a tonal shift. Blumberg groans that he's "tired of doing this story over and over again"; Davidson agrees with a deadpan mixture of hipster cool and hyperbole: "the fundamental architecture of our economy has changed, yet again. Big Whoop." Once again, the economy is rationalized as both exceptional and predictable, even boring. And Ferguson's attempt at humanizing modern financial systems as part of an ancient tradition of social relationships is subtly ironized by the musical caption.

Toxie: The Marriage of Twee Sensibility and Neoliberal Values

If "giant pool of money" represents the ur-text of *Planet Money* and the high-water mark of *TAL* storytelling, the series of episodes in 2010 on "Toxie," the toxic asset and *Planet Money* mascot, represents the show's signature contribution to the emerging podcast structure of feeling that was transforming public radio and influencing independent producers. It is a compelling example of how empathy, always a vexed term on *TAL*, transformed from a momentary spell, a magic trick, to an almost purely tonal mechanism, indicating a particular approach to storytelling. In January 2010, the *Planet Money* team collectively purchased for themselves a toxic asset as part of an experiment in understanding, in detail, the typical lifespan of the novel financial products made of sliced and bundled mortgage debts. Five members of the team kicked in to buy, what is essentially a bond, or a slice of a bond, for $1,000, with a CUSIP number of 0923434234y, which, after a write-in contest for listeners, they named Toxie, an example of what Alex Russo has called the "twee sensibility" in the first wave of podcasts to spin off from *TAL*.[22]

In short order, *Planet Money* designed an illustration of Toxie, a baby-blue fuzzy orb with a big smile and pink bow. The personification was in service of making an otherwise obscure but central financial product, a mortgage-backed security, understandable to laypeople. The amalgamation of mortgage debts from which Toxie was eventually sliced, was "born" in 2005, at the height of the housing bubble, and was composed of a mixture of solid mortgage debts and lots of mortgages that were made on much less trustworthy terms, including what previous *Planet Money* episodes had called NINA loans, a shorthand for mortgages given to borrowers with no documented income or assets.

When housing prices began to fall, the number of defaults on risky mortgages turned bundles of debt like Toxie from revenue

---

[22] Alex Russo, "'Shenanigans not Stakes': The Institutional and Cultural Production of *The Mystery Show's* 'Twee' Affect or, The World that *TAL* Wrought," Presentation at the 2016 Annual Conference of the Society for Cinema and Media Studies, Atlanta, GA, March 30, 2016.

generating machines into toxic assets, worth less than they were paid for. At $1,000, Toxie was purchased at a steep discount, about half a penny on the dollar; even so, within a year of *Planet Money* purchasing it, Toxie "died," meaning its value plummeted as more and more of the mortgages it represented fell into foreclosure and were purchased by banks. In an animated short on Toxie's life, which featured voices of actual borrowers and lenders, Toxie's doom was blamed equally on homeowners who simply stopped mortgage payments on "underwater" houses and the bankers who loaded up these bonds with NINA loans. This both-sides framing was a feature of "the giant pool of money" and represents an undercurrent to the entire project's avoidance of any critique of the inequalities baked into the systems of finance capitalism. On *Planet Money*, the profit motive of financial institutions and the homeowning dreams of individual borrowers represent parallel impulses and symmetrical moral imperatives.

Financial educators have long used mock or small-scale investments to illustrate broader themes and to make concrete what are otherwise fairly abstract processes. But *Planet Money*'s adventure with Toxie was more ambitious in ways that demonstrate picaresque narrative goals: a hero's journey through neoliberalism and cruel optimism. For Joffe-Walt, Toxie represented "our own little encyclopedia of the financial crisis." The choice to purchase debt also positioned the journalists to assume the perspective of debt collectors rather than underwater homeowners, which meant that the human faces they put on the complex crisis turned out to be the faces of bankers—and their own. The process of purchasing mortgage-backed securities, or as it turns out, a tiny fraction of one, is complicated and essentially impossible for laypeople to manage. Listeners learn about financial instruments as well as access journalism, as Joffe-Walt and her colleagues finagle their way into a larger deal with a trader named Wit Solberg in Kansas City.

Joffe-Walt and David Kestenbaum, the latter a *TAL* alumnus, are well versed in setting up narrative alcoves from which to view the play of antitheses, as outsiders rather than privileged elites. To make the deal, Joffe-Walt intones, "you kinda need to know a guy." In this case, it's Solberg whose cramped and messy Kansas City office is enlisted in the cause of irony. Joffe-Walt's entrée into finance capitalism requires travel to demi-monde; "it feels like an abandoned alley," she says of the drab, unmarked building in Kansas City. "I feel like we're buying drugs." Strewn with empty bottles and other detritus, Solberg's office signifies the chaos of the moment in the credit crisis. With computer monitors "floating on a sea of crap," the setting serves as a distancing mechanism of the hapless new traders from their position as small players in the opportunistic game of gobbling up steeply discounted debt. It also allows a novelistic approach. Glass even compares the serialized chapters of the story to a Dickens novel, to what turns into a journey across the country in search of Toxie's real-life impact.[23] Solburg's dishevelment serves as a kind of mythic warning, a troll under the bridge or the looming darkness at the entrance to the forest.

The process of figuring out how much to pay for the bond gets at the central dilemma in the credit crisis (i.e., pricing assets that nobody wants to buy). The very act of purchasing toxic debt, even at a massive discount, enacts a solution to the crisis by attaching a specific value, any value, to it. Solburg's office has a backroom dubbed "the bond cave" where a frazzled co-worker tries to compute the value of a given asset. Along the way, Joffe-Walt learns that the particular set of two thousand mortgages was packaged into a mortgage-backed security by Countrywide on terms which, in retrospect, now seem to have been somewhat fraudulent, as many of the mortgages had been made with no credit check despite assurances built into the original paperwork that set up the mortgage-backed security.

Countrywide sold it to Royal Bank of Scotland for $3.7 million, which was eventually bailed out by the UK government. It had most recently been purchased by Witt for a mere $36K, who then sold a tiny slice of it to the *Planet Money* team. Toxie is both a heuristic strategy for untangling a mystery and an example of precisely the kind of financial scavenging and realization of loss necessary for righting the ship of the global economy. In this way, Toxie provides the story with a hermeneutic code, a secret to unravel; in its indiscriminate amalgamation of thousands of mortgages, it represents a wealth of stories and a motherlode of antitheses: wealth and poverty, legal and fraudulent, lender and borrower, rise and fall.

---

[23]Ira Glass, "Toxie," *TAL*, episode #418, November 5, 2010.

As if to stoke these abstract opposites into real-life drama, Joffe-Walt, seemingly at random, cold calls Chris Hayes, owner of the Marquis of Westminster, a London pub, to ask how he, a British taxpayer, felt about paying for the RBS's unwise investment in American mortgage debts. The interview is swathed in atmospheric public background noise, as if Joffe-Walt is right there, interviewing him, vox pop style. When she explains that his tax money is bailing out a big bank, Hayes dutifully performs the role of outraged taxpayer. When he learns about some of the investors, New Jersey carpenters' union, for one, who lost money on the bailout, he manages an appropriate bit of solidarity and fellow feeling. The exchange is brief and is designed to make clear the broader stakes of the credit crisis and the hidden costs of the bailouts, through the affective performance of a representative of the petit bourgeois slice of the global ownership class.

The story of Toxie takes an odd turn when Kestenbaum and Joffe-Walt hire a private investigator to track down the individual borrowers whose defaults are now part of their bond's dwindling profitability. In a scene bordering dangerously close to parody, they track down an eighty-one-year-old retiree in Sarasota, Florida named Richard Koenig, to dun him for his late payments. "He owes us money," Koenig deadpans. As they drive up to his home, she remarks, "he doesn't look like a deadbeat." They share a similar observation with Koenig and he chuckles gamely, "I don't have horns." Koenig invites them into his home and explains that he bought a condo so he and his wife could downsize. When the market crashed, he couldn't sell his current home and keep making payments on the condo, so he walked away from the new mortgage and stayed put. Joffe-Walt jabs at him about his delinquency and its impact on their investment, prompting him to repent: "Do I feel sorry for the investors… Yes I do… I feel badly," while Kestenbaum plays good cop: "we don't hold it against you," he tells him. An abstract financial complex has been humanized through the medium of affect. Koenig is absolved, in part because of how he feels.

"And do I feel sorry for the investors? Sure I do. I'm a sensitive person, you know? I really am. I love my wife, I love my dog, I love my children and my grandkids." Joffe-Walt seems to require a bit more, however. When Koenig suggests that given his fixed income and his inability to sell either of his properties, he "didn't have a choice," she rejoins, "well you did have a choice. You could have kept paying it." The awkwardness of the moment, the sheer tonal shift from Koenig's warmth and hospitality to this venal calculus, combined with the misapplication of one set of social mores for those of another, more formal context, is worthy of a *Curb Your Enthusiasm* episode.

Later, both Joffe-Walt and Glass describe Koenig's choice to listeners as a "change in morals." "He gave his word," Joffe-Walt insists. "That's not nothing, that's what it means to be an adult, someone gives you money, you pay it back." Moving from the specific to the general, Kestenbaum adds, "There are lots of Richards these days. In fact, the moral compass of the entire country is shifting." With that, the reporters-turned-debt collectors drive off to confront other deadbeats in Florida who owe them money. Along the way, they discover that a sweet-sounding old couple in another of their homes are, in fact, involved in an ill-fated housing Ponzi scheme that contributed to the toxicity of Toxie. The discrepancy between the sweetness of their affect and their criminality is played up to great effect, part of their effort to suggest that not all the Richards out there were equally "sympathetic."

They also discover, after talking to a real estate lawyer from Arizona, that breaches of contracts are a completely unexceptional exception, a common business decision in the housing industry. "It's a business decision, not an emotional one," she tells Blumberg. Kestenbaum and Joffe-Walt, unconvinced, continue their tour of houses they partially own, peering into windows of run-down houses that nevertheless show signs of being occupied, looking for the affective traces of indebtedness.

The inevitable "death" of Toxie is played for the quirky humor of personifying a bundle of mortgage debt, but it also provides the team one last chance to assert market orthodoxy in the midst of all the exceptions to economic rules that both gave rise to the mortgage crisis and which it then required to settle. In fact, the two come together in Joffe-Walt's eulogy: "We finally figured out what you were worth, Toxie... now we have an answer to the biggest question... you were worth $449.06." This final reckoning is in fact a crucial part of resolving the credit crisis: assigning value to assets, even small ones, is necessary for restoring the creditability of credit.

In the final chapter, "Toxie in a Coma, I know: It's Serious," the team argue that Toxie's death and its poor return on investment were due in part to banks and borrowers finding compromises through loan modifications, enabling homeowners to stay put while paying less interest. Humane workarounds to impossible situations, like the elderly Koenig's "change of morals," hurt opportunistic debt collectors, like the *Planet Money* investors. Loan modifications are supposed to be "good for everyone," Kestenbaum observes, but he declares in a performative fit of pique, the *Planet Money* team is losing out. Joffe-Walt suggests to a finance executive named Samir Noriega that he petition the banks to stop making loan modifications; Noriega chuckles and says the banks don't care. Joffe-Walt rejoins, "they might care but there's a lot of pressure." Like a cartoon version of an Austrian school economist, Joffe-Walt has reserved her compassion for the banks and for Toxie, a bundle of debt. This rationalization of banks' flexible approach to contracts as an inevitable response to "pressure," paired with the "moral" critique of individual mortgage holders' decisions to walk away from underwater properties, represents *Planet Money*'s founding insight: capital is the protagonist.

Planet Money as Devil's Advocate

The early 2010s, with its sluggish growth and putatively progressive Obama administration, provided the context for *Planet Money*, now a radio and podcast institution, to explore economic questions far beyond the mortgage crisis. The Occupy Wall Street movement and the Tea Party insurgency put dueling populisms of left and right at the center of political discourse about the economy. Obama's incrementalist approach to liberal reforms, particularly in the area of health care, provided another crucial context in which the project's stridently neoliberal economic theory developed in stories about industries and policies both foreign and domestic.

In a 2009 piece for *Morning Edition*, Alex Blumberg and Chana Joffe-Walt explore Obama's new health care bill from the perspective of the health insurance industry, and in particular, the people whose job it is to deny coverage to people with pre-existing conditions. "They don't have horns, they don't hate humanity, they are nice, they have cookies in the break room," Joffe-Walt preemptively chides. "On Friday afternoon they bring their babies into the office here in Southern Connecticut." The moral hazard argument has returned to explain the microeconomics of health care, a telling frame for *Planet Money*'s coverage of what would become the Affordable Care Act. To health insurance executives, she adds, "It seems totally inappropriate that people call up already sick trying to get covered." Blumberg brings in an MIT professor of economics to warn about the unintended consequences of "publicly shaming" health insurers into covering preexisting conditions.

Government intervention is once again portrayed as an insult to the natural economic processes. "Essentially," adds Joffe-Walt, "it's trying to make health insurance as unlike insurance as possible, to make insurers do the one thing that is most unnatural to them: stop worrying about risk." Requiring healthy people to buy health insurance as a way to distribute the risk, "doesn't seem fair," to Blumberg; for Joffe-Walt it's "equally unnatural" as making insurers cover pre-existing conditions. Obama inherited a sprawling credit crisis and was in part elected over the Republican nominee because of it; this meant, however, that the free market orthodoxy now resonated differently with a Democrat and the nation's first Black president, in charge of government bailouts.

Planet Money continued to explain these bailouts using the logic of exception. David Kestenbaum got an automotive industry executive to perform this logic perfectly in an October 2009 conversation about the massive federal bailout of General Motors. When pressed, Edmunds.com CEO Jeremy Anwyl was "of two minds" about the deal, conceding he's "kind of old school" in his distaste for government intervention. Ultimately, Obama's decision to bail out the company was "the right thing." But "in principle, it's something that we probably shouldn't have done." His choice of moral language "right" and "shouldn't" belie the essentially emotional rather than logical grounds of his economic philosophy.

Human interest stories often took the perspective of those most closely allied to the interests of capital, to the giant pool of money that was, from the beginning, the show's protagonist. A January 2010 story packaged as a short spot-on *Morning Edition*, titled "Debt Collector: Tough Job but Someone's Got to Do It," explored the emotional struggles of debt collectors, whose empathy and cynicism rise and fall over the course of a shift, talking to people in desperate circumstances, eliciting a range of responses from the pathetic to the apathetic. The toll of talking to such people on the phone all day, led one part-time debt collector, John Goebel, to shut off his capacity for empathy, an outcome with which we're encouraged to empathize. Debtors often cry over the phone or tell lies; one debtor explained that his son had killed himself, his wife left him, and he lost his job. The litany of such stories, one after another, "doesn't put you in a very charitable frame of mind," Alex Blumberg helpfully explains. The emotional toll of collecting debt enforces a kind of mental discipline with the market logic that animates the program's central orientation: "You got to turn off some part of you," Goebel concludes, "you can't put yourself in everyone's shoes."

Planet Money stories set the tone for other human interest economy stories on NPR, it should be noted. A 2008 *Morning Edition* piece takes a sympathetic look at the trials and tribulations of the owner of a car repossession business. Thanks to the increased volumes of car loan defaults, Bill Johns is not getting as much money for repossessed vehicles as he used to. An economist employed by a massive auction house chimes in to add that auction prices are way down. "The overall market is very poor," Tom Webb says. "Everyone has taken a hit recently."[24] Another *Morning Edition* piece from 2010, ostensibly sympathetic to those filing for personal bankruptcy, framed the increase in such events thus: "Some who choose to file for personal bankruptcy may be deadbeats or individuals who deliberately run up debts that they have no intention of paying."[25] A 2009 *Planet Money* report, repurposed for *Morning Edition*, suggested that monkeys' grooming behavior demonstrated "natural" processes of economic laws, particularly around the relationship between status and capital.[26]

By 2010, *Planet Money* had turned its attention to the economic foibles of other countries. Alex Blumberg and Adam Davidson traveled to Jamaica to find that well-meaning government interference in propping up the devalued currency was interfering with small businesses trying to get loans. Stories filed from India identified government bureaucracy, a legacy of British colonial rule, as the main impediment to economic growth, but here too,

---

[24]Tamara Keith, "More Lenders Reluctant to Repossess Vehicles," *Morning Edition*, NPR, October 31, 2008.
[25]Wendy Kaufman, "Hard Times Lead to Dramatic Rise in Bankruptcies," *Morning Edition*, NPR, January 21, 2010.
[26]Laura Conaway, "Scientist Monkeys Around with The Economy," *Morning Edition*, NPR, October 23, 2009.

reporters zigzag their way through exceptions to market forces as they try to untangle the red tape. Kestenbaum travels to Denmark to try to understand why "Denmark thrives despite high taxes." He's incredulous when he encounters Danes who like high taxes but relaxes when a Danish economist points out the benefits of America's "killer instinct" in matters of economic competition. Having settled on Denmark as an interesting exception, Kestenbaum closes by warning that "economic miracles don't always last."[27]

Back home, Joffe-Walt and Glass drew criticism for their 2013 story, "Unfit for Work: The Startling Rise of Disability in America," with versions airing on *Planet Money*, *TAL*, and NPR's *All Things Considered*. According to mainstream and liberal sources alike, the report's inaccuracies supported many of the conservative talking points that led to cuts in welfare programs in the 1990s. Joffe-Walt argued that the system was not restrictive enough and was being abused by able-bodied opportunists who preferred government checks to work. Experts pointed out that the system rejects more applicants than it accepts and pushed back on the notion that parents keep able-bodied children on disability for the free money. Also, the high employment rate pushed workers into difficult manual labor they would have otherwise been able to avoid in times when more jobs were more plentiful and diverse.[28]

It should come as no surprise that conservative academics teaching macroeconomics flocked to these bite-sized lessons in free market principles with high production values and human-interest angles. William J. Luther, an admirer of neoliberal icon Friedrich Hayek and former Fellow at the libertarian Cato Institute (founded by the Koch Brothers), wrote a paper on the utility of *Planet Money* for helping undergraduates across nine courses and two universities understand macroeconomic principles. Rebecca Moryl, an economist formerly of the conservative Beacon Hill Institute,

---

[27]David Kestenbaum, "Denmark Thrives Despite High Taxes," *Planet Money*, Podcast, NPR/*TAL*, January 29, 2010.
[28]Brad Plumer, "Harold Pollack: What 'This American Life' Missed on Disability Insurance," *The Washington Post*, March 28, 2013; Michelle Chen, "How 'This American Life' Got Disability Wrong," *In These Times*, March 13, 2013; Dean Baker, "Planet Money Misses the Boat on Social Security Disability," *Center for Economic and Policy Research*, Blog, March 25, 2013, https://www.cepr.net/planet-money-misses-the-boat-on-social-security-disability/.

used the podcasts in her microeconomics courses.[29] A collaboration between the show and *Wired* magazine on Omaha, Nebraska's economic renaissance relies on Richard Florida's "creative class" hypothesis to argue, speciously, that the addition of gourmet restaurants was the key to economic growth.[30] David Shaywitz, of the conservative American Enterprise Institute, took to the pages of *Fortune* magazine in 2013 to share that *Planet Money* had become "a not-so-guilty pleasure" for him and his venture capital friends. In recent years, the show has added features like "Planet Money Summer School" and a portal "for educators" with links to stories by economic concepts like "Scarcity," "Economic Growth," and "Government Failure."[31] In 2018, the Federal Reserve Bank of St. Louis linked *Planet Money* stories to their "Econ Lowdown Teacher Portal." Timothy Geithner, former head of the NY Federal Reserve and Obama's Treasury secretary, was also an outspoken fan of the show.

Breathtaking: Feeling over Structure

It will likewise come as no shock to learn that financial institutions, including those whose fortunes rose and fell on the very policy and philosophy choices covered by *Planet Money*, were keen to sign on as underwriters. Ally Bank became the lead sponsor of their short pieces on *All Things Considered* and *Morning Edition* in May 2009. Formerly known as General Motors Acceptance Corporation (GMAC), Ally Bank was one of the more notorious recipients of the massive federal bailouts discussed on the program. GMAC and its officers were eventually found to have engaged in massive mortgage fraud in the years leading up to the housing crisis. Even as investigations were under way, GMAC received $15 billion starting in December 2008 and going well into December of 2009, after it had become *Planet Money*'s main sponsor through a "special series

---

[29]Rebecca L. Moryl, and Shuyi Jiang, "Using Economics Podcasts to Engage Students of Different Learning Styles," *International Advances in Economic Research*, 19 (2013): 201–2.
[30]"Creative Industries," *Jump Cut*, http://www.ejumpcut.org/archive/jc53.2011/kleinhans-creatIndus/3.html#10.
[31]Planet Money for Educators, 2024, https://planetmoney.listenwise.com/.

adjacency" in which all its stories fed to NPR shows come with the Ally Bank promotional message.

Alicia Shepard, NPR's ombudsman at the time, allowed that "cynical listeners are bound to wonder if NPR might go easy on news coverage of Ally."[32] A report for Fairness and Accuracy in Reporting, a left-leaning media watchdog organization, called it a textbook example of a "conflict of interest" of the sort every journalist is taught to shun. It was a criticism shared by more conservative publications as well; "even *Ad Age*," she reported, "the advertising industry's trade publication, was taken aback by the "close alignment of message and news program."[33] *Business Insider* was also shocked, calling the bailout and GMAC's rebrand, "the biggest scam of them all."[34]

NPR's senior vice president for news, Ellen Weiss, told Shepard that this promotional adjacency was "an experiment" that they were "being careful with." The arguments on both sides of course miss, or perhaps concede through omission, the larger point that Ally Bank is comfortable sponsoring the show precisely because of how well it represents their interests. An activist media group has gathered a long list of pro-business, anti-tax positions that co-host Adam Davidson has espoused starting long before his association with *Planet Money*.[35] Shepard closes with a reminder that NPR has a "large pool of creditability," with its listeners, an on-the-nose, if unconscious, hearkening back to "the giant pool of money."

The relationship came under additional scrutiny after Davidson's aggressive interview of Elizabeth Warren, then chair of the Congressional Oversight Panel, who was pushing for the creation of a Consumer Financial Protection Bureau, a policy that Ally Bank

---

[32] Alicia Shepard, "Ally Bank: Is it a Good Fit?" *Public Editor*, NPR, December 14, 2009, https://www.npr.org/sections/publiceditor/2009/12/14/114431256/ally-bank-is-it-a-good-fit.

[33] "NPR's 'Planet Money' Makes Deal With Rebranded GMAC," *AdAge*, June 5, 2005, https://adage.com/article/media/npr-s-planet-money-makes-deal-rebranded-gmac/137115.

[34] Vincent Fernando, "How the GM Bailout Was the Biggest Scam of Them All," *Business Insider*, Feb 27, 2010, https://www.businessinsider.com/how-the-gm-bailout-was-the-biggest-scam-of-them-all-2010-2.

[35] "Adam Davidson," *S.H.A.M.E Project*, https://shameproject.com/profile/adam-davidson/.

was lobbying hard against.[36] Davidson was forced to apologize for his antagonistic approach to Warren and her concerns, conceding, "I opened myself up to people thinking I don't care about the middle class." Even the NPR ombudsman called Davidson's attack on Warren "a meltdown." He was also criticized for giving corporate speeches paid for by the same financial institutions the show was covering while criminal investigations into mortgage fraud were underway.[37] Another Public Editor, Edward Schumacher-Matos, felt compelled to respond to criticisms of "the normally brilliant Adam Davidson" for a misleading and hostile story on the effectiveness of local and state government economic development. Such efforts by well-intentioned governments, Davidson fumed in the story, "drive me crazy." Even Davidson admitted his tone was "snarky," in a later apology. Schumacher-Matos felt that the problems were "more than tone" and that "journalism was sacrificed for style." Davidson's conclusions, he continued, in an apt summation of the dangers of the "listeners should feel," approach, were "breathtaking," but "unsupported."[38]

It is precisely in such circumstances, when shocks to the system are reverberating through the institutions of banking, journalism, and government, that style, tone, and affect does its most important work, Naomi Klein tells us. The application of "market logic" to government bureaucracy required an approach "more psychological than physical, more spectacle than struggle."[39] In their anthropological studies of neoliberal management practices in developing economies, Richard and Rudnyckyj remind us that "the transitive aspect of affect" helps present "economic relations as something more than transparent rational choices borne by self-interested individuals."

---

[36] "GMAC, Inc. Lobbying Report 2009: Consumer Protection Agency and Other Bills," https://lite.evernote.com/note/5a90fc2f-25cf-4340-bcce-293bd69bc7b8.

[37] Alicia Shepard, "Planet Money Meltdown," *Public Editor*, NPR, June 1, 2009, https://www.npr.org/sections/publiceditor/2009/06/01/104782824/planet-money-meltdown; https://shameproject.com/profile/adam-davidson/.

[38] Edward Schumacher-Matos, "Planet Money Misfires on Local Economic Developers," *NPR Public Editor*, NPR, Blog, June 22, 2011, https://www.npr.org/sections/publiceditor/2011/06/22/137349286/planet-money-misfires-on-local-economic-developers.

[39] Naomi Klein, *Shock Doctrine: The Rise of Disaster Capitalism* (New York: Picador, 2007), 356.

While some journalists balked at the conflict-of-interest problems raised by these close ties between *Planet Money* and the industry it reported on, for the most part, the program avoided any lingering or widespread criticism. Instead, the show and its on-air talent won awards and influenced a generation of journalism podcasts. "Affect is a means of subjectification," say Richard and Rudnyckyj, "that simultaneously produces those who enact it and those upon whom it acts."[40] Put another way, Lindgren observes that the show's success, even when listeners disagreed with specific concepts presented, rested on the fact that "the personality and presence of the hosts is crucial to the audience's ability to relate to the content."[41]

Planet Money has been a particularly productive finishing school for journalists and producers seeking commercial outlets for their storytelling prowess. In 2014, Alex Blumberg launched Gimlet, a for-profit podcasting network of shows that included an impressive roster of former public radio talent, including Jonathan Goldstein, Starlee Kine, and P. J. Vogt, all of whom got their start at *TAL* and Alex Goldman, one of many hired from WNYC. Adam Davidson, who started at Public Radio International's *Marketplace* before co-founding *Planet Money*, landed a column in the *New York Times Sunday Magazine*, a plum perch from which to take on the knotty economic conundrums of the day before getting his own show on Gimlet as well.

Since the demise of Gimlet, which we'll explore in Chapter 7, Davidson has taken up coaching for "business storytelling." Alex Goldman has started his own storytelling consulting shop, Enrichment Corporation. In 2024, he launched *Hyperfixed* a podcast styled as "The help desk for life's most intractable problems," which appeared to borrow from the formula used by fellow Gimlet podcasters Jonathan Goldstein ("Heavyweight") and Starlee Kine ("Mystery Show") These shifts represented the larger trend of public radio-trained journalists joining the for-profit podcasting industry. More than any other network, Gimlet exemplified this transition. "They had the public radio aesthetic—the charmingly nerdy, self-effacing demeanor, the soft-spoken inquisitive affect that

---

[40] Analiese Richard and Daromir Rudnyckyj, "Economies of Affect," *Journal of the Royal Anthropological Institute (N.S.)* 15 (2009): 61. © Royal Anthropological Institute.

[41] Mia Lindgren, "Personal Narrative Journalism and Podcasting," *Radio Journal: International Studies in Broadcast and Audio Media* 14, no. 1 (2017): 1–29.

sounds so good on the low end of the FM dial. The big difference was that unlike their former colleagues in public media, they were going to make money."[42] Podcasting about the economy was an effective way "to convert the social capital they accumulated from their public radio experience into a new entrepreneurial project."[43]

The "proleptic" mood in coverage of podcasting, starting almost from its birth but increasing in intensity in the 2010s, has presented an obstacle to historical understanding, argues Neil Verma. Something always on the verge of becoming is hard to historicize. However, certain themes have emerged. Verma points to the centrality of "obsession" in discourse about podcasting and podcast hosts in particular. It is worth noting in this context that the *Planet Money* team was an early adopter of the obsession posture. Joffe-Walt's "obsession" with Social Security's disability benefits program epitomized the nerdy, humble-brag style specific to the show, and is what drove her reporting on the system, which, as noted above, was widely criticized for "basic errors" as well as a conservative bias.

Conclusion: The Planet Money Structure of Feeling

Planet Money's birth and rise, from 2008 to 2010, overlaps with the period in which NPR and *TAL* became seriously invested in podcasting as more than a platform for sharing already broadcasted material but rather as a unique medium. One of the key affordances that the podcast demonstrated to both was its ability to "serve" content at a more tightly defined and more easily quantifiable audience demographic. Experimenting with novel funding approaches, like "special series adjacencies" for content produced for multiple platforms, was explained as, in part, a response to the austerity of the environment for public funding that the economic crisis had contributed to. As more public funds were ploughed

---

[42]Alex Sujong Loughlin, "If you love podcasts, dump Spotify," *Defector*, Blog, April 12, 2024, https://defector.com/if-you-love-podcasts-dump-spotify.
[43]Samuel M. Clevenger, "On Idleness and Podcasting," in *Podcast Studies: Practice into Theory*, eds. Dario Lliares and Lori Beckstead (Waterloo, ON: Wilfred Laurier University Press, 2024), 45.

into financial institutions, public radio broadcasters increasingly saw podcasting as a way to monetize content in ways that had previously been out of consideration.

In this way, *Planet Money* represents a broader shift that was underway in other public radio content as it moved from a defensive posture relative to internet-based media to a fulsome embrace of multiplatform media content. The show also took advantage of podcasting's temporal affordances, particularly the freedom from the tyranny of the broadcast clock. This enabled the kinds of "obsessions" that could produce open-ended projects that did not conform to the broadcast schedules and programming limitations. Like the purchase of Toxie, *Planet Money*'s 2013 creation of a T-shirt enabled an exploration of global supply chains from design to production to shipping to after-market circulation that followed the momentum of the story rather than the dictates of newsworthy "pegs." At the same time, repackaged bits from this story could be easily cut to drop into NPR's programming whenever a topical take on it presented itself. These obsessions consistently put the team in the position of owners, lenders, and producers, rather than renters, borrowers, or consumers.

Planet Money, with a business-friendly philosophy, must have seemed heaven-sent at this moment. Because of its distribution arrangements with NPR and *TAL*, it continued to produce stories that could fit onto radio platforms as standalone "stories" that qualified as both newsy and thematically rich human-interest pieces. This period was also a time when the creative energies of public radio institutions NPR, WNYC, and PRI were turning to podcasting as a platform for long-form narrative journalism. In 2010, National Public Radio officially changed its name to NPR, signaling its intention to transform into a multi-media platform for news and entertainment.

Most of the early NPR podcasts were just rebundled radio-first programs, like WBEZ's *Wait, Wait, Don't Tell Me* in 2006 and WHYY's *Fresh Air* and WAMU's *The Kojo Nnamdi Show* in 2007. Starting in 2005, NPR member stations like WNYC and WBEZ, *TAL*'s home, increased their investment in their own podcasts, shifting radio-first shows like *On the Media* and *Radiolab* into podcasting juggernauts. In 2010, *TAL* and *Radiolab* were numbers 2 and 3 respectively on the iTunes charts.[44] By 2011, almost twice as

---

[44] https://web.archive.org/web/20121214020656/http://www.apple.com/euro/itunes/charts/podcasts/top10podcasts.html.

many listeners heard *Radiolab* via podcast than on the airwaves. In 2010, NPR launched landmark shows like *Alt.Latino, Pop Culture Happy Hour*, and *Snap Judgement,* followed by *Code Switch* in 2013 and *Invisibilia* in 2015. By 2014, NPR accounted for six of the ten most downloaded podcasts in the United States.[45] In 2015, WNYC Studios spun off from WNYC, as its own podcast producing entity, launching shows like *The New Yorker Radio Hour*, signaling a shift in how legacy media organizations had begun to pool resources and branding prowess to grab market share in the crowded podcasting space.

TAL made headlines in 2010 when Glass raised over a hundred thousand dollars via "mobile giving" with a simple appeal announced at the start of each week's podcast. It was a landmark moment in the smartphone economy as both a delivery and payment system while putting local public radio affiliates on notice that they no longer had a monopoly on soliciting listeners' funds.[46] *Planet Money* won a DuPont-Columbia award in 2014 for the series on the production and circulation of a T-shirt. Audience research at that time continued to implore public radio broadcasters like NPR to super-serve its super-core, doubling down on its appeal to a "highly educated" high income and mostly white listener base. Glass's drive appeals reminded listeners that they, like him, were "middle class, with normal jobs and responsibilities."[47]

This period of radio-podcast hybridity presented editorial challenges. When presented with criticisms of problems of reporting or tone, NPR distanced itself from the podcast model. In 2009, veteran NPR editor Uri Berliner told NPR's public editor, Edward Schumacher-Matos, that "*Planet Money*'s podcast does not have the same degree of radio production or intense editing and supervision as NPR's regular shows." It was, after all, "a relatively new venture,"

---

[45] "NPR, WNYC and WBEZ Announce First Ever Podcast Upfront," NPR, April 27, 2015, https://www.npr.org/about-npr/402536549/npr-wnyc-and-wbez-announce-first-ever-podcast-upfront.

[46] "Radio and Text Donations: 'This American Life's' Experience with Mobile Giving," MobileActive, April 14, 2010, https://www.mobileactive.org/american-life-joins-mobile-giving-revolution/.

[47] "This American Life—Evergreen Fundraising Pledge Spots," https://beta.prx.org/stories/129659.

he continued. Berliner cited the explosive growth of its audience and success across platforms to explain editorial shortcomings. "There just hasn't been enough time in the day to make sure that every podcast interview is vetted by network editors." He told Schumacher-Matos that "supervisory responsibilities have not been spelled out for the blog and podcast."[48]

Schumacher-Matos conceded the "institutional pain" caused by these hybrid productions. "Such joint ventures are increasingly common as news organizations seek to manage costs, but the arrangements raise sticky issues." In response to Davidson's "snarky" coverage of state and local economic development in 2011, he insisted "*Planet Money* is distributed by Public Radio International, not NPR." He further distanced the network from the story by observing that "NPR editors were not involved." In other instances, the NPR's ombudsman pointed to "shortcuts in storytelling ... and in some cases, in reporting" or conceded that *Planet Money*'s sweeping policy arguments hung on relatively thin journalistic reeds, exacerbated by the abbreviated broadcast versions that air on NPR's news shows.[49]

From the perspective of the mid-2020s, *Planet Money*'s legacy appears to be its exemplary synthesis of myriad contradictions into a temporarily stable and highly successful formula that anticipated much of podcasting's next decade. It brought together *TAL*'s tonal intimacy and literary narrative structure with NPR's journalistic gravitas and topicality. This meant that it helped to invent nonfiction podcasting's unique temporarily—all the freedom from the tyranny of the broadcast clock and related industrial format constraints but with an ear for the cultural moment, whose contours were unpredictable and improvisational, yet somehow still quite precise. While there were a few moments of editorial disavowal, for the most part, the production logic of podcast, news program, and storytelling outlet worked well together. It also managed to reconcile public radio's reputation for liberal feeling with a relentlessly pro-

---

[48] Elizabeth Jensen, "Call The Midwives, But Ring The Doctors, Too," NPR, September 27, 2018, https://www.npr.org/sections/publiceditor/2009/06/01/104782824/planet-money-meltdown.
[49] Jenson, "Call The Midwives."

business perspective, creating an ideologically broad and enduring audience that, except for rare occasions, it has managed not to alienate.

Finally, *Planet Money* represents a watershed moment in the reconciliation of public radio's cultural cachet with a nakedly commercial sponsorship deal with Ally Bank, anticipating Glass's 2015 declaration that public radio was "ready for capitalism." More importantly, by partnering with a major financial institution, it laid out the path for Gimlet's business model, which began with raising venture capital and providing content aimed at an audience comprised largely of white male tech professionals, affectively liberal and fiscally conservative. Most impactfully, it demonstrated the durability and versatility of "surprise" and "empathy" in audio storytelling. The biggest and best surprises, for *Planet Money*'s reporters and hosts, was in empathic stories of those with money and with money itself.

CHAPTER SIX

Fellow Feeling: *This American Life* and the Gendered Voice

"We won. Our aesthetic prevailed." So declared Adam Ragusea in 2015.[1] He was talking about the predominance of the public radio sound in podcasting's first wave of hit shows, many of which were coming from public radio stations and networks. As of the year before, eight of the top ten podcasts on the iTunes Charts were public radio shows.[2] Many more of them, however, were coming from new commercial networks like Gimlet and Panoply or from commercial websites like *Slate*. It was a hard claim to refute. As mentioned in the previous chapters, the early 2010s saw a massive exodus of talent from public radio, including *This American Life*. And they all sounded like public radio, but what does that mean, precisely?

Some of the new shows borrowed production techniques from NPR: the Neumann 187 microphones with the bass-roll off button switched on so as to be heard above the rumbling of automobile tires on the road; closely miked interviews recorded with no volume

---

[1] Adam Ragusea, Presentation to the Radio Studies Scholarly Interest Group, Society for Cinema and Media Studies, Atlanta, GA, April 1, 2016.
[2] US Podcasts Top Entries, iTunes Charts, 2014, https://web.archive.org/web/20141112024105/http:/www.ituneschnarts.net/us/charts/podcasts/2014/11/09.

compression; and of course the amateur's warmth pioneered by Susan Stamberg in the 1970s on *All Things Considered*, and the conversational style that *TAL* and NPR spinoff *Planet Money* raised to house style in the 1990s and 2000s. Others approached the long-form audio nonfiction with different technical specs and with a different kind of narrative delivery, like *Snap Judgment*, Glyn Washington's story-based show, but with the same multi-act structure, thematic throughline, and emphasis on small stories of everyday life. Even *The New York Times*' flagship straight news podcast, *The Daily*, borrowed from its structure and tone.[3] Like *TAL*, it organized stories thematically; like *TAL*, it showcased small stories by and about people who are typically overlooked by news media. But it eschewed the chatty repartee between hosts and reporters or reporters and interview subjects.

Perhaps Ragusea had in mind shows like *Invisibilia*, an NPR radio show that spun off into a podcast and became an influential and durable hit until its abrupt cancellation in 2023 (following a period of tumult that will form the focus of the next chapter). Or perhaps he meant the equally influential *Radiolab* (2002–), a WNYC radio program turned podcast in 2009, which matched *Planet Money* in friendly badinage between hosts Jad Abumrad and Robert Krulwich and surpassed *TAL* in its painstaking editing style, turning conversations about science into hypnotic sound art compositions.[4] It's likely he also had in mind investigative podcasts like *Serial*, the runaway smash hit podcast that spun off from *TAL*.

By 2015, at glossy corporate conferences like Podcast Movement, held in Dallas that year, the exhibition halls were filled with vendors hawking software and hardware, webhosting support, and webinars designed to teach interview skills. One vendor and conference sponsor, Clear, offered a high-quality internet voice service that podcasters could use to conduct long distance interviews more cheaply than ISDN lines and more reliably than phone lines or VOIP, a sign that public radio's high-quality sound, rather than AM's scratchy, populist immediacy might be the emergent standard

---

[3] Sarah Larson, "Three decades into 'This American Life,' the host thinks the show is doing some of its best work yet—even if he's still jealous of 'The Daily,'" *The New Yorker*, July 7, 2024.
[4] See Glass on *Radiolab*, "*Radiolab*: An Appreciation," *Transom*, November 11, 2008, https://transom.org/2011/ira-glass-Radiolab-appreciation/.

for the new medium. Featured speakers like Marc Maron from *WTF*, Roman Mars from *99% Invisible*, and Thea Tau from *The Moth* and *Strangers*; community radio pioneers like The Kitchen Sisters, and keynote speaker Sarah Koenig, fresh from season 1 of *Serial*, all suggested that Ragusea was right: podcasting in the mid-2010s sounded like public radio.

Of course, by the mid-2010s, there were already hundreds of thousands of podcasts available for download, the vast majority of which hewed more closely to DIY production values and audio quality. If the Dallas conference attendees were a fair representation, the majority saw their podcasts as an extension of an existing small business, a form of brand extension. Many attendees noted a bifurcation in the conference between what they called "pro-casters," the high-quality, longform narrative shows featuring public radio veterans and their ilk, and "podcasters," the small-time entrepreneurs with tinny audio and big dreams.

Ragusea's assertion was partly accurate, but also partly an exercise in what Bourdieu calls "distinction," an implicit class-based value judgment that serves to define a preferred "habitus" through exclusion and inclusion.[5] As Eleanor Patterson has argued, *TAL* had by this time, expertly navigated the cross currents of commercial marketing and brand expansion from within the highbrow habitus associated with public radio. For small business podcasters, that navigation proved elusive. The attendee who first introduced the Podcasters vs Pro-casters distinction to me, himself a self-described "scrappy amateur" who hosted a men's fitness podcast, tried to describe the pro-caster sound, but the cultural referents were just out of reach: "they sound the same, you know?"

Vocal Performance, Habitus, and Social Identity

I've tried myself to capture the quality of vocal performance that distinguished the public radio sound from the rest of the audio world with only limited success. An essay I wrote in 2008 started with an

---

[5]Pierre Bourdieu, *Distinction: A Social Critique of the Judgement of Taste* (Cambridge, MA: Harvard University Press, 1984).

observation from one of my students that on NPR, "the women sound like Bea Arthur and the men sound like Truman Capote."[6] Some desultory experiments with pitch measurement software provided mixed support for this assertion. The sociolinguistic study of the human voice, and of prosody in particular, has shown a remarkable diversity in pitch across genders and cultures. To speak of a "normal pitch range" for men or women, sociolinguistics tells us, is a fool's errand.

Further, attempts to objectively measure, in wave form, features commonly associated with "gay voice" have similarly come to naught.[7] Curiously, however, the same research has determined that people are often accurate in identifying sexual identity by voice.[8] That said, it's not foolproof and, across cultures and languages, men's voices are thought to "sound gay" to the extent that they sound effeminate, itself a flexible standard with wide cultural variation.[9] By one measure of feminine vocal performance—clarity and precision of utterance—someone like Ira Glass, who was often thought by listeners to be gay, falls far short.[10] His dropped r's and swallowed l's, combined with the studied indifferent locutions of the hipster, often make him sound unclear and imprecise.[11]

Cross-cultural and cross-linguistic research has revealed that "sounding gay" varies widely as does "sounding like a woman," suggesting of course that these categories are culturally determined. Sounding both "more precise and metropolitan," seem to be features that people find to "sound gay" across multiple national-linguistic contexts, terms which are themselves highly subjective and hard to measure.[12] There is in fact an enormous body of subjective opinions

---

[6]Michael Patoka, personal communication, May 2007.
[7]Moira Lavelle and Nina Porzucki, "'Do I Sound Gay?' Researcher searches world for answer," *The World*, July 9, 2015, http://www.pri.org/stories/2015-07-09/do-i-sound-gay-researcher-searches-world-answer.
[8]For a discussion of the linguistic features that determine people's perception of sexual identity, see Erez Levon, "Hearing Gay: Prosody, Interpretation, and the Affective Judgements of Men's Speech," *American Speech* 81, no. 1 (2006): 56–78.
[9]Levon, "Hearing Gay," 57.
[10]Kathryn Berquist, "The Host with the Most," *The Advocate*, November 26, 2002, http://findarticles.com/p/articles/mi_m1589/is_2002_Nov_26/ai_95263265.
[11]"There's this whole chain of perceptual events whereby it ends up feeling that if a man is speaking more clearly that he also sounds more gay," says Smyth.
[12]Levon, "Hearing Gay," 60.

on the subject, much of it hewing closely to the hyperbolic comment of my student: public radio's men spoke effeminately; its women, not so much. This assertion, this social fact, if you will, provides an insight into ideologies of gender and language in which broadcast and podcast speech has developed over the past century. It provides an opportunity to explore the liberal feeling running through the public radio sound that, by the mid 2010s, had become the sound, or at least one of the dominant sounds, of podcasting's big decade.[13]

Perhaps overlooked in the discussion of the relative pitch of men's and women's voices in public radio is the revolutionary fact that, thanks to NPR, women's voices were so prominent in broadcast journalism from the very start of the public radio era. Susan Stamberg, the first woman to anchor a daily national newscast in the US, shaped what became an industry trend. Linda Wertheimer, Nina Totenberg, and Cokie Roberts, joined Stamberg in short order on NPR; on air and behind the scenes, they were powerful forces in shaping the upstart network into one of the nation's dominant news outlets by the end of the 1980s.

To be clear, women's voices were still *underrepresented* among correspondents and reporters for much of the network's history and *dramatically underrepresented* among expert talking-head interview subjects. The media watchdog group Fairness and Accuracy in Reporting (FAIR) documented NPR's consistent under-representation of women, people of color, and people outside elite institutions as experts, commentators and news sources in several studies during the 1990s and 2000s. And until recently, the top jobs in the network went to white men. In other words, in the roles that require expert opinions about and access to the wider world, the network continued to mirror the dismal sexist patterns of the media industry and society in general.[14] But as hosts, newscasters,

---

[13]Research has demonstrated that, on average, "the male pitch range is narrower than the female/effeminate and shows slower and less frequent pitch shifts," Sally McConnell-Ginet, *Intonation in a Man's World: Language, Gender, and Society* (New York: Newbury House Publishers, 1983), 75. Further, men typically avoid higher pitch in their speech and use "uptalk" only for special effect. See Ruth M. Brend, "Male-Female intonation patterns in American English," in *Language and Sex: Difference and Dominance*, eds. Barry Thorne and Nancy Henley (New York: Newbury House, 1975), 85–6, qtd. in McConnell-Ginet.
[14]Steve Rendall and Daniel Butterworth, "How Public is Public Radio? A Study of NPR's Guest List," *Extra!* (Fairness and Accuracy in Reporting, 2004).

reporters—those extra-diegetic voices most closely identified with the identity of the network (i.e., the network's public voice)—women's voices have been central, if not dominant.[15]

By 2012, women had begun to make progress in crucial behind-the-microphone positions as well; women held the top editorial job in five out of seven news shows and constituted almost half of all network staff. This was in stark contrast to the rest of radio journalism where women made up 18 percent of radio news directors.[16] By 2015, women made up nearly 55 percent of the NPR newsroom staff, a dramatic improvement in gender equity that for decades was lacking except in those voice-of-network roles.[17] Those roles were especially important because they represent the network itself and at the same time, as Renee Montaigne put it, hosts are "essentially a surrogate for the listener."[18]

By the 2000s, the voices and names of Ann Taylor, Korva Coleman, Snigdha Prakash, Lakshmi Singh, Windsor Johnston, Renee Montaigne, Michele Norris, Audie Cornish, and Melissa Block, to name but a few, were well known for speaking as the voice of the network on a daily basis. By the time NPR opened its podcast shop in 2005, it was fair to say that the network spoke, quite a bit of the time, in a woman's voice. Furthermore, the women hosting, anchoring, and reporting also brought racial and ethnic diversity to a staff that had otherwise struggled to diversify itself and its core audience. Michele Norris, who began to co-anchor *All Things Considered* in 2002, was the first African American woman to anchor a national news broadcast.

Hourly news updates by presenters Korva Coleman and Shay Stevens, who are African American, and Snigta Prakash and Lakshmi Singh (both from India) began presenting the hourly

---

[15]Michael Tkaczevski, "Some Things Considered, Mostly by White Men: Study of NPR commentators shows a retreat from diversity—and politics," July 15, 2015, http://fair.org/home/some-things-considered-mostly-by-white-men-2/.
[16]"How NPR Became a Hotbed for Female Journalists," *Newsweek*, March 5, 2012, http://www.newsweek.com/how-npr-became-hotbed-female-journalists-63707.
[17]Tyler Falk, "Racial diversity of NPR's newsroom stays level over three years," *Current*, December 16, 2015, http://current.org/2015/12/racial-diversity-of-nprs-newsroom-stays-level-over-three-years/.
[18]Jonathan Kern, *Sound Reporting: The NPR Guide to Audio Journalism and Production* (Chicago: University of Chicago Press, 2012), 154.

headline news in the 1990s and 2000s. Mandalit del Barco, Doualy Xaykaothao, Lulu Garcia Navarro, Ofeabia Quist-Arcton, Soraya Sarhaddi Nelson, Leila Fadel, and Yuki Noguchi, some with detectable accents revealing an upbringing outside the Anglo-US, began filing on-air reports from all over the world in the first decade or so of the century. Before taking over as *All Things Considered*'s co-anchor, Ailsa Chang began reporting on Congress for NPR in 2012. Eleanor Beardsley, her White South Carolina background accenting her English and French, reported regularly from Paris since 2004.

For Norris, coming to NPR from commercial television news was an opportunity "to sound like herself." Her TV producers had discouraged her "naturally low—and one might read, Black-register." They told her "there's so much smoke in your voice, can you lift it up a little?" At the turn of the century, NPR spoke not just with a woman's voice, but with the voices of women of color from all over the world, "women with exotic names,"[19] coming closer, at least in these roles, to Siemering's vision of "many voices and many dialects."

NPR's record, it must be said, stands in stark contrast to the history of gender equity in broadcasting. "History has many themes and one of them is that women should be quiet," says Kathleen Hall Jamieson.[20] Jamieson has traced this theme, and the associated fear that women's public speech threatened the health of both their reproductive organs and the civic order (bodies and bodies politic), across centuries of political thought. Picking up on this theme, Christine Ehrick notes that "the discomfort (or dissonance) with women's voices, especially women's voices speaking publicly and/or with authority, carried over into and shaped the history of radio," starting from the beginning, a point that has been made, in different national contexts by Michele Hilmes and Kate Lacey.[21]

---

[19] Jeff Porter, *Lost Sound: The Forgotten Art of Radio Storytelling* (Chapel Hill: University of North Carolina Press, 2016), 182.
[20] Kathleen Hall Jamieson, *Eloquence in an Electronic Age* (Oxford: Oxford University Press, 1988), 67.
[21] Christine Ehrick, "Vocal Gender and the Gendered Soundscape: At the Intersection of Gender Studies and Sound Studies," *Sounding Out*, Blog, 2015, https://soundstudiesblog.com/2015/02/02/vocal-gender-and-the-gendered-soundscape-at-the-intersection-of-gender-studies-and-sound-studies/.

More recently, Jeff Porter has concurred that, in its Golden Age, "radio was no place for the female voice."[22] While Golden Age radio dramas devised a range of creative and disturbing responses to "the problem of the speaking woman" on nighttime news and cultural affairs programs, the response was, effectively, to ban women's voices from the airwaves, especially in authoritative roles of hosts, announcers, news presenters, and experts.[23] This history might help to explain the outsized reputation for liberal bias that NPR has garnered since its debut.

In 2008, only two years into *TAL*'s podcast era, I argued that the predominance of women's voices from many backgrounds in these voice-of-the-network roles served collectively as a kind of sound effect for NPR's liberalism, articulated in Siemering's "Founding Purposes" document. Some of these liberal purposes were gradually abandoned as access journalism, political influence, and ideological tendencies within liberalism itself exerted rightward pressure. Women's voices carrying the implicit imprimatur of the women's movement, and of Siemering's pluralistic vision, continued to define the network's sound and feel. If some of the women broadcasters of public radio spoke at comparatively lower pitches, an assertion with which some sociolinguists would quibble citing an implied and specious normative pitch range, it was part of a century-long industry trend, according to Anne Karpf.[24] The age of broadcasting, of amplified human voices penetrating domestic spaces with entertainment, news, and commercial speech, Karpf argues, exerted a moderating influence on women's voices, deepening them, and on men's voices as well, raising their average pitch range.

Jamieson has likewise charted the shifts in vocal performance in the late twentieth century effected by television and in politics in ways that are instructive for thinking about public radio as well. According to Jamieson, the nature of closely miked broadcast performances privileging a "womanly" style of speech displaced an older, aggressive "manly" one, a feminization of speech that privileged men in media and politics alike:

---

[22]Porter, *Lost Sound*, 107.
[23]See Alison McCracken, "Scary Women and Scarred Men," in *Radio Reader: Essays in the Cultural History of Radio*, eds. Michele Hilmes and Jason Loviglio (London: Routledge 2002); Porter, *Lost Sound*, 127.
[24]Anne Karpf, *The Human Voice: How This Extraordinary Instrument Reveals Clues About Who We Are* (London: Bloomsbury, 2006), 175–7.

only a person whose credibility is firm can risk adopting a style traditionally considered weak. So a male candidate whose credibility is in part a function of presumptions made about those of his sex is more likely to succeed in the "womanly" style than is an equally competent but stereotypically disadvantaged female candidate. Ronald Reagan can employ a female style, Geraldine Ferraro cannot.[25]

Such an insight suggests that women needed to moderate both their pitch and their expressiveness in order to be taken seriously, even as men were rewarded for experimenting with pitch dynamics, a finding that has been supported in other research.[26]

The voices of public radio's men, I argued in 2008, demonstrated not so much a feminization as a broad spectrum of prosody. From the basso profundo of Bob Edwards on *Morning Edition* to the flutey uptalk of his 2004 replacement Steve Inskeep, public radio masculinity was broad, playful and performative; consider for example Carl Kasell's Midwestern pump-organ of a voice, shellacked to a hard brown shine. Kasell was equally successful in the deadpan seriousness of hourly news updates and in the pure camp of his *Wait Wait... Don't Tell Me* performances. The late Daniel Schorr, with a voice all jowls and music, soared cantor-like from baritone to plaintive tenor and back. Shorr was the superannuated prophet of Conventional Wisdom, a grizzled bluesman, wringing the last notes out of his indefatigable instrument. As the last standing member of the "Murrow's Boys," he hearkened back in vocal gravitas and old-school credibility, to an era of golden-voiced men.

Peter Overby, a business reporter heard almost daily throughout the 2000s, spoke in a voice that listeners have likened to any number of high-pitched cartoon characters, including Daffy Duck and Tex Avery's laconic Droopy. Robert Siegel, the longtime co-host of *All Things Considered*, sang the news in a voice full of music

---

[25] Jamieson, *Eloquence in an Electronic Age*, 89.
[26] Cara Giaimo, "An Algorithmic Investigation of the Highfalutin 'Poet Voice,'" *Atlas Obscura*, May 1, 2018, https://www.atlasobscura.com/articles/cultural-analysis-poet-voice.

and dynamics, sliding between baritone and tenor.[27] Disciplined in volume and meditative in attitude, his voice in terms of pitch is all over the place, a virtuosic performance of the new flexibility of the male NPR voice. Importantly, Seigel's vocal discipline includes a mostly successful suppression of accent and the traces of ethnicity (Jewish, of Eastern European descent), and region (New York) that it carries.

Together, these voices, women's and men's, arguably lower for one and all over the place for the other, struck me in 2008 as representative of an asymmetrically distributed gender liberation. To be taken seriously as newscasters, on public radio and elsewhere in broadcasting, women had to lower their voices. Men on the other hand, had been liberated from narrow constraints of gendered vocal performance. When Stamberg says that men's use of language has changed since 1990 and that "feminism has done a lot for men!" she may well have had the careers of these male reporters and hosts in mind.[28] Both aural signifiers had been recruited, it seemed to me then, to signify a kind of atmospheric liberalism in tune with the "metrosexual" androgyny celebrated elsewhere in the culture of the turn of this century.[29]

These changes in vocal performances served the interests of NPR's founding commitment to "affective education." The theme of authenticity runs through the NPR's relationship to vocal performance training. The network, which has trained many of public radio's most well-known voices, even those like Ira Glass who have gone on to fame outside of network programming, has long relied on voice coaches to help hosts and reporters bring out "the most emotive, evocative and distinctive qualities" in their voices. "Host Whisperer" David Candow worked with hosts,

---

[27]"A tone consists of a fundamental tone plus related higher tones—overtones or harmonics. The quality of any particular tone is the result of the relative intensity and amplitude of the fundamental tone and its associated harmonics." Anne McKay, "Speaking Up: Voice Amplification and Women's Struggle for Public Expression," in *Technology and Women's Voices*, ed. Cheris Kramarae (London: Routledge, 1988), 204, n.5.

[28]Deborah Ross, Jewish Women's Archives, Washington DC Stories, Interview with Susan Stamberg, March 28, 2011, http://jwa.org/exhibits/dc/stamberg-susan. Accessed November 13, 2015.

[29]This paragraph adapts passages from Jason Loviglio, "Sound Effects: Gender, voice and the cultural work of NPR," *The Radio Journal—International Studies in Broadcast and Audio Media* 5, no. 1&2 (2007): 67–81, https://doi.org/10.1386/rajo.5.2-3.67_1.

anchors, and reporters to improve their pitch, pace, volume, and rhythm starting just as *Your Radio Playhouse* debuted. Getting on-air talent to "sound more like themselves" was the mantra for most of these coaches. Ira Glass echoed this injunction, urging his colleagues to speak in the natural, conversational styles that make for better storytelling, greater empathy, and a more intimate connection with listeners.[30] Avoiding the "cookie-cutter" sound of other radio performers, public radio offers listeners an "artisanal" approach to voice.

For Candow, improvements in sound technology opened the field of radio to a wider range of voices. "Now cars are quieter, and most car radios produce CD-quality sound. Ergo, radio can tolerate a wider range of voices." Another factor to be considered as we chart the change in vocal performance in recent years is that many of public radio's new voices were coming to broadcast news from the world of print journalism, which was haemorrhaging jobs in the 1990s and the 2000s. According to Candow, "the school of beautiful voices [was] dead" by 1995 when he began working at the network.[31] Peter Overby, Jim Zarroli, David Greene, and David Folkenflik, among others, all came to NPR from newspapers during this period of contraction in print journalism. In other words, the shift in voices, from sonorant baritones to squeakier ones, speaks to a set of changes not driven purely by technological advances, aesthetic considerations, or social change.

Even so, one of the cultural hallmarks of the rise of neoliberalism was the embrace of precisely this kind of identity-based social liberalism as cover or "legitimating cloth," in Wendy Brown's words, to mask not just an increasing stinginess in liberal programs like welfare and educational funding, but instead a shift toward a new governmentality in which the state's legitimacy resides in its ability to support capitalism.[32] NPR, it seemed to me, sounded liberal because women were allowed to speak and some of the men spoke in voices "liberated" from a traditional voice-of-god baritone style associated

---

[30]Ira Glass, "Mo' Better Radio," *Current*, May 25, 1998, https://current.org/1998/05/mo-better-radio/.

[31]Paul Farhi, "When This Guy Talks, NPR Listens," *Washington Post*, August 31, 2008, https://www.washingtonpost.com/wp-dyn/content/article/2008/08/29/AR2008082900683.html.

[32]Wendy Brown, "Neoliberalism and the End of Liberal Democracy," *Edgework: Critical Essays on Knowledge and Politics* (Princeton, NJ: Princeton University Press, 2005), 48.

with the heyday of broadcast news. As Lisa Duggan has argued, the neoliberal transformation achieved a rare bipartisan success in the United States during the 1990s and 2000s in part because of the way it "relied on identity and cultural politics" in ways that have not been adequately explored.[33] The cultural institutions, like public radio, propping up the idea of liberal democracy, historically existed "in tension with capitalist political economy." For theorists like Brown, neoliberal governmentality, bipartisan and dynamic, rendered such institutions of liberal democracy "increasingly void of substance."[34]

I'd like to take another listen to the vocal performances of radio and podcasting's last two decades in the hopes of catching other traces of meaning, and to situate the cultural politics of voice into the contemporary context on this side of the various moments of reckoning, grievance, and tumult that have rocked American politics and media in the twenty-first century. In particular, I want to attend to the ways that vocal performance on *TAL* extended public radio's gender liberalism into a gender queer version of masculinity. Ira Glass, David Sedaris, David Rakoff, and Dan Savage, along with other frequent contributors, made a significant intervention into the sound of twenty-first-century masculine cultural authority. For much of its run, the show was preoccupied with masculinity, with extending the limits of its possibility, and with exploring, often with layers of ironic distance, the performative nature of gender.

From vocal performance to narrative preoccupations with absent fathers to queer travesties of masculinity, *TAL* centered men's voices as both culturally vital and also somehow queer enough to speak as if from a narrative alcove, a critical perch from which to observe modern life from just outside of it. Taken together, the gender queer masculinity of its storytellers evoked a robust and infectious fellow feeling that informed the signature public radio sound that could be heard taking over podcasts in the 2010s. It was this fellow feeling that shaped the appeal of many of podcasting's early hits. It also shaped some of the limits of those shows and their production cultures in ways that became starkly and painfully obvious by 2017 when the overlapping contradictions and tensions in the audio industries came to a head.

---

[33]Lisa Duggan, *The Twilight of Equality? Neoliberalism, Cultural Politics, and the Attack on Democracy* (Boston, MA: Beacon Press, 2003), xii.
[34]Brown, "Neoliberalism and the End of Liberal Democracy," 47.

In the first section of this chapter, I will focus on *TAL*'s gender queer masculinity with a close listening to a 2002 episode called "Testosterone" and several other stories that feature a variety of "sissies," the show's word for men—straight, gay, and trans—who endlessly narrate their own problematic relationship to masculinity. After that, I will explore an occasional multiplatform NPR series from the mid-2000s called "Vocal Impressions," which resonates with the preoccupation with vocal performance and social difference that animated the public radio structure of feeling.

Public radio's curatorial approach to "great voices" in the 2000s was an extension of the empathy work that characterized its approach to social difference and cultural democracy. Featuring the sounds of difference carried by the human voices became a crucial part of public radio's traffic in liberal feeling. It was part of a playful approach to sound editing, story themes, and social difference that podcasting's big hits of the 2010s drew inspiration from. I also document some of the ways this approach to sound and social difference backfired, igniting backlash from resurgent white male grievance politics. These vocal performances set up expectations for social justice and workplace democracy that proved difficult to live up to. The chapter will conclude with an exploration of the first wave of "reckonings" to hit the empathy machines of public radio and podcasting, in which the question of voice and difference becomes quite a bit more vexed.

Sissies on Testosterone

A key feature of *TAL*'s embodiment of the public radio structure of feeling was its neoliberal celebration of the elasticity of gender and sexual identities. But it was the limits of that elasticity, and the surprising *snap!* that occasionally followed, that drove the narratives' inevitable epiphany and whimsical (and at times, retrograde) humor. In these stories, the expanded terrain of masculine performativity emerged as the central artifact of neoliberalism's post-feminist sensibility. In the years leading up to the MeToo Movement, the show slyly celebrated the ironic windfall of feminism's critique of gender: an expansion of possibilities for the public performance of masculine identity, which sometimes meant a relatively narrower spectrum of possibilities for women's voices in public.

At the same time, *TAL*'s version of feminism pushed back insistently on the historical and political claims of feminism that extended beyond the liberated male voice. Liberal notions of gender roles have been one of the most compelling features of NPR since its inception in 1970. These unusual vocal performances signal some kind of social formation, a post-feminist generation and social class, eager for the sound effects of social change but ambivalent about progressive social change itself. In the media landscape of the 1980s, when the backlash against feminism spread across journalism, film, radio, and social science, NPR's feminist credentials may have been guaranteed by its lack of outright hostility to women's voices in the public sphere.[35] NPR's flexible attitudes towards gender roles and inclusion of women's voices on the air obscure the ways in which the network has come to epitomize the neoliberal gesture towards feminism and progressive politics in general.

Nowhere have the sound effects of this liberal post-feminist social formation been more elegantly performed than on *TAL*, where mostly white men's voices explored the landscape of their inner lives with a nonchalance and confidence that embraced a broad range of masculine identities and simultaneously repositioned sexual difference and the performance of masculinity from the social and the political to the aesthetic and the personal. Consider the off-kilter interiority on display in "The House on Loon Lake," in which narrator Adam Beckman deadpans, "I was 13 years old and I was in love with a house," an example of the ways in which the confessional audio mode expanded the possibilities of masculine narrative expression. Everyone, *TAL* reassured us in its early years, is queer, especially during puberty. This kind of universalizing masculine queerness was a narrative alcove from which the best stories could be told.

Perhaps this is the reason for the show's studied insouciance regarding sexual minorities: the gay white educated men who grace the microphones eschew any particularistic identification as gay, performing inclusion while denying politics. Michael Warner has called this kind of move equal parts "utopian universality" for the tolerance it represents, and a "major source of domination," because

---

[35] Susan Faludi, *Backlash: The Undeclared War Against American Women* (New York: Crown, 1991).

of the narrow and apolitical parameters it places around who gets represented and what they get to say about themselves.[36] Scholars in queer theory have examined the power of homonormativity to recuperate queer liberalism to its own conservative ends. *TAL*'s early embrace of David Sedaris, David Rakoff, and Dan Savage epitomized this recuperative move, as did its narrow focus on nuclear families, fathers and sons, and laddish camaraderie.[37] Beckman is straight, but he's queer for houses. Likewise, Scott Carrier's stories of alienation, like "The Friendly Man," evoke queer relations; he experiences his professional communication as coercively false, as "whoring in radio news," and authentic human connection lingers at the margins of his job as forbidden desire.

A 2002 episode, entitled "Testosterone," another favorite of listeners and *TAL* staff alike, presented a set of stories that exemplify the trademark preoccupation with the personal search for identity leavened with a jolly helping of heteronormativity in the guise of post-feminist masculinity, trans-cis camaraderie, and parental love.[38] The episode highlights the way in which elastic assumptions about gender and sexuality that are the uncontroversial starting point for *TAL* revert back into surprisingly traditional patterns. In the second act of "Testosterone," producer Alex Blumberg interviews Griffin Hansbury, a transgender man, who recounts the changes to his libido, his intellect, and his attitude about women and feminism after taking large doses of testosterone.

In his introduction, Glass introduces Hansbury as "a woman [who] gets pumped up with several times the testosterone that most men have," a framing that signals the story's reductive approach to gender, which can be attributed partly, but not wholly, to the fact that in 2002, mainstream journalism had not developed clear policies for acknowledging and respecting people's chosen gender, pronouns, and names. It is also a framing that allows for a literary play of opposites like those Barthes discovered in Balzac's story of a castrato, in which things are not what they seem. The episode

---

[36]Michael Warner, *Publics and Counterpublics* (New York: Zone Books, 2002), 165.
[37]Some gay men featured on *TAL*, particularly Rakoff, have explored more radical perspectives on queerness in other venues than on *TAL* stories, which can be seen as more evidence of the recuperative power of the show's homonormativity or perhaps, as an opening into a more critical discourse on sexuality.
[38]Ira Glass, "Testosterone," *TAL*, episode #220, August 30, 2002.

includes a scene in which the *TAL* staff have their testosterone blood levels tested—a friendly competition. The episode also employs Barthes's proairetic code, the slow agglomeration of events in sequence, as in this exchange between Blumberg and Hansbury exploring the early effects of testosterone treatments:

> AB: What were some of the changes that you didn't expect?
> GH: (*laughs*) The most overwhelming feeling is the incredible increase in libido and change in the way that I perceived women and the way that I thought about sex... Testosterone... flooded my mind with aggressive pornographic images one after the other... and I could not turn it off. And it made me understand men and it made me understand adolescent boys a lot... Before, it was cool... I was a butch dyke and that was very cutting edge and that was very sexy and raw. And now I'm just a jerk, you know.
> (*Blumberg laughs*)
> And I've gotten into a lot of arguments with women friends, coworkers who did not know about my past as a female. You know, I call myself a post-feminist. And I had a woman say, you're not a post-feminist, you're a misogynist. And I said, no, that's impossible, I can't be a misogynist... And to her I was just a misogynist. And that's unfortunate because it's a lot more complicated than that.
> (*Blumberg laughing*) I'll say. Wow. Testosterone didn't just turn you into a man, it turned you into Rush Limbaugh!
> GH: I know. (*laughs*) That I was not expecting... So I had to relearn how to talk to women... And I'm not very good at it so I get into trouble.
> AB: That is so fascinating. Because... as a man I think, from the time I went through puberty that's something I've been learning to do in certain ways. To say things without getting myself in trouble. (*laughs*)
> GH: Yeah, yeah.

A laddish camaraderie springs up between these two men. The ease with which Blumberg embraces Hansbury's experience, voice, and subjectivity as a trans man is of a piece with the program's general attitude toward gender and sexual diversity. Indeed, this casual enlightenment was one of *TAL*'s most striking features, though it was tempered by flashes of reductive, even mean-spirited irony, like referring to Hansbury as "a woman... pumped up with testosterone." The people whose subjectivity we are asked to empathize with happen quite often to be queer. Queerness or, and this is crucial, a certain voice associated with educated white male homosexuals (and in Hansbury's case a laddish straight transman), functions as a privileged subject position from which many of the stories are told.

Blumberg is as sanguine with Griffin's trans identity as he is about Griffin's essentializing narrative of gender difference. The latter is redeemed, in some ways, by the former. Griffin's experience of testosterone provokes gales of laughter as if at a subversive truth that saves us from other, drearier ways of looking at the world. The point of convergence for these two men is acknowledging how difficult it is for men to "say things without getting in trouble." Like most everyone else in this episode of *TAL*, their voices and their stories are funnily, quirkily, even queerly at odds with feminism—and women.

Whether this posture of casual acceptance of sexual minorities is more striking as a political act itself or as a rejection of the fact of the bitter politics of sexuality in the United States at the turn of the century is an open question. And this openness, this un-pin-down-ability, is a hallmark of the program's style, its "tone." Does this diffidence towards the political debates of the moment represent the appeal of a collectively experienced "as if" moment of queer acceptance? Or does it represent the insouciance of the apolitical, privileged elites for whom political debates are not fights over scarce resources and physical well-being, but interesting abstractions? Or does the recuperation of certain privileged queer masculinities represent the politics of homonormativity, a concept we'll explore below.

Hansbury's journey from edgy feminist lesbian to leering "jerk" maps the terrain that most amuses Blumberg. In fact, Blumberg admits that the concept for this episode began with his own uncomfortable encounter, in puberty, with a "man-hating" work of feminist fiction (Marilyn French's 1977 bestseller, *The Woman's Room*), which he read accidentally, whilst looking for porn. His anguish over this novel's critique of male desire sets the stage for the entire episode, which has as its central theme the power of testosterone as the driver, not only of male sexual desire but of masculine—indeed human—identity itself. Blumberg comes to the project of understanding testosterone with a particular concern.

> AB: My testosterone and how it affects me and how I react to it, I think about on a daily basis, all the time. It often feels like there's something in my body giving me instructions that I probably shouldn't follow.

This hyper-awareness of the perils of masculine sexual desire and of its inevitability is the fundamental contradiction that Blumberg explores in each of the program's four acts. It is the source of the humor too, and of the laughter and conspiratorial boyish snickering, associated with a guilty pleasures and politically incorrect truths. The elision of problematic masculine behavior with testosterone is critical to the premise of the episode. Glass explains in the introduction that while the concept seemed "dumb-headed and simplistic" at first, they decided to plow forward with it, perhaps on the ironic postfeminist logic observed by Rosalind Gill, that "the extremeness of the sexism is evidence that there is no sexism."[39]

Cordelia Fine, following Richard Francis, has explored the durability of this reductive hormonal essentialism across contemporary gender debates. Despite its popularity, "Testosterone Rex," the myth that a single hormone is responsible for male social behavior across species, is wildly unscientific. But it is precisely in their wrongness that the stories in "Testosterone" cohere. As Susan Douglas has noted, the postfeminist media of the 2010s assumed a "knowing smirk" shared between producers and audience, that enabled hyperbolic displays of sexism, enjoyed from the safe distance of irony.[40]

> AB: Other than the visual, and other than the libidinal, are there other ways that you feel like testosterone has alternated the way that you feel or perceive.
> GH: I became interested in science. I was never interested in science before.

---

[39]Rosalind Gill, qtd. in Susan Douglas, *Enlightened Sexism: The Seductive Message that Feminism's Work Is Done* (New York: Times Books, 2009), 13.
[40]Douglas, *Enlightened Sexism*, 14.

AB: No. Way. Come on! (*laughs*) Are you serious?
GH: I'm serious! I'm serious.
AB: You're just setting us back a hundred years, sir. (*laughs*)
GH: I know I am. I know... Again, I have to have this caveat in here. I cannot say it was the testosterone. All I can say is that this interest happened After T. There was BT and AT and this was definitely After T. And I became interested in science. I found myself understanding physics in a way I never had before. (*laughs*)
AB: (*guffaws incredulously*)
GH: It's true, it's true...
AB: (*with relish*) Wow.

This singular account and the laughter it evokes represent another exemplary *TAL* moment: a hilarious universal truth, upsetting political expectations, revealed through first-person testimony. Masculinity's secrets become the grounds for a forbidden queer connection. The show's dedication to finding surprises, solving for a different Y, as it were, achieve an almost perfect formal state in this piece. Like the narrator of Balzac's *Sarrasine*, discussed in Chapter 4, Blumberg's glee comes from his proximity to the bundle of sexual antitheses at play all around him.

Blumberg is likewise telling a story as part of a larger seduction. His seduction of Hansbury, from queer subject to lad, parallels the seduction of *TAL*'s audience from default assumptions about the plasticity of gender and sexuality to a deliciously retrograde ("dumb-headed") exploration of the deterministic power of testosterone through linked anecdotes. What is sedulously avoided is any acknowledgement of the role of political or social factors, like the power relationships between men and women. The power of the personal, the anecdotal, is a powerful rebuke to the political, another ironic reversal of the feminist "personal is political" strategy at the heart of the public radio structure of feeling since the days of Susan Stamberg.

Taken together, the stories in "Testosterone" provide a brief on behalf of an ironically crude biological determinism. Blumberg begins the program with an interview with an unnamed man whose body, due to an undisclosed physical ailment, temporarily stops producing testosterone. We never learn anything else about the man, his age, his illness, his family life, or the circumstances that led to his eventual medical recovery and return to normal levels of testosterone. His account of life without testosterone, in which he loses not only sexual desire but all desire, and ultimately, all subjectivity, is so stark, so stripped of the context in which lives have meaning (e.g., relationships, jobs, conflicts) that it has the quality of a parable. It's a story about the loss of the self, an almost existential meditation on the dissolution of identity.

> Everything that I identified as being me, my ambition, my interest in things, my sense of humor, the inflection in my voice, the quality of my speech even, changed in the time that I was without a lot of the hormone. So yes, the introduction of testosterone returned everything.

One might expect an episode of a documentary-style program whose quarry is the very nature of identity itself to include a range of positions on the controversial notion that testosterone is responsible for men behaving badly, as well as their entire subjectivity and humanity—perhaps some kind of counterpoint. But *TAL* piles on, so to speak, by getting the entire staff tested for their testosterone levels in the third act. Suspense builds as the stakes in the testosterone context are raised. The test will reveal some kind of "truth" about manliness of the staff; a voyeuristic reveal, unraveling a gender enigma, similar to the plot of *Sarrasine*.

Predictions are shared. Anxieties are shared. Men worry about having too little; women about having too much. The results are read on the air. The men with the least feel bad; the woman with the most feels bad. And the others try to sympathize and not gloat. And Surprise! The gay guy, in this case David Rakoff, described for maximum irony as a "gay Canadian Jew living in Manhattan," wins with the highest testosterone level. It is a triumph of both surprise and homonormativity. The entire episode can be understood as a brief in the struggle for men to talk "without getting in trouble." At one point, Rakoff protests that "Canadian" should not be included in the list of ironic signifiers that attach to the "surprise" winner of the testosterone game. What wasn't up for debate, however, was the ironic play with regressive notions of masculinity, now embracing them, now laughing them off—all of it reassuring "evidence that there is no sexism."

On the one hand, we have a program that promised, expansively, to tell us stories of "this American life," and that inclusively featured the voices and stories of men gay and straight, trans and cis. If queerness was a privileged subject position from which American life can be observed, perhaps *TAL* was dedicated to troubling dominant categories of identity. On the other hand, we have a collection of stories whose insistent theme is the confrontation of male pornographic desire with feminism, of the "man-hating" and lesbian varieties. The episode's origin story is, quite literally, a backlash against 1970s feminist literature and its traumatizing impact on a straight white adolescent boy who grew up to make radio for a living. Testosterone functions thematically across the stories to smooth away the conflict. Taken together, the stories provide Alex Blumberg that thing he's sought since his puberty in the 1970s: a way to talk without getting into trouble. That he confesses to thinking "all the time" about testosterone's effect upon him echoes the intense inward-focus of so many of the program's male storytellers and journalists.

The fear of "getting in trouble" which Hansbury and Blumberg share, and its deleterious effect on art in general and storytelling in particular, has long been an anti-feminist talking point.[41] Figuring out how heterosexual (and white) masculinity gets to speak in a world that has absorbed, or at least encountered, some of the lessons of the movements of feminism, civil rights, and others, has been central to the cultural work of public radio. *TAL*'s gleeful embrace of the centrality of traditional gender roles in these stories resonates as part of a larger disavowal of, withdrawal from, and abandonment of a set of specifically political commitments. There is a sense in which, to return to Susan Douglas, the role of women's voices have to be both "accommodated and contained" (which is how she describes the cultural work of AM talk radio) in the post-feminist public sphere that public radio constitutes for its audience.[42]

---

[41] For a roughly contemporaneous account see Katie Roiphe's complaint about the sexual ambivalence of contemporary male novelists whom she contrasts unfavorably to the pornographic and misogynist brio of Updike, Mailer, and other "Great Male Writers" who came of age before women's studies courses became a *de rigeur* rite of passage for sensitive male authors. Katie Roiphe, "The Naked and the Conflicted," *New York Times Book Review*, December 31, 2009.

[42] Susan Douglas, *Listening In: Radio and the American Imagination* (New York: Times Books, 1999), 285.

It is worth noting that the only woman's voice heard on this episode, other than the anxious voices of *TAL* staff as they await the testosterone levels, is in the last segment, which is titled "Learning to Shut Up." It features a mother of a teenage boy who learns to stop talking to her son about his masculinity, to honor his silence about sex and sexuality. Unlike men's fascination with testosterone, which is endlessly entertaining, women's interest in and critiques of testosterone's role in the workings of masculinity are intrusive, embarrassing, and ill-placed.

It can be argued that the laddishness of a sympathetic heterosexual man chatting with a trans man is so subversive of sexism as to render it harmless, even radical. This argument is tempting for a show whose entire sound signature seems designed to appeal to the young, urban, educated audience for whom traditional debates about gender and sexuality may have been considered largely settled. This episode's delight in the surprising power of testosterone depends upon the inoculation of the aural performance of "soft masculinity," epitomized in the host Ira Glass. Glass's version of masculinity turns on its simultaneous proximity and distance to a certain kind of queerness: "I'm a Jewish guy who doesn't like sports and is into musicals," he explained in an interview with *The Advocate*, when asked about rumors of his gayness. "He's a total fag but he's not homosexual," David Rakoff deadpanned.[43] Together, they manage to claim masculinity as a queer identity while fending off homosexuality.

Glass's adenoidal, nerdish delivery—l's and r's melting into his hard palate, voice cracking into an adolescent uptalk—is a sound effect of the post-feminist masculinity at the heart of the public radio structure of feeling. Blumberg's vocal performance is strikingly similar to that of Glass's, perhaps more so than any other *TAL* alum. Central to this project, at least for *TAL*, is the mainstreaming of certain kinds of men's voices and experiences. This aesthetic is echoed in the show's early embrace of indie pop band They Might Be Giants, whose quirky lyrics, twee sensibility, and vocal stylings all reinforce a kind of soft, ironic version of masculine performance. But while some of the voices may be queer, the lessons learned seem to re-enforce fairly traditional notions of gender. In other words, the fascinating stories don't necessarily "queer" the deal; they

---

[43]Berquist, "The Host With the Most."

don't challenge traditional gender assumptions behind the stories of this American life. "We're pussies," Glass conceded to a listener in an online forum, exemplifying the way that performative soft masculinity centered men at the expense of women.[44]

This apparent contradiction hearkens back to Judith Butler's prescient observation that the subject of queer theory is not fixed but instead will change in different social and historical moments.[45] More recently, Jasbir Puar has noted a liberal "realignment" by which "certain queer subjects" are privileged and normalized and "disaggregated" from "queer racialized populations."[46] The advent of "homonormativity," she argues, reinforces heteronormativity and replicates narrow racial, class, gender, and national ideals.[47] For Lisa Duggan, this kind of homonormativity brings about the "neoliberal privatization of affective as well as economic and public life." "'Freedom,'" according to Duggan in this context, means "impunity for bigotry and vast inequalities in commercial life and civil society." Further, "the 'right to privacy' becomes domestic confinement, and democratic politics itself becomes something to be escaped."[48] This seems an adequate description of the structure of feeling on display in Blumberg's conversation with Hansbury. Their shared search for "a way to speak without getting in trouble" is in fact the expression of a desire for "impunity for bigotry." The pleasure of Blumberg's mock condemnation that Hansbury is "setting us back a hundred years, sir," evokes the notion of democratic politics as "something to be escaped."

---

[44]Ira Glass, "I am Ira Glass. Ask Me Anything," Reddit, r/IAmA, 2013, https://www.reddit.com/r/IAmA/comments/1197d7/iam_ira_glass_creator_of_this_american_life_ama/ September 8.

[45]Judith Butler, "Critically Queer," in *Bodies that Matter: On the Discursive Limits of "Sex"* (London and New York: Routledge, 1993); Judith Butler, "Sexual Inversions," in *Discourses of Sexuality: From Aristotle to Aids*, ed. Donna Stanon (Ann Arbor: University of Michigan Press, 1992); Haley D. O'Shaughnessy, "Homonationalism and the Death of the Radical Queer," *Inquiries Journal/Student Pulse* 7, no. 3 (2015), http://www.inquiriesjournal.com/a?id=1003.

[46]Jasbir K. Puar, *Terrorist Assemblages: Homonationalism in Queer Times* (Durham: Duke University Press, 2007).

[47]There has emerged a robust debate about the role that anti-normativity plays in queer theory. Sharif Mowlabocus, *Homonormativity: Gay Men, Identity, and Everyday Life* (New York: Springer, 2021).

[48]Duggan, *The Twilight of Equality?*, 165.

It is in the context of homonormativity, understood broadly as a conservative and masculinist set of constraints on the political meanings of sexual and gender difference, that *TAL*'s sexual politics can be most fully understood. The seeming nonchalance about queer voices is not subversive of norms so much as it is a recognition of normativity's dynamic capacity for liberal appropriation. As Drew Daniel put it, "The celebration of gay and lesbian difference offers no real alternative to a dominant neoliberal capitalist democratic culture that is only too happy to reinforce, include, and cater to them all as a dutiful rainbow coalition of subject-consumers."[49] Nonchalance can be read instead as the recuperation of a normative relation to gender and sexuality, one in which the prerogatives of race and gender hierarchies have been affirmed around the question, "Who gets to speak?"

In "Testosterone," we have an unusually clear and articulate answer. In Hansbury's conversation with Blumberg, we can also detect heterosexuality's recuperative normative power. As Butler puts it, "Heterosexuality can augment its hegemony through its denaturalization, as when we see denaturalizing parodies which reidealize heterosexual norms without calling them into question."[50] The humor at the heart of Hansbury and Blumberg's shared concern with "getting in trouble" with women and Hansbury's newfound scientific prowess is at once a travesty of heterosexual (and masculine privilege) and a restoration of it. Hansbury's rueful confession that he's been accused of misogyny parallels Blumberg's own concerns about his testosterone and "what it makes him do." The humor, irony, and fear associated with getting in trouble is the shared ground of masculine identity, rather than a particular level of sex hormones.

The entire episode functions according to the logic of gender performance, as defined by Butler, not as a "voluntarist exercise of choice" (an interpretation of her work she is at pains to reject). But instead as an example of a "compulsory repetition of prior and subjectivating norms, ones which cannot be thrown off at will, but which work, animate, and constrain the gendered subject." Further, the hyperbolic celebration of testosterone is legible through this logic

---

[49]Drew Daniel, "All Sound Is Queer," *The WIRE*, London, Issue 333 (November 2011).
[50]Butler, "Critically Queer," 22.

as a melancholic performance associated with the drag queen, but whose proper object, Butler argues, is the heterosexual, performing an identity that must be rejected as love object (a rejection that cannot be grieved, following Freud's notion of melancholy).

Testosterone as a way to talk about essentializing masculine subjectivity, sexuality, speech, and importantly "trouble," performs several remarkable moves at once. It demonstrates a virtuoso command of the codes of gender norms by mocking and reinforcing them simultaneously. The episode also organizes the entire production staff around their relationship to testosterone and thus, to masculinity, and notions of masculinity. It is a move that enables them to mark a variety of comfortable distances from an external norm and thus to disavow and recenter a biological understanding of gender. Finally, the program once again realigns "certain queer subjects" as privileged and normative, an act that is performatively inclusive and implicitly exclusive.

From the start, the voices of articulate white queer men have become a critical sound effect and affective touchstone on *TAL*. They have co-mingled easily, and in the case of Hansbury, ecstatically, alongside voices of straight cis men of a similar demographic. Perhaps no soundmark better epitomizes this queer relation than– David Sedaris's stories on *TAL*. Glass and Sedaris set the tone that Blumberg and Hansbury follow. Consider Sedaris's sullen elf in "The Santaland Diaries," in which his queerness is expressed in terms of his alienation from retail work and its campy theatricality. It is not so much the specifics of his costume with its "perky green cap" as it is the artifice inherent in all wage work that resonates.

Sedaris's "gay voice" is not the transgression; rather, it is the medium through which we are able to understand and forgive the larger transgression of anti-social behavior in a department store (i.e., threatening harried customers and mocking Christmas). Gayness, a particular kind of gayness: white, male, literary, is in the service of something else, a useful alcove from which to tell on the rest of us. Listeners to *All Things Considered*, where the first version of "The Santaland Diaries" debuted and *TAL*, where a longer version aired, must have found in Sedaris's cheeky elf an apposite performance of 1990s-era performative disaffection from consumer culture.

We can hear a similar ventriloquism of homonormativity as a narrative alcove from which to lay claim to outsider status in the stories gathered under the theme of "Sisses" (December 1996). "Though being gay no longer has much of a stigma in parts of the country," the episode begins, "being a sissy still does—even among gay men." In one of the stories, John Conners, a white gay performance artist from Chicago, reads from a 1942 self-help book for men on how to avoid effeminacy in word and deed. Conners adopts a mocking voice to read the advice about which words and gestures are too effeminate for men to safely manage. He chooses a breathy declamatory style and an accent that approximates the Middle Atlantic of Golden Age broadcasters, round vowels and all. It manages to sound both fussy and authoritative, a kind of drag performance of old-world hypermasculinism, rendering it queer.

Conners is particularly interested in the passages that police men's voices for signs of femininity. He ends with a passage on the proper sound of laughing: giggling is discouraged in favor of big booming haw haws, both of which he acts out in hyperbolic fashion. After this performance, Glass and Conners try to imagine a sissy circa 1942 earnestly reading the book for tips, a conversation which devolves into giggling jokes about "homosexual battalions" of GI's liberating Parisian lingerie shops during World War II and rendering the city "too foofy" for the Germans to tolerate. The silliness and retrograde homophobia of the jesting only makes sense according to the logic that "the extremeness of the homophobia is evidence that there is no homophobia." Or perhaps to the giddiness associated with a certain kind of laddish camaraderie confirming a shared distance from both normative masculinity and from the historic otherness that once relegated queers to violence and repression. It is, in any case, two men, one queer and one straight (but a sissy), who have figured out how to talk without getting into trouble.

This is a popular formula and Glass uses it to introduce the episode, enlisting his friend Michael, "who is constantly talking about what is manly and what is not manly." The surprising twist? "He's gay. He's *a gay*," the repetition and odd construction ("*a gay*") another fending action—as if the very idea of naming it is a bit naff for someone like Glass. In episodes like "Testosterone" and "Sissies," manliness is something that can only be spoken about, and only from just outside its traditional boundaries. Glass speculates that for Michael, "it's almost like being a gay has made him more obsessed with manliness than any of my straight friends. If you're a man, I think it just comes with the territory." These last two contradictory observations, manliness is a special preoccupation of "the gays" and "if you're a man... it comes with the territory," succinctly captures the triumph of *TAL*'s homonormativity. To talk about manliness is both queer and quintessentially the fate of all men. Masculinity, to the extent that it can be spoken about, is both queer and urgently ubquitous.

In the act that follows, Glass talks with Dave Awl, another gay white man, who recalls his youth as a sissy. After a childhood of taunts and violence, Awl turns for protection to the punk aesthetic—mohawk, safety pins, leather, and painted nails. It works—he's still an outsider, but no longer a sissy, or rather, he realizes, he hides his sissy side, behind a protective shell. Glass notes, "It's interesting that you were able to accept the fact that you were gay long before you were able to accept the fact that you were a sissy." Awl agrees that his gayness is less a threat to him on the streets than his relationship to the feminine.

The recuperation of white homosexual identity into a normative masculinity continues in the final act featuring frequent contributor Dan Savage, a gay man whose advice column *Savage Love* made him a bit of an authority on sexuality and identity in the 1990s and 2000s. Savage shares his campaign against internalized homophobia in the personal ads in alternative city weeklies. Men looking to hook up with "straight-acting" men seemed to Savage both ridiculous and an erasure of gayness itself. The real sissies, he concludes, are the men who are afraid to be sissies—or afraid to be seen with one. The category sissy, a man who lacks a certain kind of privileged masculinity, remains, but has been shifted so that confident, self-actualized, and sex positive men—real men—like Savage are no longer implicated. Masculinity and a certain kind of queerness are again narratively elided. While Savage has been an outspoken critic of a narrow, hetero-model for gay relationships and sex, in this story, he seems to leave in place, even solidify, the real man/sissy binary.

If some stories told by gay men were framed in such a way as to imply a narrow race-and class-bound homonormativity, there's an overlapping sensibility running through *TAL*'s straight white male storytellers, who also identify as sissies of one kind or another.[51] The contributions of Jonathan Goldstein in the 2000s were something of a master class on the narrative authority of the gender-queer-but-straight white man, particularly in scenes of laddish intimacy with another man. Goldstein was one of the staff who anxiously submitted to the testosterone test in 2002; he expressed this anxiety in an interview with Glass through a continuous series of moments of regret for having spoken. "Even as—even now. Even now. Even now." Because the speaking of regret itself engenders fresh regrets, it is also a way to never stop talking, which is part of the joke and part of the point. It is another way to talk without getting in trouble. In a 2011 show, Goldstein tells a story from the perspective of the man who Lois Lane dates on the rebound after things don't work out with Superman.

In a hyperbolic scene of abject humiliation, the man becomes Superman's doughy sidekick. In his encounters with Lois or Superman, he never seems to get a word in edgewise. Superman cuts him off by saying "silence." He never dares to contradict Lois when she's excited about something. It's only when he runs into Clark Kent, another man adjacent to masculinity proper, that he finds his voice: "I told Clark all about [it] and he listened to me. That was all I really needed just then, to be listened to."[52] Of course the joke's on him, because he's pouring his heart out to his rival in disguise. The failure of words is central to Goldstein's performance of masculinity. "Life isn't about saying the right thing anyway," Goldstein concludes at the end of another episode. Instead, life seems a series of miniature dramas of beset manhood, in which talking proves disastrous or elusive but always necessary.

Elsewhere, Goldstein notes a generational difference in how his father manifested masculinity compared to himself. But he renders his father's admired machismo so ironically that we understand that

---

[51]Savage's 2010 "It Gets Better" campaign, aimed at helping suicidal gay youth, while inspirational and well-intentioned, has been criticized for its class-bound assumptions about upward mobility. See Laurie Phillips, "A Multimodal Critical Discourse Analysis of Race, Class, Gender, and Sexual Orientation in the 'It Gets Better Project,'" *AoIR Selected Papers of Internet Research* 3 (October 2013), https://doi.org/10.5210/spir.v3i0.8390.

[52]"Father's Day 2011," *TAL*, episode #438, June 17, 2011.

we're comparing men who linger in an alcove just off mainstream manliness. "Growing up, I watched him charm his way out of traffic tickets, whistle for cabs, and say things like 'va-va-voom,' all things I cannot now help but think of as being a part of a manliness from a bygone era."[53] His prediction about testosterone levels reinforces the relationship between voice and manliness: "It's like who can yell the loudest, right?" Central to masculinity in these stories, is talking—how one talks, how much, how loud.

In these stories, talking is acknowledged as compensatory—speaking speaks to a lack, but it is also constitutive of masculinity. Speaking to others and over them ("silence!") is part of being a man. "Let me start by saying I have pretty much regretted every single thing that's ever come out of my mouth," Goldstein begins another story, an echo of his confession to Glass in "Testosterone" that was essentially, a rolling monologue of instantaneous regrets. His solution: tape record everything for one day and then go back and edit in snappy repartee that eludes him in the moment. "Lord knows saying the right words to people's faces would be far too frightening, but right here, behind their backs on the radio, that would be just right."[54] In this moment of reflexivity, Goldstein properly captions one of the animating principles of the program: recording, editing, and broadcasting are the necessary affordances for proper masculine vocal performance—to speak without getting in trouble.

The insight calls to mind the relentless preoccupation with manliness at the heart of the Torey Malatia gag, mentioned in a previous chapter. At the end of (almost) every episode, Glass repurposes a bit of tape, a bit of speech, and creates a brand-new context for it, in which Malatia, formerly his boss at WBEZ, is ventriloquized, often as a girl, a woman, or in a context which suggests that he is some sort of sissy. Matthew Murray has traced the historical origins of "going swish" back to early radio and vaudeville. On golden-age radio comedy, ostensibly straight comics put on a "gay" voice in order to titillate mature audiences with a risqué acknowledgement of queerness while also distancing themselves, through their hyperbolic performance of swish, from

---

[53]"Goldstein on Goldstein," *TAL*, episode #294, August 5, 2005.
[54]"What I Should Have Said," *TAL*, episode #257, January 16, 2004.

any hint of effeminacy.[55] This having it both ways, it seems to me, is at the very heart of broadcasting's century of masculine vocal performance: ways to talk—and talk and talk—without getting in trouble.

Vocal Impressions: Curating Sound

Part of *TAL*'s charm in the 1990s and 2000s was its coy insouciance about its remarkable stable of unusual voices. Sarah Vowell, Jon Ronson, David Sedaris, David Rakoff, Alex Blumberg, Jonathan Goldstein—their voices carried so much information about their personalities, perspectives, and habitus. They are confidently ironic; privileged and marginal; cosmopolitan and uneasy. Men's voices were the object of frequent mockery, both within and without the industry. "Dweeb with gayvoice rising," is how one listener put it to describe Ira Glass and his ilk.[56] "Your somewhat nerdy friend," rather than an old-fashioned broadcaster "with an impossibly low voice" is how John Biewen put it when describing quintessential NPR reporter, Daniel Zwerdling.[57] Glass, for his part, describes himself as "slightly femme-y" in the "Testosterone" episode. Part of the show's hetero- and homonormativity was precisely this confidence that these voices represented an emergent cultural authority. It was also an acknowledgement about the affective power of voice. As Dario Llinares put it, "personality, identity, experience and knowledge are intrinsically articulated, not just through the words spoken or ideas expressed, but within the very sound of the voice in all its textures, complexities, uncertainties and emotional valences."[58]

At the same time, *TAL* along with the other big institutions of public radio, were developing a new self-consciousness about sound, and particularly the curation of voices. This was a period

---

[55]Matthew Murray, "Going Swish," in *Radio Reader: Essays in the Cultural History of Radio*, eds. Michele Hilmes and Jason Loviglio (New York and London: Routledge, 2002).
[56]"Most Annoying Voice," DataLounge, Reply #6.
[57]John Biewen, *Reality Radio: Telling True Stories in Sound* (Chapel Hill: The University of North Carolina Press, 2010), 1–2.
[58]Dario Llinares, *Podcasting: New Aural Cultures and Digital Media*, eds. Dario Llinares, Neil Fox, and Richard Berry (New York: Springer, 2010), 141.

in which the term curation had migrated from museums to an entire swath of cultural practices, in part because easy access to digital copies of art forms of all kinds had created a deluge of media stimuli. The hipster exhaustion with media overload, discussed in Chapter 3, inculcated a desire for new tastemakers and coolhunters to help cut through the noise. Of course, the end of the century and millennium encouraged the curation of best-of lists across broad swaths of human endeavor including lists of "great voices," as we'll explore below.

Finally, audience research was reinforcing the notion of public radio as part of a coveted "psychographic" taste culture: highly educated and "societally conscious."[59] Public radio was adept at bringing together the highbrow connotations of curation with the massive digital archive and popular appeal of the internet and web content. This combination of carefully curated content, made digitally accessible, represents one of the chief affordances of early podcasting. This shift seems crucial for understanding how the public radio structure of feeling shaped early podcasting. Learning the right ways to talk and knowing the right things to talk about both required an aesthetic disposition.

Ashley Sterner, a local *Morning Edition* host for WYPR, a Baltimore NPR member station, told me in a 2016 interview that she programs sonic "easter eggs" in the bumper music leading in and out of national stories with allusions, resonances, and puns aimed at a tiny fraction of the audience. "We make musical references that we don't expect everyone to get. One percent will get it and ninety-nine percent won't, but they won't miss out on anything." Sterner assumed that for any given reference, there would be different groups of listeners comprising that one percent. The benefits of distinction, of being "one who understands," is thus mass-distributed, one percent of the audience at a time, and is consciously built into the programming at the local and national level.

Marketplace, a public radio program nationally distributed by American Public Media, an NPR competitor, "does this all the time,

---

[59]Michael McCauley, *NPR: The Trials and Triumphs of National Public Radio* (New York: Columbia University Press, 2005), 92–3; Jack Mitchell, *Listener Supported: The Culture and History of NPR* (Westport, CT: Praeger, 2005), 24–6.

too," Sterner assured me.[60] Playlists, icons, classics, brief musical "buttons" in between segments, "tiny" concerts: these features provided an apt soundtrack for the tricky maneuver of managing popular culture and elite taste at once. Audience curiosity about the name of songs and artists played in these brief buttons was the inspiration for *All Songs Considered,* which began in 2001 as an online list of song titles and credits and expanded to become a robust web-based program and NPR's first proto-podcast. It was in this context of public radio's growing curatorial role, that books by *TAL* contributors David Sedaris, Jon Ronson, and Sarah Vowell hit bookstores. In his curatorial mode, Glass edited a volume of essays called *The New Kings of Nonfiction* in 2007, the term "kings" making clear the gendered assumptions at work in the curation, as the collection features twelve male authors and only two women.[61]

Around the turn of the century, public radio developed a fascination with distinctive literary voices—its own and those of its listeners and of iconic singers and speakers of the past century of recorded audio technology. This seems to have taken place just as its dominant institutions were shifting their aesthetic and technological priorities away from "sound-rich" field recordings to emphasize human voices in studios. Several industry practices and innovations combined to produce this shift, starting with the inclusion of musical buttons in between stories and musical beds *within* feature stories, running *under* the narration. These practices, attributed by many to NPR producers Joe Frank and Keith Talbot, were taken up by Ira Glass, who was trained by Frank and Talbot, and it formed one of the sonic signatures of *TAL* storytelling.

These shifts privileged the use of pre-recorded music and sound, replacing in some sense the sonic pleasures of the high-quality field sound recording and editing. This was partly an economic decision, due to high costs associated with sending engineers out with producers and reporters on assignment. Throughout the 1990s, this practice became increasingly rare, thanks to belt-tightening and union tensions around the division of labor between producers and

---

[60]Personal communication, Ashley Sterner, June 19, 2014.
[61]David Sedaris, *Me Talk Pretty One Day* (New York: Little, Brown, 2000); Jon Ronson, *Them: Adventures with Extremists* (New York: Simon Schuster, 2001); Sarah Vowell, *Radio On: A Listener's Diary* (New York: St. Martins Griffin, 1997); Ira Glass, *The New Kings of Nonfiction* (New York: Penguin, 2007).

engineers. This meant that stories increasingly relied on a "talking head" and the use of pre-recorded music, rather than expertly recorded field sounds.

The shift from analogue to digital recording and editing exacerbated these tendencies. In the early 1990s, NPR chose the Dalet software, which was ideal for allowing multiple editors to work on sound files in real time (even during the recording process), but which, at the time, was not nearly as good as other systems in producing high quality, "sound-rich" tape. "It simply didn't allow for the kind of fine tuning" that had been possible with reel-to-reel tape, veteran NPR sound engineer Flawn Williams told me.[62] The high cost of sound engineers, and the limited affordances of early digital editing software, combined with the network's growing reputation for curation of recorded music, moved the network toward features like *All Songs Considered*. Of course, the network's shift to "hard news" and inside-the-beltway dominance can be considered an important part of the context in which talking-head interviews replaced sound-rich feature stories associated with the arts and culture desk. But it is not a sufficient explanation, as we can hear in the emphasis on curation, voices, interviews, and recorded music buttons in arts and culture programs like *TAL*.

In the 1990s and 2000s, public radio networks, shows, and stations transformed from alternative public sector broadcasters to a competing group of transmedia brands with increasingly market-based business models.[63] As public radio blossomed into multiple competing networks and spread into digital platforms, the relationship of producers to listeners became more self-conscious and more opaque at the same time. As cultural and technological trends pointed the way towards "interactive" media, public radio producers leaned into curation to encourage engagement around questions of habitus. Inviting listeners to participate in public radio's "curatorial" relationship to sound was one key strategy for developing a new transmedia brand and a compelling feature of the public radio structure of feeling.

---

[62]Flawn Williams, personal communication, May 20, 2024.
[63]Mike Janssen, "NPR rallies system to jointly build 'trusted space,'" *Current*, July 17, 2006, https://current.org/2006/07/npr-rallies-system-to-jointly-build-trusted-space/.

Sound curation enabled public radio programs to draw attention to unique programming and to invite and train listeners to think of themselves as sound curators as well, and thus, as co-producers of public radio's distinctive sensibility. NPR pursued this with several initiatives, like *Fifty Great Voices*, a 2010 series of appreciations for iconic and obscure singers; *All Songs Considered*, arguably its first podcast, has become an enduring music curation initiative.[64] In 2006, NPR introduced an occasional series called *Vocal Impressions*, that brought together curation, preoccupation with vocal performance of social difference, and the impulse toward greater listener "engagement" in interesting ways.

Vocal Impressions: Sound Curation as Distinction

"She sounds like diamonds dipped in caramel." That is how the university-bound daughter of the NPR reporter, Brian McConnachie described the voice of Ella Fitzgerald on the radio as they drove through the New England hills to the first of many visits to university campuses.[65] If the college visit weekend, including the all-important interview, is a rite of passage for the upper-middle class American family, then this moment of precocious synesthesia provides a glimpse into the *habitus* of this class and this ritual, and offers as well an insight into the cultural work of NPR in the 2000s. The reporter, Brian McConnachie, is of course bragging about his daughter, whose attention to the sound of the voice, rather than merely the meaning of the lyrics, bodes well for her weekend of college interviews. "Wow," he enthuses. "I guess high school worked!"

Vocal Impressions was featured on *All Things Considered* from 2006 to 2008, the early years of the network's development of podcasts, web-first content, and its curatorial turn. Listeners were invited to play along with the McConnachie family's road-trip

---

[64]"All Songs Considered," NPR, https://www.npr.org/podcasts/510019/all-songs-considered.

[65]Brian McConnachie, "Vocal Impressions," NPR, http://www.npr.org/templates/story/story.php?storyId=5617413&ps=rs.

game, producing their own metaphorical descriptions of famous voices. Those belonging to Morgan Freeman, Marilyn Monroe, Truman Capote, and Patsy Cline were selected in the segment's first episode, a catalogue reflecting the network's promise of cultural breadth and the class and generational limits of that promise.

In its invitation to listeners to collaborate on vivid analogies about iconic voices, *Vocal Impressions* also exemplifies one of the great and vexing conceits of a great deal of public radio programming: that listeners are "pretty much just like us" in terms of education, values, and taste, and therefore share an aesthetic fascination with social difference. In its curatorial mode, public radio producers were very much in sync with audience researchers who exhorted them to lean into the notion of their listeners as part of an educational elite.[66] For this audience, listening—a certain kind of listening (what Bourdieu called an "aesthetic disposition")—is a critical marker of the *lifestyle* or *habitus* exemplified by the leafy university campuses that McConnachie and his daughter are exploring.[67]

Listeners proved every bit as adept at playing this game and eager to do so. The voice of African American actor Morgan Freeman evoked front porches, rocking chairs, brandy, sandpaper, the fireside, walnuts, grandfathers and "the voice of our conscience." One listener said his voice sounded simply like "the earth." Author Truman Capote reminded listeners "of kazoos, a whole menagerie of small animals, sock puppets and ingesting helium." One listener said his voice was "the twisted thoughts of barbed wire." The blues singer Odetta was likened to "Mother Earth moaning another rain forest downed," and "A call and response in the church of the wild bear." Norah Jones sounded like "that pebbly mud that feels good squished between your toes," "Silk brushing your cheek as chocolate melts in your mouth," and "Crunchy peanut butter—mostly smooth with a bump here and there." And chocolate—lots of chocolate.

What was most striking about this catalogue of famous voices is the over-representation of racial and sexual minorities and immigrants or foreigners. Nearly three-quarters of the voices

---

[66] *Audience 88; Audience 98.*
[67] Bourdieu, *Distinction*, 11–17.

selected for the series were either Black or foreign-born or queer-coded in some way, or some combination thereof. More than 40 percent were either openly gay like Truman Capote and Harvey Fierstein; rumored to be gay (Bobby Short and Cliff Edwards); or well-known as drag icons (Mae West, Cher). Of those remaining white, straight, US-born voices, many, like Patsy Cline, Johnny Cash, and Elvis Presley, bore traces of class and regional otherness: namely the white working class of the South. This series reflects the network's tropism for the sounds of alterity—but framed, curated, that is—for an imagined audience of public radio listeners.

For these listeners, the voices of racial and sexual minorities evoke earth, mud, chocolate, nature, and the pleasure of sensual contact between the body and the world. The fascination with "strangers" as a literary trope and as an ideological tool for connecting to the public radio audience as familiar objects of empathy is a key feature of the public radio structure of feeling, as I've discussed in previous chapters. Rendering perceived others as objects of empathy involved a process of distancing that hardened the cultural boundaries that were performatively crossed. In 2009, some wag observed that NPR only played "Black music" if the artist was Dead, Old, Retro, or Foreign, dubbing this phenomenon "the DORF Matrix."[68] A website kept track. This pattern made perfect sense for an imagined audience for whom Black music represents a kind of cultural competency, rather than a lived experience. Of course, this was before the rise to mainstream influence of the *Tiny Desk Concert* and the profound changes wrought on public radio in the 2010s by multiple forces, which we will explore in the next chapter.

Then there is the matter of the listeners' poetry. Like McConnachie's daughter, their synesthetic metaphors seem like emblems of the educational and cultural mastery that is at the heart of public radio's construction of its audience. Audience participation on weekend quiz shows, weekly audience emails in response to stories on *All Things Considered*, and the endless psychographic research conducted by NPR (explored in Chapter 2) speak to the ways listeners' voices

---

[68] Jody Rosen, "The DORF Matrix: Towards a Theory of NPR's Taste in Black Music," *Slate*, https://www.slate.com/blogs/browbeat/2009/10/12/the_dorf_matrix_towards_a_theory_of_npr_s_taste_in_black_music.html.

shaped public radio's sonic identity. Likewise, *Vocal Impressions* demonstrated to listeners and underwriters alike the markers for education, class, and consumption that make public radio both a self-conscious community and a unique upscale market segment. Rendering Capote's voice as "the twisted thoughts of barbed wire" neatly conveys several forms of cultural competence, literary, biographical, and ideological, into one elegant phrase.

Like McConnachie's daughter, listeners' analogies are deliberately multi-sensual, mixing oral pleasures of taste and texture and cultural stereotypes of racial and sexual difference with the social consciousness that McCauley has identified as central to the audiences' connection to NPR. Thus, Morgan Freeman sounds like "our conscience" and Odetta is our environmentalist anguish apotheosized into "Mother Earth" herself. These metaphors draw attention to the relationship between the voice, the body that produced it, and the bodies that hear it in ways that recall Jennifer Stoever's critical concept of "the sonic color line," an often-overlooked historical, cultural, and technological process by which sounds, and the politics and aesthetics of sound, have become racialized in the service of a dominant whiteness. For Stoever, voices of African Americans have been represented in literature as especially fraught, evoking layers of embodied sexual desire and threat.[69] Michele Hilmes has also observed the cultural anxieties around race and gender inspired by radio's disembodied voices, and the industrial and narrative practices that were enlisted to police them and to metaphorically, re-embody them.[70]

The aggregation of listeners' synesthetic analogies provides a set of easy-to-decode signifiers of the sonic color line and the cultural work it performs. Morgan Freeman's voice was also likened to "a silk trombone," which is a nonsensical concept unless you understand the words synesthetically and as part of a collection of signifiers that have gathered in the mainstream (white) consciousness around Freeman's exceptionalist cinematic personae and their reliance on a selective set of tropes around African American masculinity that

---

[69]Jennifer Stoever, *The Sonic Color Line: Race and the Cultural Politics of Listening* (New York: The New York University Press), 180–228.
[70]Michele Hilmes, *Radio Voices: American Broadcasting 1922–1952* (Minneapolis: The University of Minnesota Press, 1997), 130–50; 230–70.

easily collide with a broader commodification of Blackness and the high-brow recuperation of jazz.[71] Metaphors like, "What rich river-bottom soil feels like," drew from a common set of sonic color-line signifiers linking African Americans to the "lower" and more sensual elements of the natural world, particularly earth, dirt, and mud. For another listener, Freeman sounded like "A lion gargling with pebbles."

Black men and women often sounded to these NPR listeners like forces of nature. The booming baritone of Paul Robeson's voice was likened to "the morning fog in the redwood forest" and "rolling thunder sweeping away everything it encounters." For another listener, he sounded like "blood running away from your heart," another metaphor that seeks to find in the disembodied voice traces of embodiment, though in this case, passion and violence intermingle in unsettling ways. Others, like the one who likened Morgan Freeman to "Darth Vader's brother," signaled an awareness of the always-mediated nature of Black voices while betraying the vast undifferentiated Otherness of the sonic color line. Some of these responses leaned into dominant meanings of specific celebrities and their voices as in the description of Barry White: "what you'd expect to hear when you put your ear up to an empty bottle of Viagra." Still another called White's voice "a deep need on a Saturday night."

Such metaphors were perhaps overdetermined by the series' focus on voices like White's that came swathed in well-known sexual resonances, especially for Baby Boomers and Gen Xers, who in the mid-2000s made up the lion's share of public radio's listenership. Over-the-top metaphors for White's voice, like "a roller-coaster ride in a hot tub through the tunnel of love," gestured to this overdetermination. "If syrupy, thick erotica owned a vehicle, [White's] voice would be the car alarm," and "satin sheets caressing a velvet leisure suit on a waterbed" likewise indicate that piling up familiar signifiers for late-twentieth-century sensuality trumped any attempts at describing sonic or musical elements of these iconic

---

[71]Perhaps this image is inspired by crooner Mel Tormé's nickname, "The Velvet Fog"; Tormé's voice, and crooning in general, produced anxieties about race, gender, and sexuality in the early twentieth-century crooner era. Allison McCracken, *Real Men Don't Sing: Crooning in American Culture* (Chapel Hill, NC: University of North Carolina, 2015).

voices. Describing Samuel L. Jackson's voice as "like the subject of several Jim Croce songs" or "Mr. Ed recast as a rottweiler" again pairs Boomer media savvy with a reductive Otherness. Perhaps these evocations are influenced by the snippet of Jackson's voice played in the episode's prompt, a scene from *Pulp Fiction* (1994), Quentin Tarantino's edgy, post-modern valentine to genre violence in which anti-Black racism and Blackness' "cool" factor weave together playfully, as if to say, "the extremeness of the racism is evidence that there is no racism."

Ray Charles sounded like "honey on a bruise" and "chocolate melted in the heat of summer," metaphors that link the sensuality of sweets to different kinds of embodied ambivalence. For another, he sounded like "Brown sugar Cream of Wheat sliding down your throat on a winter morning." These impressions underscore the preoccupation with the body. As Hilmes has argued, the effort to re-embody the disembodied voices of radio (and other sound technologies of the twentieth century) often seemed more vital when those voices signaled social difference, particularly along the axes of race, class, and gender. Thus, the iconic voices of Black celebrities are rendered in bodily metaphors again and again.

Of course, the physicality of the human voice has been a central concern for a long time, animating debates and insights across several fields. Roland Barthes has described the "grain of the voice" as a physical thing: "the 'grain' is the body in the voice as it sings, the hand as it writes, the limb as it performs." Christine Ehrick adds, "If the voice is not the body, what is it? Even when it travels over long distances (via telephone or radio, for example) and/or if its source remains out of sight, the body is there, present via the sound vibrations it produces." More bluntly and queerly, Daniel reminds us that voice sounds come from "air pushed thru a tube of meat."[72] He is referring in particular to the orgasmic moaning in a gay club anthem, but the point extends to all vocal sounds. At the same time, his example, and the ones I've drawn from *Vocal Impressions* above, suggest that some voices evoke embodiment more, and more sensually, than others. This point puts pressure on Barthes's understanding of "the grain of the voice" as "outside the

---

[72]Daniel, "All Sound Is Queer."

law of culture," which is to say, beyond semiotic understanding. But Barthes also says, paradoxically, that the grain of voice is "equally [outside] that of anti-culture."

In fact, Barthes describes the "the grain of the voice" in pairs of contradictions: it is "the paradoxical state at once totally abstract... and totally material." It is individually experienced "but in no way subjective."[73] In other words, it's complicated. The attention to the voice as a material force that exceeds "word meaning" is critical to understanding the affective power of vocal performances and the impressions they make, and to understanding the curatorial approach to voice and sound. But getting "outside the law of culture" is easier said than done. *Vocal Impressions* seems more usefully understood as part of a Bourdieuan system of distinctions, designed to fix voices into socially legible and hierarchical place.

At the same time, this system fixes two publics—Williams's elite "minority culture" and those minoritized voices who are both politically marginal and culturally central. *We* are the college-bound, who make metaphors and *they* are the earth-bound, who make sounds. Or as Bourdieu puts it, "two antagonistic castes, those who understand and those who do not." Mastering such distinctions is akin to making chocolatey metaphors—the "ability to perceive, classify, and memorize"—what Bourdieu would call the "unintentional learning" taught by institutions of the family and the school.[74] Such distinctions represent a kind of fluency with cultural signs and with sensory input and a facility of moving back and forth between them. Thus, when Anne Karpf says "Humphrey Bogart... is his rasp," she's making a point about sound and sensibility, the physical body and the social body.[75]

For Jeff Porter, focusing on both the literal meaning of words spoken and the unpredictable meanings conveyed by their sound, requires close reading and close listening. Sound has the potential to "upset the semantic order of language."[76] And voices in particular have the potential to "exceed speech" and "transcend its referent." NPR's sonic history, according to Porter, has been marked by

---

[73]Roland Barthes, *Image, Music, Text* (Mariposa, CA: Fontana Press, 1977), 188.
[74]Bourdieu, *Distinction*, 28, 31.
[75]Karpf, *The Human Voice*, 11.
[76]Porter, *Lost Sound*, 8.

competing waves of "sound-meaning and word-meaning."⁷⁷ *Vocal Impressions* seems an attempt to resurrect the avant-garde cachet of "sound-meaning" in the service of conspicuous distinction-making. For Barthes, the very physicality of the voice, its embodiment, is what queers it, putting it beyond the realm of signification. The grain of the voice *is* its embodiment; it needs no further language to sort it out. The grain exceeds culture and language. The *plaisir* of this exercise is the sensuality and variability of the collected metaphors—each one individual yet anonymous, each one abstract yet physical—rooted in the sensuous contact between the vibrating flesh in the speaker's throat and the membrane of the hearer's eardrum.⁷⁸

And yet the muddy, bumpy earthiness heard in NPR's curated voices of minorities (Blacks and queers and southerners and foreigners) clearly are not invitations to some Barthesian opening out into the unpindownability of meaning. The grain of the voice for Barthes is that which might escape the reductive interpretation of symbolism. *Vocal Impressions* brings these embodied sounds and the bodies that produced them back into language as metaphor, and back into a logic of distinctions that is central to the cultural work of public radio. Like the reporter's college-bound daughter, for these NPR listeners, the voices of racial and sexual minorities evoke earth, mud, chocolate; nature; and the pleasure of sensual contact between the body and the world. Their analogies form a one-way bridge over the compelling boundary between two distinct groups: public radio listeners and producers on one side and the sensuous, earthy voices of difference on the other.

If iconic Black men's voices like Barry White's inspired a cascade of metaphors for sensuality, then voices belonging to men coded as gay set loose an entirely different set of associations and embodiments, which provides broader contexts for understanding the relationship between vocal performance and hetero- and homonormativity in play in the public radio structure of feeling. Along with the highbrow associations with the curation and distinction-making, there is a slightly more vernacular form of "reading" at work in calling out queerness in vocal performances. At once inside and outside,

---

⁷⁷Porter, *Lost Sound*, 13.
⁷⁸Barthes, *Image, Music, Text*, 181, 188.

these readings perform otherness in the process of identifying the sonic traces of queerness in men's voices.

Thus 1960s pop singer Bobby Short sounds like "your sixth-grade music teacher on spring break in Paris," an impression which manages to straddle condescension and tenderness. "The yellow smell of books forgotten in a closet," is another metaphor that places Short within a scene of quaint nostalgia and queer melancholy. The homonormativity in these impressions is evoked through metaphors of social scenes rather than sonic elements, echoing the larger industry shift away from sound-rich content to mass-media curation of signifiers. Even impressions that gesture towards a focus on sound, like "he's the sound of a valet brush whisking a tuxedo," evokes class, and particularly homonormative versions of upper-class masculinity represented by Short's visual appearance and upscale cabaret act rather than anything to do with Short's smooth baritone singing voice. "He's the last bluebird over the rainbow high-fiving Dorothy," a cartoonish image but one that economically gathers together several mainstream cultural signifiers evoking a particular gayness, though again, nothing specific about his voice or its *grain*. These readings also obscure Short's Blackness, echoing a theme of his career and of the longer racialized erasure of the "American Songbook" tradition in which he worked.

Actor Harvey Fierstein's distinctive voice, bearing traces of his native New York along with a gravelly quality caused by an overdeveloped vestibular fold in his vocal cords, inspired a range of responses limning awkward, even painful embodiment and queer masculinity. His voiced recalled "My bunions begging me to stop wearing stilettos," offered one. Some likened the voice of Fierstein, one of the few openly gay men on the list, to a retrograde transvestitism, complete with a seedy urban milieu like, "A 70-year-old, chain-smoking barfly named Margo." Perhaps because of the colliding codes of class, ethnicity, and sexuality that Fierstein's voice evokes, listeners embraced a wider field of metaphors, moving beyond nostalgic scenes of queer masculinity. Fierstein sounded like "blowing bubbles in the mud," and "a burning rose bush"—images that manage to evoke aesthetic paradoxes and abstract antitheses.

Women's voices come in for even more dramatic metaphors of queer embodiment in *Vocal Impressions*, perhaps in part because of the provocative selection of iconic women. In response to a brief audio file of Cher covering "It's in His Kiss," listeners were explicit in their hostility to a voice and a woman that seemed to violate codes of acceptable femininity. Cher sounded like "the aunt you liked much more when she was your uncle," and "The masculine voice Wayne Newton always wanted." A current of hostility runs through these reads, which not only attempt to "queer" Cher's voice, but to enact scenes of humiliation, disappointment, and discovery. "The pulse in your head when you figured out that the hottie you were told was a transvestite really is a woman after all."

What's most striking in these images is not the melancholy projected onto trans identity, but the strenuous assertion of straight masculinity from the listener-poets. "A vibrating mixture of molasses and testosterone," suggests, with some ambivalence, the power of sexual and racial ambiguity in Cher's voice. Sexuality is evoked in these impressions, but hostilely, as a form of distancing. "A middle-aged woman on a roller-coaster giving birth," directs the hostility toward Cher's femininity, but along a different axis of difference or non-conformity: her age. "An air raid siren warning of a dance party about to invade the city," harnesses ambivalence about her voice in a gentler, still humorous manner.

Like McConnachie's daughter, listeners understood that their ability to render voices socially and sensually was part of a valuable cultural skillset. One listener posted online that Totenberg was

> her absolute favorite, with her black-coffee-&-a-cigarette voice and those long, compound-complex sentences, nougat-rich in detail, sending out their tendrils, peppered with excerpts of conversation, each longer than the longest paragraphs of the mere mortals who share her airspace.[79]

This is of course praise that reflects back upon the giver of praise, her use of figurative language and her capacity for discrimination. By the 2000s, the names and voices of public radio hosts, anchors, and reporters became part of a repertoire of taste-making among a certain sector of the professional managerial class. This preoccupation took two distinct forms: enthusiastic identification with the names and voices—a kind of hyperbolic fandom—and a

---

[79] "Most Annoying NPR Voice," DataLounge, https://www.datalounge.com/thread/9224876-most-annoying-npr-voice.

withering criticism, bearing some of the misogyny and homophobia found in the *Vocal Impressions* project.

Together these popular and highly performative responses to public radio's voices demonstrated an untapped cultural surplus, an excess of signifiers circulating around a limited output of broadcast content. Public radio's curatorial impulse had turned on itself and found a treasure trove of iconic voices, distinctive names, and vocal performances that had become useful markers of class distinction. Loving Nina Totenberg's "black-coffee-and-cigarette voice" was itself part of finding a particular kind of voice of one's own. The self-conscious fandom took off just as Facebook, Twitter, Reddit, and Tumblr made curating a page, wall, feed, or blog both easy and compelling and for many young people, a key part of finding one's voice. Early web culture, along with the comments sections underneath online articles on mainstream journalism sites like *Slate*, intensified the performativity in fan cultures.

Articles curating the "exotic" and euphonious names of public radio voices began to appear in this period as well.[80] "I love the sound of Sylvia Poggioli when she says her name!" declared one fan. Another common object of affection was the name and voice of Ofeabia Quist-Arcton. "I just love writing/saying/conjuring her name," swooned another.[81] By 2009, NPR's *Pop Culture Happy Hour*, a weblog and proto-podcast with a curatorial mission, was circulating the "What's Your NPR Name" game: "Take the first letter of your middle name and insert it anywhere you'd like in your first name. And then your last name is the smallest foreign town you've ever visited."[82] An automatic NPR name generator was also in circulation. (I got "Deepak Overby-Adelstein").[83]

---

[80]Deirdre Mask, "Why Do NPR Reporters Have Such Great Names?" *The Atlantic*, May 16, 2013, https://www.theatlantic.com/national/archive/2013/05/why-do-npr-reporters-have-such-great-names/275493/;
https://www.fluther.com/55346/what-are-your-favorite-npr-hostscorrespondents-names/; https://ask.metafilter.com/119586/Why-All-the-Weird-Names-at-NPR.
[81]DataLounge.
[82]Linda Holmes, "Pop Culture Happy Hour," *NPR*, Blog, April 15, 2009, https://www.npr.org/2009/04/15/103124731/whats-your-npr-name.
[83]Mask, "Why Do NPR Reporters Have Such Great Names?"

Gayvoice Rising

Public preoccupation with the names and voices of NPR's diverse corps of international correspondents eventually developed into a Rorschach test for cultural competence in an era of globalization, immigration, and cosmopolitanism. At the same time, a different kind of distinction-making was at work circulating among those who saw in public radio's diverse names and voices signs of cultural decline. Not surprisingly, the critics zeroed in on race, gender, and sexuality as the markers of this decline. GOP speechwriter and rightwing AM talk radio host Tony Snow epitomized this attitude and the politics of racial grievance animating it in a 2006 interview. "One of the problems with NPR is that there is so much political correctness that if you've got a name that looks like it was made up by Rudyard Kipling, you've got a better chance of getting hired. I'm a white guy named Tony Snow for heaven's sake. That's as white as it goes."[84]

One doesn't have to be a postcolonial scholar to pick up on the hostility, distancing, and exoticization at work in this comment. Snow's implication of discrimination against white men with Anglo-Protestant names tells us much about his ideological position. Also, his cultural touchstone for recognizing "foreign-sounding" names is Rudyard Kipling, the nineteenth-century British novelist of imperial Britain with all the racism and condescension that it brings, which tells a great deal about the vast chasm of the US culture wars in which NPR's evolving sound was developing. For many, public radio was the sound of performative racial, gender, national, and sexual difference, at odds with its implied nationalizing mission, discussed in Chapters 1 and 2.

Public radio voices were vulnerable to cultural criticism from mainstream liberal circles as well, as in Sarah Vowell's 1997 comment that all NPR women sounded "so alike in their sober nasal condescension." The network's sound and feel, she opined, in a critique that pairs a gender and class critique, would benefit from "more dirt bikes."[85] By the turn of the century, the network faced complaints about its tight control over pronunciation, cadence

---

[84]Tony Snow quoted in *The Washington Post*, July 7, 2006.
[85]Vowell, *Radio On*.

and accent, especially for women and people of color. WBUR anchor Irene Doyle recalls, with incredulity, how her boss once informed her she was mispronouncing *her own name* on the air (the station wanted Doil instead of Doy-al).[86] In a 2005 *Nation* piece, Scott Sherman noted the widespread criticism, even among NPR staffers, about "the NPR drone," which he says is how "some staffers describe the network's overall sound." "If you listen to a lot of NPR," Brian Montopoli averred in *The Washington Monthly* in 2003, "you realize how similar it all sounds: no matter who is talking or what they are talking about." Sherman also quoted music critic Greil Marcus's 1998 complaint that the network "imparts a sense of boredom with the world." Even Stamberg said in a 2010 interview that one price of NPR's success was that listeners weren't "hearing great voices anymore."

NPR voices became culturally resonant at the same moment for sounding too boring and alike and at the same time, far too exotic. This tug-of-war over what public radio should sound like is an echo of the wider culture wars that have raged since the 1980s, in which mainstream liberal institutions have been buffeted from all sides for capitulating to a permissive political correctness, failing to more fully represent the nation's diversity and failing to appeal to younger, more fickle listeners. Cutting across political lines, public radio's sound, and its voices in particular, became the site of a battleground over something larger, a referendum on its status as a representative public institution. Not surprisingly, given what we know about the gendered origins of the public sphere, the public/private distinction, and the historical silencing of women and minorities in putatively democratic systems in the West, critics zeroed in on gender, race, and sexuality as markers of undesirable or illegitimate voices.

Groups sprung up on Reddit and other sites dedicated to discussions of "the most annoying voices on NPR." Together they provided a platform for the kind of performative distinctions that animated *Vocal Impressions*, albeit in a forum that encouraged and even celebrated trollish commentary. Not surprisingly, such trolling often bore out homonormative attitudes about masculine

---

[86] Mark Jerkowitz, "Jane Christo: The aloof, devoted, intense, insensitive, shy, tyrannical perfectionist behind WBUR," *The Boston Globe*, December 4, 1997, D1-D5.

vocal performance and sexist vitriol about women's voices on the airwaves. Glass was described as a "dweeb with gayvoice rising," on DataLounge, an online forum for discussion of news and celebrity gossip from a gay perspective.

On a thread that ran for *three years* dedicated to criticizing annoying public radio voices, starting in 2009, posters complained that "the Ira Glass affectations have become kind of standards for public radio wannabees."[87] Another poster sniped "Does Ari Shapiro have ovaries and a vagina? Sounds like it." Shapiro, *All Things Considered* host, was one of the few high-profile NPR staffers to have come out as gay. He was the subject of most of the criticisms aimed at men for "sounding too gay." The performative nature of celebrity gossip on DataLounge may have contributed to the irreverence and graphic nature of some of these criticisms.

Zoe Chace, longtime *TAL* and *Planet Money* reporter and host, came in for the worst of the criticisms aimed at the women of public radio, perhaps due to the "vocal fry" (what linguists call "creaky voice,") attributed to her. "Zoe Chace, the most offensive voice in all radio," cried one poster. On the r/NPR sub-Reddit from 2014, another listener couched their complaint about Chace in BDSM imagery: "I imagine her to be a master in an S&M situation, whipping her slave with a leather whip while reading her report."[88] Such elaborate fantasies joined aesthetic displeasure with fears of women's power. The alarm raised about "vocal fry" in the early 2010s, approached the level of a moral panic and focused particularly around the phenomenon in young college-educated women and in the vocal performances of women in public radio. It was explored on public radio programs at length as part of the medium's reflexive approach to sound and to its own uneasy relation to the public for which it stood.

Lexicon Valley (2012), NPR's health blog *Shots* (2014), *Fresh Air* (2015), *This American Life* (2015), and others explored the generational change represented by vocal fry while touching only

---

[87] "Most Annoying NPR Voice," DataLounge.

[88] Ashwinmudigonda, "Need to get this off my chest," *Reddit*, March 10, 2014, https://www.reddit.com/r/NPR/comments/202shv/need_to_get_this_off_my_chest/.

very lightly on the sexism inherent in much of the criticism.[89] This hostility was not confined to the anonymous posts on websites dedicated to snark. Public radio veteran-turned-podcaster Bob Garfield called vocal fry "repulsive" and "mindless." When challenged on this on feminist podcasts and blogs, Garfield doubled down, sarcastically declaring "I hate women" and asserting that the episode of *Lexicon Valley* in which he criticized vocal fry received a record 700,000 responses, most of whom "loathed" the sound, which he described as "a human record scratch."[90] Garfield, it should be noted, was fired from WNYC in 2021, for "incidents of bullying and harassment that were not reported to senior management out of fear of reprisals."[91]

I share this extravagantly trollish commentary to make clear the extent to which public radio's sound effects of liberalism, of feminism, and of openness to diversity cut both ways, not just as an alibi for the organization's creeping centrism politically and corporate mindset in its business practices. Featuring voices other than those of men with a traditional middle American broadcast baritone also provoked a growing politics of backlash and grievance bubbling up across the country and increasing in intensity and vulgarity in the years after Barack Obama's election, a period that followed the housing crisis of 2008 and overlapped with the Great Recession. The cultural politics involved in feeling liberal became increasingly vexed as public institutions found it difficult to evoke shared cultural meanings through sound.

---

[89]Mike Vuolo, "Get Your Creak On," Lexicon Valley, Podcast, *Slate*, https://www.slate.com/articles/podcasts/lexicon_valley/2013/01/lexicon_valley_on_creaky_voice_or_vocal_fry_in_young_american_women.html; Selena Simmons-Duffin, "Talking While Female," *Health Shots*, Blog, NPR, October 24, 2014, https://www.npr.org/sections/health-shots/2014/10/24/357584372/video-what-women-get-flak-for-when-they-talk; Terri Gross, "From Upspeak to Vocal Fry: Are We 'Policing' Young Women's Voices?," https://www.npr.org/2015/07/23/425608745/from-upspeak-to-vocal-fry-are-we-policing-young-womens-voices; Ira Glass, "Freedom Fries," *TAL*, episode #545, January 23, 2015, https://www.thisamericanlife.org/545/if-you-dont-have-anything-nice-to-say-say-it-in-all-caps.

[90]Bob Garfield, "'Old Fart' Responds to the Great Vocal Fry Outcry of 2013," *Slate*, January 11, 2013, https://slate.com/human-interest/2013/01/young-women-and-vocal-fry-slate-podcast-wars-continue.html.

[91]Katie Robertson and Ben Smith, "WNYC Fires Bob Garfield, Co-Host of 'On the Media'," *The New York Times*, May 17, 2021.

The seeming impossibility of representative voices to speak to and for a national public stands in stark contrast to the claim at the start of this chapter that public radio's sound had "prevailed." Public radio's sound prevailed as the dominant aesthetic of early podcasting, perhaps for the same reasons it was so contested on late 2000s radio: it didn't sound particularly public anymore, if it ever had. Instead, it sounded increasingly intimate, niche, and private. It was this private sounding, intimate feeling quality, evoking embodiments queer and otherwise, that seemed to be provoking irritation across the ideological spectrum. At the same time, this traffic in feelings moved so easily, precisely because of the public, or social, nature of the media of commentary. "I googled Terry Gross annoying and found this treasure!" exulted one DataLounge commenter. "I have read every single post, and I feel such deep cathartic relief."

The circulation of hateful associations linking women's and queer voices to grotesque embodiments and to threats to the body politic recalls Sarah Ahmed's affective economy, through which powerful affects, like hate, do not reside in or on a specific voice or network or sound. "Hate is economic," Ahmed reminds us. "It circulates between signifiers in relationships of difference or displacement."[92] Much of the vitriol online centers the affective impact of certain voices, while also making clear the pleasure of being part of the circulation of these affects. The frustration and anxiety produced by disembodied female and queer voices, resolve themselves in participation in the social movement of hate—the affective economy.

Conclusion

Rather than "escape the tyranny of meaning," public radio's approach to the sound of the human voice in the first decade or so of the century was engaged in a complicated project of negotiating an impossible contradiction: the public as the site for democratic access across historic boundaries of race, class, and gender *and* as a

---

[92]Sara Ahmed, "Affective Economies," *Social Text* 22, no. 2 (Summer 2004): 117–39.

site for the maintenance of the distinctions that re-impose some of those very boundaries.[93] Looked at from this perspective, the sixty-year history of modern public radio was not about an ideological shift to the right; rather, it was the working out of an inevitable tension in its founding mission. The 2025 defunding of public media seems, from this perspective, less an example of Donald Trump's often unconstitutional departures from precedents and norms and more of an historic inevitability.

The differences celebrated and policed in *TAL*'s stories on masculinity, on NPR's *Vocal Impressions*, and in the online discourse around public radio voices provide insights into the curators, metaphor-makers, and other cultural workers that form the network's habitus much more than they tell us about the voices themselves. Like a college education, NPR had become a credentialing institution for entry into the shrinking, anxious, and increasingly status-conscious professional managerial class. A curatorial approach to sound had become another handy tool for the aspirants to that class.

The question of voice (who speaks) is central to the struggle for democracy. In the United States, the neoliberal transition was powered by a politics of voice; but it was a politics not of democracy or even populism, but of their sound effects and the affects they put into circulation. In the neoliberal politics of the time, "voice," with its layers of evocative meanings (modern subjectivity, universal access, "active constructive participation") gives way to the affective terrain of "vocal performance." And the public airwaves, like the rest of the public spaces in the US, have become another site where distinctions and hierarchies masquerade as freedoms.

At the same time, voices carry their own affective power that exceed and disrupt attempts to delimit dominant intentions. Tiffe and Hoffmann, for example, have explored "feminized vocality" on podcasting in terms of resistance, "an immaterial element of the material body, that can take up space" in ways that resist dominant gender ideologies.[94] Chenjerai Kumanyika has likewise explored

---

[93]Barthes, *Image, Music, Text*, 185.
[94]Raechel Tiffe, and Melody Hoffmann, "Taking up Sonic Space: Feminized Vocality and Podcasting as Resistance," *Feminist Media Studies* 17, no. 1 (2016): 115–18, https://doi.org/10.1080/14680777.2017.1261464.

the power of voice on radio to convey identities that challenge the default whiteness of entrenched aesthetic, ideological, and stylistic performances associated with public radio and "the public" in general.[95]

In the next chapter, new challenges to public radio's precarious identity politics crystallize as podcasting finds in its intimate voices a powerful new affordance. Podcasters navigate the seemingly impossible challenge of speaking in public to an audience of private listeners in ways that echo and modify public radio's sound. The coming together of these two distinct media forms, in an overlapping aesthetic sensibility and a largely shared approach to vocal performance, provides ways to think about the evolving challenge of feeling liberal in an increasingly divided society with myriad channels of public and private discourse. It also provides insight into the relationship of broadcast media to internet-based media as the two commingle and struggle to (re)define themselves. Finally, the rise of nonfiction narrative podcasting, in the aesthetic image of public radio, tells us something about the evolution of empathy machines, the concept that media technologies can bring us together across vast divides in fellow feeling.

---

[95]Chenjerai Kumanyika, "Challenging the Whiteness of Public Radio," *All Things Considered*, NPR, January 29, 2015, https://www.npr.org/sections/codeswitch/2015/01/29/382437460/challenging-the-whiteness-of-public-radio.

CHAPTER SEVEN

Feeling Uncomfortable: The Politics and Aesthetics of Cringe

There's an awkward moment in the opening minutes of season 3 of *Serial* (2018) when host Sarah Koenig shares a scene from inside an elevator at the Justice Center, a cluster of towers in downtown Cleveland, Ohio comprising offices, a courthouse, and jail, whose manifold problems form the subject of the entire season. The elevator, stuffed with people from all walks of life, serves as a metaphor for an ideal American pluralistic democracy.

> Koenig: When I'm feeling optimistic, I appreciate that an elevator car in a government building is one of the few places left in our country where different kinds of people are forced into proximity. I like to think that we can all stand so close to one another, with our sensible heels, and Timberland boots, and American flag lapel pins, and fake eyelashes, and Axe cologne, and orthopedic inserts, and teardrop tattoos, and to-go coffees. And when the elevator doors open up, spilling us out onto our floor, the fact that no one is bloodied or even in tears, it's a small, pleasing reminder that we're all in this together.

It's a lovely passage, and one that captures, I think, an often merely implied fantasy that operates in our modern experience of public institutions like schools, libraries, courthouses, and of course, media organizations like WBEZ and *This American Life*, who developed *Serial* in 2014. In the next moment, Koenig's sentiment collapses under its own weight. "Other times, the shoulder-to-shoulder closeness only magnifies the obvious—we're not the same, not at all." She demonstrates this by sharing a self-deprecating bit of audio from an elevator ride in that same building. In the audio clip, she awkwardly attempts to bridge the racial divide by lamely joking to a fellow passenger, a young Black woman, that her portable speaker blaring hip-hop from her bag is "quite a soundtrack for the elevator." Motivated in part by her own keen discomfort that the white people in the elevator had been exchanging looks about the music, she now bows her head in embarrassment "to avoid the looks the black people are probably giving each other." She hasn't bridged the divide so much as called attention to it by "saying the lamest thing I possibly can." The vignette serves as an entrée into an important observation about the Justice Center, and by extension, the state of the justice system in the United States: "This place is primarily black and white."

> The majority of the courthouse staff is black. Clerks are mostly black. Most of their managers are white. In the sheriff's department, most of the security guards are black. Most of the deputies are white. Most of the attorneys are white. Almost all the county judges are white, and their bailiffs are white. Most of the defendants and crime victims are black.[1]

It also serves as a way for Koenig to acknowledge that the awkward perch from which she tells this story is not merely an obstacle but a necessary condition. Rather than simply "reporting" the story, Koenig's presence in the Justice Center elevator becomes an uncomfortable element of the story. Rather than plumbing human tragedy for universal truths, she lets us know from the start that in the Cleveland Justice Center, and in podcasting circa 2018, estrangement, rather than empathy, is built in.[2] More precisely,

---

[1] Sarah Koenig, "A Bar Fight Walks into the Justice Center," *Serial*, Season 3, Chicago Public Media, September 20, 2018, https://serialpodcast.org. Retrieved January 20, 2019 from https://serialpodcast.org/season-three/1/transcript.
[2] Koenig, "A Bar Fight Walks into the Justice Center."

scenes of public disappointment—embarrassing and awkward—had replaced moments of empathy as the narrative "surprise" built into the radio and podcast structure of feeling.

It's hard to think of a voice more closely associated with the "golden age" of nonfiction narrative podcasting than Koenig's.[3] A longtime producer at *TAL*, Koenig's role as the dogged, if ambivalent—some have said "obsessed"—reporter and host of *Serial*'s first season launched her into global celebrity in 2014.[4] Hers was the voice—lush and confiding, at times; briskly professional, at others—that, more than any other, demonstrated the contention that public radio's sound aesthetic had prevailed on the new medium.

Of course, *Serial* was the result of a collective effort, as Koenig herself has been at pains to acknowledge.[5] And it was more than her vocal performance at work in making the first season such a massive hit. *Serial* debuted only weeks after Apple's new iPhone was released with a built-in app for podcasts, removing some of the technical difficulties of access. Further, by focusing on a true-crime cold case, with a beautiful young victim, Hae Min Lee, buried in her cheerleading outfit, *Serial* was leaning into genres, tropes, and serialized release strategies whose popularity had already been tried and true across media and generations. Finally, public radio institutions like *TAL*, *Radiolab*, along with a handful of cult public radio shows like *Love+Radio*, had demonstrated by 2014 that podcast versions of radio content were reaching new and growing audiences.

Even so, Koenig along with producer Julie Snyder, and their boss Ira Glass, brought distinctive narrative structures and plotting, along with masterfully dexterous tonal shifts to *Serial* that made for irresistible listening, with some scholars hailing it as the start

---

[3]Tiziano Bonini, "The Second Age of Podcasting: Reframing Podcasting as a New Digital Mass Medium," *Quaderns Del CAC* 41, no. XVIII (2015): 21–30; Richard Berry, "A Golden Age of Podcasting? Evaluating Serial in the Context of Podcast Histories," *Journal of Radio & Audio Media* 22, no. 2 (2015): 170–8, https://10.1080/19376529.2015.1083363; Steven Goldstein, "The Three Eras of Podcasting," *Amplifi Media*, September 19, 2023, https://www.amplifimedia.com/blogstein-1/the-three-eras-of-podcasting.
[4]Neil Verma, *Narrative Podcasting in an Age of Obsession* (Ann Arbor: The University of Michigan Press, 2024).
[5]Sarah Koenig, Podcast Movement, Conference, Dallas, Texas, 2015.

of "a new human journalism."⁶ Perhaps most important for understanding the opening anecdote, and its role in this chapter, the *TAL* team brought to *Serial* a mastery of making "a scene"—to be specific (if redundant), a "public scene." *TAL,* with its penchant for surprise and magic, had long since mastered the art of the moment.

This chapter traces the evolution of the "public scene" as a trope in narrative nonfiction podcasts to understand how these story-based shows managed the contradictions inherent in feeling liberal at the heart of the public radio structure of feeling. Throughout the 2010s, podcasts that spun off from public radio shows (or that drew talent from public radio veterans) explored art, design, science, politics, and the small dramas of everyday life. But many of them also explored, often as an undercurrent, the public and its problems. By this I mean the conceptual and structural challenges that highly specialized and unequal societies present to democratic theory and practice, which was the theme of John Dewey's prescient 1927 book, *The Public and Its Problems.*⁷

Narrative nonfiction podcasts also used public scenes to explore their own relationship to the competing notions of the public associated with its ancestral broadcasting medium and with kindred web-based communities. Finally, the entire audio industry, including public radio stations and for-profit networks, found themselves in the middle of uncomfortable and unwanted public scenes during a decade in which "reckonings" with entrenched racism, sexism, and hostile workplaces intersected with national crises and flashpoints. The public scene proved an exceptionally versatile conceit for representing and navigating a period of tumult, in part because of how deftly it centered the experience of disappointment in public institutions. Thanks to Sara Ahmed, we know how affects move across the surface of objects and subjects, rather than residing in them, gathering intensity through association and circulation. Podcast stories featuring these moments of public disappointment, humorous or tragic, generated moments of cringe, producing

---

⁶Martin Spinelli and Lance Dann, *Podcasting: The New Audio Media Revolution* (London Bloomsbury, 2019), 188.
⁷John Dewey, *The Public and Its Problems* (New York: Henry Holt, 1927).

currents of rueful empathy and a bit of *schadenfreude* for the storytellers and for podcasting itself.

Koenig's public scene in the elevator of the Justice Center was a canny bit of authorial framing from a podcaster famous for constructing a fifteen-year-old murder investigation around her own shifting perspective in *Serial*'s first season. Koenig introduces the new season as, in some ways, a corrective to seasons one and two which focused on exploring the guilt or innocence of individual men (Adnan Syed and Bowe Bergdahl, respectively) regarding single criminal acts. To understand the criminal justice system, one ought not to extrapolate from a single extraordinary case, Koenig cautions. One ought instead to tell the stories of ordinary criminal cases, 98 percent of which are pleaded out before trial.

In a narrative departure for *Serial*, Koenig sets out to tell the story of structures, rather than merely structure a story. The first character she introduces is the Justice Center, a name we'll quickly come to understand as ironic, if not Orwellian. She describes the structure as "hideous but practical." Roughly speaking, the building functions like most hierarchies—vertically. In this case, from the bowels up.

> The main court tower is 26 stories high, so the elevator really runs the place. If a person's arrested in Cleveland, they're coming into the Justice Center from the basement. Weary cops escort suspects from the underground parking garage. They get booked, go up a few floors to the jail.

What's novel here is the assertion, from the start, that structures (social, political, and physical) are a necessary context in which to understand audio stories. This represents a departure from the US public radio structure of feeling, in which intimate voices and unique stories provide the warp and weft for the universal fabric of human experience.

If the message of *TAL* in its first fifteen years could be pared down to a single motto, it would be this: "See? We all feel." *Serial* season three weaves together stories from across a year of reporting in Cleveland to find a different moral. Even as it does so, Koenig gamely reminds herself and her listeners that the impulse to find common ground, tempting though it may be, is a fool's errand. In the final episode, Koenig summarizes thus: "let's all accept that something's gone wrong. Let's make that our premise." This may sound tame, but it marks a significant journey for the US public radio structure of feeling, which has preferred the complacency implied in the ethos of empathy as the goal of narrative. US incarceration rates are "wildly out of whack and unprecedented in our history," she continues. And "every joint in the skeleton of our criminal justice system is greased by racial discrimination," a line that joins structural analysis to the embodied nature of oppression with impressive economy. Balancing dramatic tension with the codes of journalistic objectivity often leads nonfiction audio work to conclude with mealy-mouthed equivocation, or worse, with a Rashomon-like shrug at the infinite varieties of perspective.

The shift to a more openly political approach to storytelling, and to a more structural approach to politics, did not happen all at once. *Serial* one and two, in their focus on single crimes and their imbrication into traditional and even stereotypical "scenes" of sociality, explored complexity at the microscopic level, a tacit assumption, perhaps, that the macroscopic conditions are neutral, unexceptional, or simply not accessible to journalistic exploration and critique.[8] It is not an inherently conservative move, though its use in police procedurals and true crime narratives traditionally have allowed deeply conservative themes to fill the void where political and structural analysis could have gone.[9]

For instance, season one's focus on the murder of a teenaged girl and subsequent conviction of an allegedly jealous ex-boyfriend mobilized a whole set of tropes in which the violence against young women gets sensationalized in fictional and nonfictional

---

[8]Susan Douglas, "The Turn Within: The Irony of Technology in a Globalized World," *American Quarterly* 58, no. 3 (2006): 619–38.

[9]Tanya Horeck, *Justice on Demand: True Crime in the Digital Streaming Era* (Detroit, MI: Wayne State University Press, 2019); Stella Bruzzi, "Making a Genre: The Case of the Contemporary True Crime Documentary," *Law and Humanities* 10, no. 2 (2016): 249–80; Mats Hyvönen, Maria Karlsson and Madeleine Eriksson, "The Politics of True Crime: Vulnerability and Documentaries on Murder in Swedish Public Service Radio's *P3 Documentary*," in *Vulnerability in Scandinavian Art and Culture*, ed. Adriana Margareta Dancus, Mats Hyvönen, and Maria Karlsson (New York: Palgrave Macmillan, 2020).

treatments.[10] Likewise, the story of Bowe Bergdahl, in its empathy for his ordeal as a POW, balanced with empathy for his comrades whose search for him put them in grave peril, inevitably obscures a critical focus on US foreign policy and leaves little room for empathy for the tens of thousands of Afghan civilians killed during the US invasion and occupation.

Serial's 2018 decision to investigate the criminal justice system represents a departure and an acknowledgement that small stories about empathy for strangers were no longer sufficient. Such a shift can seem, in retrospect, inevitable, given the congenital paradoxes at work in the public radio structure of feeling in which empathy for some was produced in stories told by others. It can be understood, narrowly, as a direct response to the shocks of the Trump election in 2016, the Charlottesville hate rally of 2017, the #MeToo and #BlackLivesMatter movements, and the well-documented political fragmentation and polarization encouraged by social media sites and their engagement algorithms. In addition to its significance as a measure of white male grievance against the gradual advances of liberal pluralism, Trump's election, in particular, mobilized and intensified a hostility to journalism from the right. From the left and center, mainstream journalism seemed feckless and even complicit in Trump's 2016 win (perhaps even more so in 2024).[11]

Audio journalism at NPR, *TAL*, and in the growing field of politics podcasts across commercial and public networks, responded with searching explorations of the electorate, which suddenly seemed inscrutable, and of its own assumptions, some

---

[10]Neroli Price, "Can True-Crime Podcasts Make Structural Violence Audible?" in *The Routledge Companion to Radio and Podcast Studies*, eds. Mia Lindgren and Jason Loviglio (London and New York: Routledge, 2022), 358–67; Ryan Engley, "The Impossible Ethics of Serial: Sarah Koenig, Foucault, Lacan," in *The Serial Podcast and Storytelling in the Digital Age*, ed. Ellen McCracken (London and New York: Routledge, 2019), 87–100.

[11]Neal Gabler, "Five Ways the Media Bungled the Election," *Columbia Journalism Review*, January 24, 2017, https://www.cjr.org/criticism/media_election_trump_fail. php; Duncan J. Watts and David M. Rothschild, "Don't Blame the Election on fake news. Blame it on the media," *Columbia Journalism Review*, 2017, https://www.cjr. org/analysis/fake-news-media-election-trump.php.

of which now appeared blinkered and complacent.[12] The role of narrative surprise had shifted from delightful epiphany to moments of performative cluelessness and uncomfortable self-analysis. In addition to these contexts, the audio nonfiction industry was in the midst of a painful and shocking set of scandals, calling into question its long reputation for empathy and social consciousness, even among supporters. By 2018, in other words, Serial Productions had lots of reasons to foreground politics and the political aspects of podcasting itself.

This web of interrelated contexts provides some background for understanding the approach to investigative journalism in podcasts like *Serial*'s third season. It was part of a wave of hard-hitting investigative audio journalism that took on systems of criminal justice, incarceration, health care, and education with an attention to character, plotting, and ethical reflexivity, including Madeleine Baran's *In the Dark* (American Public Media, 2016–20), Al Letson's *Reveal* (Center for Investigative Reporting/PRX 2013–), and *Stolen: Surviving Saint Michaels* (Gimlet, 2021–2). These and other podcasts won prestigious awards and inspired conferences, review journals, university classes, and other podcasts.

They also provide some context for the gold rush in podcasting that followed. Podcasts that were evaluated, celebrated, and marketed as empathy machines very quickly became targets of investors seeking profits, exacerbating the tension between market logic and public service much less subtly than public radio's funding model had. After public radio talent was gobbled up by small for-profit networks, platforms like *The New York Times* and Spotify acquired these small networks, transforming journalism and storytelling into "content" and the public first into listeners, then into valuable marketing data. What followed was a predictable spasm of cuts, cancellations, and contractions rationalized in the language of efficiencies, exacerbated by the pandemic, and

---

[12] Corey Hutchins, "NPR's Kirk Siegler on covering America's Urban-Rural Divide," *Columbia Journalism Review*, March 30, 2017, https://www.cjr.org/united_states_project/qa-nprs-kirk-siegler-on-covering-americas-urban-rural-divide.php; Pete Vernon, "9 Podcasts Keeping up with Trump," *Columbia Journalism Review*, June 15, 2017, https://www.cjr.org/politics/donald-trump-podcasts.php.

immediately recognizable as an extension of Cory Doctorow's notion of "enshittification."[13]

Tuning into the current of discomfort and disappointment rippling through the texts and contexts of longform nonfiction podcasts in this period provides insights into bigger questions about the fate of "feeling liberal" at the twilight of the neoliberal era.[14] One of the arguments of this book has been the role of public radio's approach to empathy in providing an "affective education" in a particular form of liberalism, in which educated *haves* stage scenes of empathy on behalf of the *have-nots*. Listening in to this kind of audio narrative and feeling empathy, I have tried to argue, became part of the cultural work of the highly educated, socially conscious Americans that comprised public radio's imagined audience. As the emotional and material forms of comfort promised by this brand of liberalism proved increasingly unsustainable and unsatisfactory, and as the makeup of the imagined audience proved more various, public radio and its podcast progeny had to navigate an increasingly complicated set of contradictions.

This chapter explores these challenges and tries to discern the outlines of new emotional and political currents in the audio storytelling as podcasting enters its third decade and US public radio enters its seventh. Once again, so much of this story cannot be told without *TAL* and its spinoffs. Podcasters proved unusually adept at staging scenes of public disappointment and discomfort in part because so many of them had learned how to stage scenes of ambivalence about liberal feelings on *TAL*. If Richard and Rudnyckyj are right, and affect is the medium in which subjects are formed, then the longform audio narratives of the last fraught decade and the messy scenes they depict may tell us a bit about the kind of listening subjectivities that are emergent.[15] And we may also

---

[13] Cory Doctorow, "Social Quitting," *Medium*, November 15, 2022, https://doctorow.medium.com/social-quitting-1ce85b67b456.

[14] Louis Menand, "The Rise and Fall of Neoliberalism," *The New Yorker*, July 17, 2023, https://www.newyorker.com/magazine/2023/07/24/the-rise-and-fall-of-neoliberalism; "America's 'Neoliberal' Consensus Might Finally Be Dead," *The New York Times*, May 25, 2023, https://www.nytimes.com/2023/05/25/opinion/neoliberal-consensus-china-trade.html.

[15] Analiese Richard and Daromir Rudnyckyj, "Economies of Affect," *Journal of the Royal Anthropological Institute (N.S.)* 15, no. 1 (2009): 55–77. © Royal Anthropological Institute.

glean some of the "specifically affective elements of consciousness," providing insight into the emerging political and social structures of our young century, which as of this writing remain very much "still in solution."[16]

Telling Stories Out of School

When the definitive history of *TAL* is written, it will no doubt focus on its decades-long interest in the state of public education in the United States and the terrible impact of racism and inequality in the life chances of students unlucky enough to be born poor and Black or Brown in poorly funded school districts. Taken together, the dozen or so episodes on schools, produced over thirty years, provides a master class in the empathy machine approach to nonfiction storytelling in its least ambivalent and most well-intentioned form. Perhaps because the condescension of adults to children partially occludes other dynamics in the empathy machine model, *TAL*'s explorations of inequality in schools have been unusually successful. The exploration of daily life and an epidemic of gun violence at Harper High School in Chicago (2013) has been praised by the Peabody Award for "masterfully localizing a national crisis in a vivid, unblinking, poignant, and sometimes gut-wrenching manner."[17]

From the start, there has been an element of tragic irony built into these stories about school reform and their seemingly inevitable collapse. The collaboration with *New York Times* journalist, Nicole Hannah-Jones, "The Problem We All Live With" (2015) won a George Polk Award for its deeply reported story on the re-segregation of US public schools in the decades after the *Brown v Board of Education* ruling ended de jure segregation.[18] "Two Steps Back" (2004) explored the rapid decline of a Chicago public school only months after a round of promising reforms.[19] "Is

---

[16] Raymond Williams, *Marxism and Literature* (Oxford: Oxford University Press, 1977), 172.
[17] "Harper High School: WBEZ Chicago's *This American Life*," Peabody, https://peabodyawards.com/award-profile/harper-high-school-wbez-chicago-91-5/.
[18] Ira Glass, "The Problem We All Live With, parts 1 and 2," *TAL*, episodes #562 and #563, July 21, 2015.
[19] Ira Glass, "Two Steps Back," *TAL*, episode #275, October 15, 2004.

This Working?" (2014) explores the unintended consequences of disciplinary innovations like restorative justice in public schools.[20]

Like "Is This Working?" "Three Miles" (2015) features Joffe-Walt exploring another educational reform fiasco.[21] Another Peabody winner, "Three Miles," tells a very small story of the disastrous emotional impact of a well-intentioned "exchange program" in which students from one of the country's poorest schools in the Bronx visit one of the nation's most elite schools, a private one, only three miles away. Making listeners empathize with poor Black and Brown Americans was likely an easier trick to pull off when the subjects of empathy were children. US schools also provided an unusual site for revealing the shockingly structural nature of inequalities of opportunity, making it an ideal "meta trope for liberal journalism," per Neil Verma.[22]

Such stories, in *TAL*'s hands, captured the ideological ambivalence at the heart of the public radio structure of feeling better than stories on any other topic. On the one hand, they seem to reprise an essentially conservative argument about the ironic futility of liberal reforms and high-minded, but foolish, "do-gooders," which make up so many of the stories analyzed in Chapters 3 and 4. On the other hand, they also represented the most artful and compelling version of the liberal approach to audio journalism, in their humanizing of forgotten Americans. *TAL*'s embedding in Harper High for five months, for instance, provides a deep and intimate exposure to the conditions of poverty and violence in a racially segregated public school, harkening back to the best and most vital muckraking style of journalism animating historic progressive reforms.[23]

By 2020, when longtime *TAL* producer Chana Joffe-Walt explored the NYC public school system as a reporter and parent for *Serial*'s "Nice White Parents," it was a return to very familiar subjects and themes. It was also a fitting topic for *Serial*'s first podcast since its acquisition by *The New York Times*. Joffe-Walt and Koenig continued to produce *Serial* shows, but now with

---

[20] Ira Glass, "Is this Working?" *TAL*, episode #538, October 17, 2014.
[21] Ira Glass, "Three Miles," *TAL*, episode #550, March 13, 2015.
[22] Personal communication, June 8, 2021.
[23] See Jack Mitchell on the Progressive spirit in public radio journalism, Jack Mitchell, *Listener Supported: The Culture and History of Public Radio* (Westport, CT: Praeger, 2005), 3–10.

the vast research and circulation resources and legacy media imprimatur of the *Times*. *TAL*, for its part, got another tendril of its content production out of the public radio ecosystem and into the commercial market, part of a long-term process of franchising we explored in Chapters 4 and 5. Joffe-Walt's approach to NYC schools, informed by her role as a parent of school-aged children, proved an ideal vessel for narrative ambivalence.

The season's title, "Nice White Parents," telegraphed its operating assumption: that the default empathy of people like Joffe-Walt for the Brown and Black denizens of New York City's public school system merits scrutiny. The season represented in some ways a capstone for the decades-long project of *TAL*'s exploration of inequality in US public education. But it also serves as a corrective to that project in its focus on the disastrous consequences of the white liberal empathy that was its quarry. It is a joke that *Serial* is telling on itself, one that eschews easy answers for hard questions and heavy realizations. Public scenes provide dramatic moments and comic relief to leaven this heaviness, and "Nice White Parents" stages a lot of them, each more awkward than the last.

Public scenes of the kind Koenig created in the courthouse elevator in Cleveland, like Driveway Moments, remind us of the affordances of sound-and-time-based media: sonic snapshots that capture the affective impact of being in a specific place and time. Unlike Driveway Moments, *Serial*'s public scenes don't speak to the pleasures of stolen moments of live listening. They're moments of regret, frozen on tape, and returned to as to the scene of a crime. They're distillations of conflict that open out into the world outside the podcast, confessions designed to seduce listeners into empathy for storytellers.

"Nice White Parents" is a five-episode podcast that begins with Joffe-Walt's own search, as a white parents of means, in the New York City borough of Brooklyn to find a public school for her own children. Her way into the story of NYC's tortured relationship to diversity in public education is to recognize herself as a part of the historical bloc implicated in the city's bad-faith relationship to equity in schooling. Joffe-Walt developed her informal narrative voice during her years on the staff of *TAL* and *Planet Money*. Like Koenig, she places a public building at the center of this narrative; instead of a courthouse, it's a middle/high school in Brooklyn, known in various eras as the School for International Studies, IS #293, and The Nathan Hale Middle School, among other incarnations. Instead of looking at a range of stories that take place within one building over the course of a year, this is the sixty-year story of "an utterly ordinary, squat, three-story New York City public school building," and the many schools it has housed; each one, it turns out, shaped in ironic ways by white parents' ambivalent relationship to racial diversity.

The season's first public scene takes place in the school's library during a Parent-Teacher Association (PTA) meeting. Joffe-Walt marvels, for a brief doomed moment, at the diversity and democratic spirit of the gathering. The group of parents, sitting "around a wooden table in a library in a public school," evokes the same kind of democratic hopes Koenig felt in the elevator (i.e., a coming together in common purpose of people from disparate backgrounds). "People with vacation homes in Sonoma County and people who live in public housing," as she puts it; a class distinction that enfolds the larger racial gap that is the focus of the season. She soon learns that there is little common purpose or even mutual understanding around that table.

The scene's central drama is the revelation that the school has two competing parent groups working at cross purposes: one group, newly arrived at the school, of wealthy white parents raising big bucks for a French language immersion program; and the original PTA, comprised of mostly working-class Black, Latino, and Middle Eastern parents whose children had been attending the school for years. Much like Koenig's awkward elevator ride with everyone suddenly acutely aware of racial difference, this scene ends with two groups of people shooting knowing looks amidst awkward silences.[24] Before one of the parents asks Joffe-Walt to turn off her recorder, we hear the surprise ripple through the room when the mostly Black and Brown parents on the PTA hear that a new parent, a white dad named Rob who works as a fundraiser, has taken it upon himself to raise $50,000, earmarked for the French program that none of the PTA officers know much about. Maurice, another one of the PTA officers, still clinging to the tone of civility

---

[24]Parts of this analysis first appeared in Jason Loviglio, "Nice White Parents and the Phantom Public School," *RadioDoc Review* 7, no. 1 (2021), doi: https://doi.org/10.14453/rdr.81.

and decorum, hazards a gentle observation to try to unfreeze the conversation and deal with the growing tension: "I think a lot of us feel that there's two different groups." Another parent adds, "it's easy to feel steamrolled." With building frustration, Maurice continues, "what about the rest of your school? Where's all this money going? We have no answer. We don't know."

Joffe-Walt has an ear for difficult scenes, and her microphone has already captured cringe-inducing comments like when Chris, another new white parent, says, "I love diversity." There's an *Alice in Wonderland* quality to the upside-down world she explores, in part because she captures people at their least self-conscious. "Diversity" for the white parents, she explains, means Black and Brown kids at the school, which they feel is culturally enriching for their own kids. For the Black and Brown kids and parents Joffe-Walt spoke to, "diversity" meant something else: here come the white kids and "gentrification," a word that nobody uses, but it's clear that the new families feel like a threat.

Democracy represented another topsy-turvy concept. For Rob, the democratic process of a PTA meeting poses a problem for him as a fundraiser, who needs to honor "donor intent." The principal, Jillian Juman, is "thrilled and relieved" at the influx in white parent interest in the school. PTA president Imee Hernandez is blunt: the money of "white people with high socioeconomic backgrounds" scares her. Joffe-Walt whisks us from the PTA meeting to another scene, this one literally on stage, where the children are rehearsing a play they've written in French as part of the new immersion initiative. A girl who is fluent in French corrects a girl named Maya for making a mistake in delivering her line. "Language learning is not new to Maya," Joffe-Walt tells us in a voiceover. We hear a tape of Maya explaining to the principal "my dad speaks Arabic and my mom is Turkish … so it gets confusing." The principal is cheerfully brisk: "And now you're learning French!"

The episode ends with another excruciating scene at the French Cultural Services building, a gorgeous Italian Renaissance-style "palazzo" in Manhattan with a "huge marble staircase" where the wealthy white parents have staged a lavish gala fundraiser for the school's French program. The working-class PTA members have showed up gamely for a night of condescension, opulence, and "seventeen different cheeses." Joffe-Walt intersperses audio recordings of the noisy event with her own sardonic narration with deft comic timing. The audio has been edited for maximum cringiness, and profound cognitive dissonance. At one point, a wealthy Manhattanite donor, her voice straining above the noise of the cocktail party, lectures one of the Hispanic parents (listening with a mixture of forbearance and incredulity, in Joffe-Walt's narration) on the importance of knowing a second language, urging her to visit France in the spring, her "*saison preferée.*" The scene ends with Susan, the PTA co-president, sitting dumbstruck on a bench. "It's just hard to explain how this is a public-school fundraiser," she says.

The 2010s was a decade in which the staging of uncomfortable scenes took on a new kind of public centrality across popular culture, particularly for characters and celebrities who represented marginalized groups. Issa Rae's web series *Awkward Black Girl* debuted in 2011, and quickly went viral, landing her a Netflix show, *Insecure*. A similar kind of humor, featuring scenes of comic humiliation, was central to Ilana Glazer and Abbi Jacobsen's *Broad City* (2009–11) web series that was picked up by Comedy Central (2014–19). Donald Glover's *Atlanta* (FX, 2016–22) staged humorous scenes of social discomfort within a landscape of muted rage and alienation. Shows in the "cringe core" category can be seen as descendants of Larry David's *Curb Your Enthusiasm*, which debuted in 2000, and also gained popular and critical acclaim and cultural relevance throughout the 2010s, perhaps buoyed by youth-oriented shows dedicated to a new generation of awkwardness.[25]

In this decade, "cringe" replaced "cringeworthy" and migrated from verb to noun and adjective as a way to explain not just behavior but the larger, shared affective environment in which awkward and creepy feelings circulated, joining subjects and subjectivities in a shared structure of feeling.[26] Understood as the experience of "vicarious social pain," this "odd emotion," has been studied by social psychology for its alignment with "empathy pathways."[27]

---

[25] Selome Hailu, "Curb Your Enthusiasm Season Finale Hits Season High of 1.1 Million Viewers," *Variety*, April 10, 2024, https://variety.com/2024/tv/news/curb-your-enthusiasm-series-finale-ratings-viewers-1235966425/.

[26] Rebecca Jennings, "When ordinary people go viral, where's the line between comedy and cruelty?" *Vox*, November 20, 2020, https://www.vox.com/the-goods/21575707/tiktok-cringe; "The Makings of Cringe Making," *The New York Times*, August 27, 2023, https://www.nytimes.com/2023/08/27/insider/the-makings-of-cringe-making.html.

[27] S. Jesus, A. Costa, G. Simões, G. Dias Dos Santos, J. Alcafache, and P. Garrido, "THAT'S SO CRINGE: Exploring the Concept of Cringe or Vicarious Embarrassment and Social Pain," *European Psychiatry* 65, no. 1 (2002): S669–S670, doi:10.1192/j.eurpsy.2022.1722.

Unlike cringe-inducing forebears *Curb Your Enthusiasm* and *The Office* (NBC, 2005–13), the heroes of these awkward scenes were not exclusively white men; and in some cases, the consequences of disastrous scenes lingered on after the end of single episodes, driving main characters further into absurdity and isolation, as in *Atlanta*, a show that was not afraid of getting into the darkly absurd side of public alienation.[28]

The grammatical structure of online memes in the late 2000s/ early 2010s depended upon the irreconcilability of expectations and realities. Memes worked like self-deflating balloons with the top line of text undermined by the text beneath an image, as in the durable "expectation vs reality" format that has organized meme grammar starting in the late 2000s.[29] In affective terms, the look of performative anger and disappointment on the face of the girlfriend in the ubiquitous "distracted boyfriend" meme serves as an exemplar for the cultural moment. Its ubiquity spoke to the endless variations on similar moments of mortification, or "public failures."[30] Such scenes of discomfort eschewed epiphanies and narrative closure for lingering unease—part of a larger cultural shift away from empathy rituals and towards something that increasingly became referred to, with varying degrees of seriousness, as "reckoning."

Reckonings

The cultural politics of the decade seemed to require an acknowledgement of disappointment, or ironic reversal, as a keystone affect. The juxtaposition of soaring expectations and brutal realities built into internet meme culture seemed, in retrospect, an apt anticipation of the shift from Obama's stylized politics of Hope into the brutal "American Carnage" of Trump's 2017 inaugural address. The political schism that Trump's election represented

---

[28]Lea Palmieri, "Atlanta is Reinventing the Cringe Comedy," *Decider*, Blog, March 16, 2018, https://decider.com/2018/03/16/atlanta-reinventing-cringe-comedy/.
[29]Alli Hayes, "Expectation vs. Reality: How '500 Days of Summer' inspired a reaction meme that perfectly encapsulates disappointment," *Daily Dot*, Blog, March 29, 2024, https://www.dailydot.com/pop-culture/expectation-vs-reality-meme-explained/.
[30]Jesus et al., "THAT'S SO CRINGE," S669–S670.

was paired with a seemingly bipartisan agreement on the rhetoric of crisis to explain American life. Strategists like Steve Bannon counseled Trump to dominate news cycles through the affective power of the shock doctrine, piling up scene upon scene, crisis upon crisis.[31] The opposition to draconian and racist immigration policies and incompetent, unqualified cabinet and judicial nominations also relied on popular mobilizations steeped in the rhetoric of crisis and a politics of defiant empathy.

NPR seemed preoccupied with understanding the perspectives of the Trump voters who had taken the rest of us by surprise. In 2018, Dave Isay, founder of StoryCorps, started an initiative called "One Small Step," modeled on the StoryCorps formula of an intimate, edited conversation between two loved ones, substituting two strangers from opposed political orientations placed in a recording booth for a moment of shared understanding "to remind the country of the humanity in all of us."[32] Empathy became an even more explicit strategy and methodology for feeling liberal.

Of course, Trumpism was animated by its own emotional politics. Contempt for liberal feelings did not require stoicism, but instead a whole set of emotionally charged affinities and narratives. Sara Ahmed has modeled the economic circulation of affects like hate through emotionally charged language used to describe immigrants, for example. In fact, it was on the terrain of affect that public radio sought to understand the mysterious Trump voter in 2017 in any number of now-parodied interviews in proverbial Heartland diners with ostensibly average Americans, exploring the nature of their economic and cultural grievances.[33]

---

[31]Michael Wolff, *Fire and Fury: Inside the Trump White House* (New York: Henry Holt and Co., 2018), 88–98.

[32]https://storycorps.org/discover/onesmallstep.

[33]See, for example, Ari Shapiro, "In Rural N.C., Trump Supporters Eagerly Await a Different Kind of Change," *All Things Considered*, January 17, 2017, https://www.npr.org/2017/01/17/510301308/in-rural-n-c-trump-supporters-eagerly-await-a-different-kind-of-change, and criticism of this approach, SemDem, "NPR's profile of one undecided voter might be the dumbest story you'll read today," *DailyKos*, Blog, March 9, 2024, https://www.dailykos.com/stories/2024/3/9/2228401/-The-dumbest-news-story-you-ll-read-is-NPR-s-profile-of-one-undecided-voter; Erin Anderson, *Cement City*, Independent Podcast, episode 1. 11 min. 20 seconds, https://soundcloud.com/erinand/cement-city-ep1/s-Qd76vEU9Xra?in=erinand/sets/cement-city/s-GqBdHFHzAus&si=e7b1019e6de4434e968c40cdf149558e&utm_source=clipboard&utm_medium=text&utm_campaign=social_sharing.

Uncomfortable Entanglements

If public scenes of local governance and the awkwardness and anxiety they occasion became irresistible to nonfiction podcasters in the years after Trump's election, so did debates about the role of science in understanding our world. This trend only intensified in 2020, when the science of the novel coronavirus and the remarkably rapid production of vaccinations created new rifts, supercharged old conspiracy theories, and gave rise to new ones. Perhaps no podcast was better suited to engage in this difficult moment than *Invisibilia*, the NPR radio show-turned podcast that, more than any other show, seemed dedicated to exploring the science of empathy.

It was also one of the most successful radio-turned-podcast shows and was heavily influenced by *TAL*'s tonal warmth, playful banter, and narrative plotting, stringing bits of anecdote together suspensefully and paying off with grand statements about humanity. It became an influential show itself, a departure from and extension of *TAL*, and an early successful model for podcasts featuring women hosts. Introduced as "a show about all of the invisible things that shape human behavior: our assumptions, beliefs, emotions," it paved the way for podcasts like *Hidden Brain* (NPR, 2015–), *You're Wrong About* (Independent, 2018–21), and *Revisionist Histories* (Pushkin, 2016–) that offered entertaining, erudite, and chatty correctives to conventional wisdom with an emphasis on understanding the unseen. In this, it owes something perhaps to *99 Percent Invisible* (2010, Radiotopia, SiriusXM).

Alix Spiegel, a founding producer of *TAL* and Lulu Miller, a former producer for *Radiolab*, joined forces to tell long-form stories that drew from their previous shows' emphases on narrative journalism and scientific inquiry. After the first season, Hanna Rosin, from *Slate*'s Double X blog and *The Atlantic*, joined the team, replacing Lulu Miller. One of the show's hallmarks was its investigation into extremes of human experience with an eye on yielding universal human truths. It was explicit that "empathy is the way" from the very beginning. It drew on enthusiasm for science from *Radiolab*, as well as its expressionistic sound design.[34] This

---

[34] Spinelli and Dann, *Podcasting*, 17.

approach to sound leaned into the hypermediated style, which drew attention to the complicating and limiting factors of the show's curiosity and perspective. From *TAL*, it took the formula of anecdote-reflection, building theories of human universals from tiny case studies of one, in the service of building empathy via scientific inquiry. *The New Yorker* found the show "reassuring" in the way it trafficked in "ideas we are happy to believe in," particularly those that point to ways all of us are invisibly linked.[35]

This approach is similar to that of women-hosted shows like the memoiristic *No Feeling is Final* (Australian Broadcasting Company, 2018), a Third-Coast International Audio Festival winning podcast hosted by a young Australian woman diagnosed with "too many feelings—about four times as many as the average person."[36] Lea Thau's extraordinary and mostly independently produced *Strangers* podcast (KCRW/Radiotopia /StoryCenteral, 2012–) predated *Invisibilia* and set the tone for podcasts centering the intensely personal landscape of the podcaster's emotional life. It took a few years for the bravery and emotional impact of *Strangers* to be recognized, perhaps a belated acknowledgement that the power of podcasting's intimacy was something that women podcasters might know a thing about.[37] By then, even Thau, who had started at *The Moth*, had shifted to the political, staging her own series of awkward scenes in one-on-one encounters with Trump supporters in 2017.[38] Here too, however, the goal was not conflict but "the beauty and limits of human connection," the kind of radical empathy that *Invisibilia* had made mainstream.

---

[35] Sarah Larson, "Invisibilia and the Evolving Art of Radio," *The New Yorker*, January 21, 2015, https://www.newyorker.com/culture/sarah-larson/Invisibilia-evolving-art-radio.

[36] Britta Jorgensen, "The Feelings Frontier," *RadioDoc Review*, December, 2019: 2, https://ro.uow.edu.au/cgi/viewcontent.cgi?article=1089&context=rdr.

[37] Mike Roe, "Strangers wins the 2015 KPCC Public Radio Bracket Madness," *Southern California Public Radio*, April 15, 2015, https://laist.com/news/kpcc-archive/strangers-wins-the-2015-kpcc-public-radio-bracket; David Haglund and Rebecca Onion, "The 25 Best Podcast Episodes Ever," *Slate*, December 14, 2014, https://www.slate.com/articles/arts/ten_years_in_your_ears/2014/12/best_podcast_episodes_ever_the_25_best_from_serial_to_the_ricky_gervais.single.html.

[38] https://www.storycentral.org/strangers/; Stacey Leasca, "A Podcaster Reveals a New Side of Trump Supporters in America," *Good*, May 30, 2017, https://www.good.is/articles/trump-supporters-podcast-strangers.

One of *Invisibilia*'s earliest shows, "Entanglement," began with an account of a quantum physics experiment in which two atoms, separated by several feet, are made to move identically when one is pelted with photons. It's a scientific phenomenon that host Lulu Miller is quick to extrapolate from: "Like, there could be one particle of you right now entangled with a person that you just passed on the street." Quantum entanglement, we learn, has been theorized to be occurring naturally to atoms all around us, "the ways we are all invisibly connected to each other." Next, we meet a woman named Amanda with a rare disorder called "mirror touch synesthesia," which causes her to feel the physical and emotional stimuli inflicted on those around her. Amanda feels loved when others are hugged and experiences physical pain when someone bumps their head. The mysteries of quantum theory serve as a nifty metaphor for the equally puzzling mysteries of the human emotions. Both are invisible forces shaping how isolated things come together.

Amanda's entanglement with others, a kind of radical empathy, presents obstacles to intimacy and to self-knowledge as we learn in interviews with her and her family. The *Radiolab*-esque sound design reinforces the weirdness of this condition and the universality of compassion, which Dowling has called "the controlling metaphor of the episode," though perhaps it could be said of the entire podcast.[39] A pleasantly relatable brain scientist joins the conversation to explain the condition as "a lack of gray matter" in the "temporal parietal junction," a part of the brain "that helps us to distinguish between the self and somebody else." Back at Amanda's, we learn that her oldest daughter suffers from similar sensitivities and from loneliness due to her mother's self-imposed emotional exile from everyone around her.

In a gloss on Amanda's predicament, and perhaps for *Invisibilia* itself, Rosin muses, "I began to wonder if maybe this is the danger of empathy—when you think you know what someone else is feeling, when you're pretty sure you've got a handle on it, you don't bother to ask." This is the *TAL*-style big takeaway, the moment of reflection, and it provides a caption for the discomfort that animated the podcasts under review in this chapter. What if empathy isn't "the

---

[39] David Dowling, *Podcast Journalism: The Promise and Perils of Audio Reporting* (New York: Columbia University Press, 2024).

way"? From Amanda, it's an easy and reassuring leap to behavioral science, demonstrating the power of social conformity. The movement (I'm tempted to say entanglement) between particular and universal is not subtle; it works so well (*Invisibilia* was one of the highest rated podcasts on iTunes for most of its run) perhaps because it traffics in ideas we are happy to believe in.

Like many stories on *Invisibilia*, this one seeks out examples of exotic human specimens to make a larger point about some universal human quality. Stories of the socially marginal, told from the perspective of the socially powerful, will humanize these others in the hearts and minds of affluent and socially powerful listeners. Likewise, *Invisibilia* argued, this process, rather than a trick of narrative enchantment, was central to the unseen natural world all around us and only needed elucidation. Other outliers, such as Allie n Steve, a single person whose gender identity is so fluid as to change minute by minute, serve as an ambassador to the quicksilver lability of identity for us all. Or more prosaically, per the *Invisibilia* website, Allie n Steve "has lessons to teach us about the beauty of not retreating to black and white."

SM, whose amygdala is completely calcified due to a rare disorder, is neurologically speaking, utterly fearless. She provides an entrée into a meditation on how we can all banish our fear. Daniel, a blind man who navigates on his bicycle using his own version of clicks and sonar, provides a way to suggest that we're all just prisoners of artificial barriers and labels. "Do you believe," the hosts ask each other, with the brio of faith healers, "that the blind can be made to see?"

There is a strong vein of "self-improvement" to these stories that is in tune with the neoliberal goals of individual freedom from constraints, unmoored to any larger social contract. Emotions (i.e., their lability, their seeming idiopathy, their mystery, and finally, their mastery) form the center of the program's appeal. In 2017, its website promoted the show thus: "Listen. Feel Different," an echo perhaps of Apple's 1990s "Think Different" campaign, a motto in tune with corporate marketing of individualism and creativity. It's also a promise of emotional succor, one that NPR has made for other podcasts, like *Hidden Brain*, which includes self-improvement tips for feeling less alone, cultivating greater empathy, and connecting with others.

The End of Empathy

The late 2010s and early 2020s brought a new urgency to the challenge of empathy and heightened sense of the proximity of the "others" in our midst, especially for the journalists working in public media and its podcast diaspora. But the fractures and contradictions within public radio's highly educated, socially conscious "us," always unstable and prone to crisis, became increasingly unstable and uncomfortable. *Invisibilia*'s 2019 episode, "The End of Empathy," provides an instructive example of the limits of empathy for managing these fractures, and enacts, in its small way, a public scene of the collapse of the public radio structure of feeling. Like *Nice White Parents*, "The End of Empathy" seems to go out of its way to make a scene. It is structured out of disparate parts whose misalignment is made clear from the start. Guest producer Lina Misitzis uses excerpts from an interview with a man who calls himself "Jack Peterson" to briskly summarize a disastrous encounter with a woman he'd been involved with, which ends with violence, police, and Jack in a mental hospital. Misitzis steps out of the story and out of the show itself to address listeners:

> This tape you're hearing of Jack Peterson, I didn't record it. The journalist Hanna Rosin did for an NPR show called *Invisibilia*. I'm in the running for a job there. They've given me this interview as a test, to make a story out of. And though it wasn't explicitly stated, it does feel obvious that I should craft it to sound like an *Invisibilia* story, which is to say empathically – a thorough and thoughtful look at Jack Peterson's brain, how his character traits were born and what we can learn from them.

As anecdote-reflection goes, it's a pretty unusual one. Misitzis addresses listeners as if from outside of the show itself, a distancing move that goes beyond glib reflexivity to something a bit more radical. We aren't listening to *Invisibilia* yet, just voices that have yet to be properly organized. At this point, Rosin and Spiegel cut in to tell us that Misitzis's version of the story shocked them into a bit of a crisis. In its impatience with "the empathetic way," Misitzis's version sounded like "almost the opposite of what we created." This last bit, "what we created," could refer to Rosin's rough-cut version of the Jack Peterson story or more broadly, *Invisibilia* itself.

Misitzis, "like much of the world, seems to be losing patience with that way," Rosin continues. "In the post-#MeToo, vigilant, polarized Trump-era world, showing empathy for your so-called enemies is practically taboo." Misitzis's version of Peterson's story "felt like a reckoning" to Rosin and Spiegel. They decide to play both versions of the story, first Rosin's, then Misitzis's, in order to determine Peterson's right to empathy and perhaps, *Invisibilia*'s as well. Rosin's version begins with her discovery of Peterson's YouTube videos, in which he excoriates himself as ugly, a loser, and undatable with a small penis. Rosin is intrigued. "I couldn't help but wonder—what's driving you?" she asks. She tells us that Peterson is an incel (involuntary celibate), "guys who hate women and in rare cases kill them because women won't sleep with them."

This disturbing bit of information comes early in her story, making it clear that this will be a test-case for the way of empathy. Rosin recounts his story from lonely, unpopular twelve-year-old to an online chat romance with an older teenaged girl, M. The story takes an ugly turn when he admits to Rosin that he sent revenge porn photos of M to her family, professors, and friends. The fraught relationship culminates in a bizarre scene in a parking lot near her house, where he has shown up unwanted and unannounced and is arrested and committed to a mental hospital for suicidal ideation.

The incident plunges him into an abyss of depression and despair. At his lowest moment, he finds the online incel community that embraces losers and who "hate empathy, sympathy, and comfort of any kind—too feminine." Peterson attracts media attention as an outspoken incel activist and encounters kind women in the media (e.g., reporters, producers, hosts). This kind treatment, combined with the horrific violence committed in Canada by a self-proclaimed incel in 2018, caused Peterson to have a change of heart. A reformed incel, he asked to be blocked from the incel site, joined online dating sites and embraced a more hopeful future.

Misitzis breaks in and calls this version of Peterson's life "a lie." She breaks down the same tape Rosin recorded into bits, each one demonstrating hostile, threatening behavior. Ignoring M's desire to be left alone; crossing state lines to surprise her; scaring her enough that she asked him to prove he wasn't carrying weapons. Misitzis also introduces new tapes from Peterson's many media appearances and from her own interviews with him. In them, he demonstrates little remorse for his actions and little empathy for

M. In a bit of radical editing and juxtaposition of contexts, Misitzis then intersperses the clips of his recorded statements about M with audio from an interview of another woman, J, telling of her own experiences with other men. Threats of self-harm, showing up at her house after being asked not to, his contempt and desperation, her fear: the stories match up—until they diverge, but Misitzis has made her point. She has "a whole choir of women's voices in her address book," who can narrate the same scenes of intimidation and coercion. Misitzis also admits that her version requires context.

> I'm usually one of those people-are-mostly-the-same types – someone who tries to find overlap in just about everything – but this *feels different*. I think it's because this week *feels different*. This week, despite multiple allegations of sexual assault and despite a clear unwillingness to entertain the possibility that he might have anything to answer for, Brett Kavanaugh was confirmed to the U.S. Supreme Court.[40]

This historical context helps to explain the explicit nature of the disagreement between Misitzis and her potential employers. This strange episode, with its multiple layers of contradictory and self-undermining narratives, speaks to a crisis at the heart of liberal journalism, which since 2016, had been undergoing paroxysms of self-doubt.[41] It also speaks to the way that movements like #MeToo and #BlackLivesMatter surfaced implicit political conflicts into nonfiction narrative audio in ways that provided newly urgent, if uncomfortable, context. Unlike Rosin, however, Misitzis doesn't present the context of Kavanaugh's confirmation as evidence that she is not to be trusted. Rather, like many of the newer voices entering audio storytelling in the years since 2016, she seems to be arguing that new realities produce new feelings that require new stories and perhaps, new storytellers.

The stakes are not lost on Rosin, who describes her response to hearing Misitzis's version on the way into work.

---

[40]Emphasis added.
[41]Kyle Pope, "Here's to the Return of the Journalist as Malcontent," *Columbia Journalism Review*, November 9, 2016, https://www.cjr.org/criticism/journalist_election_trump_failure.php?utm_content=buffer3da1f&utm_medium=social&utm_source=twitter.com&utm_campaign=buffer.

I rode the elevator up and down—six to lobby, lobby to six—feeling embarrassed, and annoyed, and called out, and taking it all very personally and thinking, Hanna, you're an idiot. Hanna, are you an idiot?

Another awkward elevator scene in which the expectations of liberal journalism and fellow feeling encounter intractable and painful divisions. Rosin calls Misitzis to hash it out: "does Jack deserve our empathy or not?" Rosin admits that understanding Peterson, "getting into his head," was her original goal and that it is "literally always my goal" when reporting stories. Misitzis asks "why," which leads Rosin into a brief history of empathy, an emotional philosophy that she says, was a foundational part of public education in the 1960s and 1970s. She brings on a cognitive scientist to co-sign. He charts the decline of "all dimensions of empathy" beginning around the turn of the century. Misitzis tells Rosin that extending empathy to abusers like Peterson does violence to his victims, leading Rosin to explore "himpathy, the tendency to empathize with men in power over vulnerable women."

There have long been reasons to be circumspect about the social and policy benefits of empathy. In his book *Against Empathy*, Paul Bloom cautions that empathy often doesn't serve the object of empathy. An empathetic response to a child terrified of needles might not be in the best interests of that child who needs you to draw on the broader and more utilitarian impulse of compassion. Also, empathy tends toward the particular, the familiar, rather than the general or the strange. Or as "himpathy" suggests, it can privilege the powerful over the vulnerable, which explains the racialized and gendered empathy gap we see in so many sectors of society from criminal justice to foreign policy.[42] Nakamura reminds us of the work of recent Black feminist theorists like Courtney R. Baker and Alisha Gaines, that reminds us that "the desire to experience empathy for the sufferings of black people while leaving structural

---

[42]Samantha J. Dodson, Rachael D. Goodwin, Jesse Graham, and Kristina A. Diekmann, "Moral Foundations, Himpathy, and Punishment Following Organizational Sexual Misconduct Allegations," *Organization Science* 34, no. 5 (2023): 1938–64.

racism in place has long underwritten pleasurable forms of cultural appropriation and projection."[43]

The episode shows Rosin considering the notion that "empathy is not an infinite resource," as if for the first time. The cognitive scientist returns to agree, but cautions that while indiscriminate empathy is misplaced, and highly partial empathy leads to "dangerous tribalism," the absence of empathy risks the death of civil society and "90 percent of what our life is all about." Resolving the high-stakes binary choice presented at the start of the episode into a balancing act at the end, this starts to sound more like a typical *Invisibilia* story. Rosin admits to being changed by the challenge, though she suspects that Misitzis was not. At the very end of the piece, after the production credits, Rosin tells us "By the way, Lina Misitzis—she's working at *This American Life* now."

Within two months of the airing of this episode, *Invisibilia* had changed its roster of hosts, Kia Miakka Natisse, an African American woman, and Yowei Shaw, an Asian American woman, replaced Rosin and Spiegel, two white Jewish women. The show veered into politics and sociology and away from pop psychology and empathy. Within two years, NPR announced the show's cancellation, part of a broader swath of cost-cutting measures in response to the post-pandemic drop off in listenership, and broader decline in podcast revenue across the industry. It is difficult to see these events as merely chronological, and not, in some ways, linked to a bigger story about the career of public radio's structure of feeling.

The second half of the decade saw the rise of podcasts with many of the personnel and stylistic influences of public radio, but which centered public scenes of conflict and discomfort over empathy. Shows like *Resistance* (Gimlet, 2020), *Not Past It* (Gimlet, 2021–4), *No Compromise* (2020), *Rough Translation* (NPR, 2017), and *White Lies* (NPR, 2021), signal, in their titles, the extent to which conflict and irreconcilability were becoming a

---

[43]Lisa Nakamura, "Feeling good about feeling bad: virtuous virtual reality and the automation of racial empathy," *Journal of Visual Culture* 19, no. 2 (2020): 47–64, https://doi.org/10.1177/1470412920906259; Courtney R. Baker, *Humane Insight: Looking at Images of African American Suffering and Death* (Champaign: University of Illinois Press, 2017); Alisha Gaines, *Black for a Day: White Fantasies of Race and Empathy* (Chapel Hill, NC: University of North Carolina Press, 2017).

kind of genre. *Marketplace*, the venerable public radio "business show for the rest of us," known for its political complacency and the unctuous sardonic vocal performance of its longtime host Kai Ryssdal, spun off podcasts, *The Uncertain Hour* (APM, 2016), *This Is Uncomfortable* (APM, 2019), and *How We Survive* (APM, 2021) that centered conflict and discomfort. "It seems that no matter the locus of funding, social-justice-minded podcasts are now mainstays at every major network," one observer noted dryly, implying a possible disconnect between the message and the political economy driving the messenger.[44]

Frustration and uneasiness were explicit ingredients of *Reparations: The Big Payback* (2021). Hosts Erica Alexander, a Black woman, and Whitney Dow, a white man, "talk about race, slavery, and America inside the conversation of reparations." Their relationship and their very different positionalities stand in for the larger irreconcilability of the emotional and policy gaps around their subject matter. When they come upon a neglected plaque in lower Manhattan marking the site of an old slave auction site, Alexander doesn't hide her anger from Dow (or us), nor her tears. Alexander explains that Dow "is a proxy for all the things I can't say to White men and White women." Dow acknowledges the dramatic importance of playing his role, though he too, gives voice to a frustration and uneasiness, a residue perhaps of that obsolete voice of white chivalry.

On the popular *Code Switch* (NPR, 2016–), hosts Eugene Demby, a Black man, and Shereen Marisol Meraji, an Iranian/Puerto Rican American woman, bristled at the constraints of doing a show on race and culture for an audience populated with millions of legacy NPR listeners (i.e., older, whiter, well-meaning liberals). On the 2021 episode on "Karens," a term for typically white women who use their privilege to harass people of color in public spaces often creating scenes, Demby steered into the anger instead of away from it. They did two episodes on the problem of "the explanatory comma," the background information about minority culture and language necessary for white listeners but off-putting

---

[44] Jess Shane, "Towards a Third Podcasting: Activist Podcasting in an Age of Social Justice Capitalism," *RadioDoc Review* 8, no. 1 (2022), doi:10.14453/rdr.108.

for others (i.e., their target audience of people of color).⁴⁵ Longtime NPR listeners wrote in to protest the idea that journalists shouldn't do everything possible to bring everyone along, invite everyone in, another explicit philosophical schism that made it plain that the very idea of these podcasts was itself under investigation. In their first episode of *Invisibilia*, "Eat the Rich," Natisse and Shaw explore the idea of forcing white people to pay reparations. "This is going to make some of our listeners uncomfortable," Kia says with an uncomfortable laugh. "Well, let's get to it," Shaw responds.

"Yes to pedagogical discomfort!" audio documentarian Jess Shane declared in 2022 in *RadioDoc Review*, an online journal of criticism. "No to explanatory commas!" she added. *RadioDoc Review* instructs potential contributors to consider "Emotiveness and Empathy: (identification with talent/characters, affect, evokes visceral response)" in their reviews. The decision to publish a document calling for an end to "handholding through complex ideas," and less focus on "individual traumas" speaks to the tectonic shifts in the public radio structure of feeling.⁴⁶ If a new generation of soundwork artists were game to take on discomfort in nonfiction audio storytelling in the era of Trump, others expressed a bit more reluctance.

We can trace the current of unease as it worked its way into *Reply All* (Gimlet, 2014–22) and the lively banter of its two hosts, Alex Goldman and PJ Vogt, two white men with public radio credentials. Perhaps along with *Invisibilia*, no podcast better exemplifies the extension and apotheosis of the public radio sensibility in the new medium than *Reply All*, an award-winning and massively popular podcast with a focus on internet technology and culture. Its very public, very awkward implosion in 2021 seemed to mark a new phase in the collapse of the empathetic way. Its special four-part series, "The Test Kitchen," on the exploitation of women and people of color in the test kitchen of *Bon Appetit*, the glossy food magazine, only made it through two episodes.

---

⁴⁵Kelly McBride, "NPR's Code Switch is an Overnight Sensation 7 years in the Making," *Poynter*, Blog, 11 December, 2020, https://www.poynter.org/ethics-trust/2020/nprs-code-switch-is-an-overnight-sensation-7-years-in-the-making/.

⁴⁶"Submitting to RadioDoc Review," *RadioDoc Review*, 2014, https://ro.uow.edu.au/rdr/policies.html.

Former Gimlet staff took to Twitter to point out the uncomfortable similarities between the conditions at *Bon Appetit* and those they encountered at Gimlet. Young women and people of color took the lowest paid and least secure jobs at the magazine and the podcast network. Eric Eddings, a former Gimlet colleague, tweeted: "I had been avoiding listening to ['The Test Kitchen'] but once I did, I felt gaslit. The truth is *RA* ..... contributed to a near identical toxic dynamic at Gimlet." The producers had anticipated some, but not all, of the fallout. An outside consultant said they had asked "should we be the ones to tell this story?"[47] Eddings agreed: "It's damaging to have that reporting and storytelling come from two people who have actively and AGGRESSIVELY worked against multiple efforts to diversify Gimlet's staff & content."

In an era of heightened demand for accountability, the critics were heard. The remaining episodes of the *Bon Appetit* investigation were abruptly shelved and co-host PJ Vogt and producer Sruthi Pinnamaneni, both of whom had spoken out against the younger union organizers on staff, left the show. Emmanuel Dzotsi, a Black reporter brought on to co-host *Reply All* only a few months prior, returned to the microphone for the show's awkward damage control reboot. He reflected upon his own position on the show as a symbol of the larger crisis. Die-hard *Reply All* fans had been taking to the internet to criticize Dzotsi's stories for a while, he acknowledged. His stories, the ones he liked to tell, the ones he was hired to tell, concerned as they were with race and racism, "didn't sound like *Reply All* stories." He agreed and saw it not simply as a criticism of him, but instead an observation about the show and the listeners and sound it had cultivated. This insight, paired with the criticism that *Reply All*'s old guard did not sound credible telling the story of "The Test Kitchen," spoke to the larger impasse rippling through the public radio/podcast structure of feeling.

Dzotsi was also hinting that blithely adding another voice and perspective, that just so happened to belong to a younger Black

---

[47]Katherine Rosman and Reggie Ugwu, "What Really Happened at '*Reply All*'?" *The New York Times*, March 12, 2021, https://www.nytimes.com/2021/03/10/style/reply-all-test-kitchen.html.

man with a British accent, failed because the whiteness and white-man-ness of the original pair of hosts was crucial to the show's appeal. Beyond this, Goldman and Vogt also shared a background at WNYC and a set of privileges that enabled the explorations of shared neuroses and pet peeves that required no explanatory commas for listeners trained in public radio journalism. The show's formula for success was often described using language that spoke to privileges accrued by white men seen as non-political, like the "relative innocence of the time" of the show's debut (2014!) and the "self-conscious nerdy sensibilities of the hosts."[48] Goldman admitted that "the continuity of PJ and I's relationship is part of what keeps people interested," acknowledging the difficulty that Dzotsi would have when he joined in 2021.

With the show's reputation in tatters and two key figures gone, the future of the show depended on what the new version would sound like and whether Dzotsi's voice could ever sound like it belonged. In the first show after the collapse of "The Test Kitchen," Alex Goldman, known for his frank self-disclosures, kept mostly quiet, as if in this context, even his voice no longer sounded like it belonged. The musical bed under Dzotsi's voice lurks way down in the mix, a low, slow bassoon-y vamp, as if awkwardly trying to reassert *the idea* of entertainment, that this is, after all, still *a show*. At that moment, however, it is a show in which a Black man is telling his co-host and listeners that the reason his voice doesn't sound like it belongs is because of the show's irreducible inaccessibility to Black voices, experiences, and perspectives. The discomfort is the emotional truth of the moment and so Dzotsi and Goldman have little choice but to lean into it a bit. Unlike the lively badinage between Vogt and Goldman, punctuated by bursts of laugher, these exchanges are difficult, bordering on cringe.

The new vibe involved less ribbing, less hilarity, and more of them responding to each other with the judicious phrase, "that's fair." *Reply All* gamely moved forward for a couple of weeks; Dzotsi and Goldman, still working on an emotionally plausible

---

[48] Reggie Ugwu, "A Song No One Remembered. A Podcast that Is Hard to Forget," *The New York Times*, March 19, 2020, https://www.nytimes.com/2020/03/19/arts/reply-all-podcast.html.

version of the laddish rapport enjoyed by Goldman and Vogt. This awkward period, before the show's inevitable cancellation later in the year, presented an opportunity to reflect on what kinds of stories, voices, and perspectives *sound-like-they-belong* in the rarified world of podcasts like the ones promoted by NPR and Gimlet. And to anticipate perhaps, a shift in the aesthetics of affect on shows "in the wake of 2020's cocktail of racial reckoning and pandemic-induced introspection," as audio producer Jess Shane put it. "Many media organizations across North America have pivoted to cater to an audience allegedly ready to 'speak truth to power.'"[49] Shane's use of "allegedly" points to a sense that the shift away from "handholding" and driveway moments had not yet been realized, and that more discomfort was needed.

Some defenders of Vogt and Goldman argued that the criticisms were overblown, a tempest in a teapot, sparked by a single tweet by a disgruntled former Gimlet employee.[50] But it's worth noting that the episodes that immediately preceded the "Test Kitchen" debacle featured some awkward moments that in retrospect, sound prophetic. The episode entitled "Account Suspended" is a response to the Jan 6 insurrection and the subsequent deplatforming of Trump from social media. Alex Goldman shouts his anxiety and despair over recent events and PJ laughs; a familiar pattern that sounds newly discordant. The episode ends with Gimlet CEO Alex Blumberg taking Goldman to task for his climate despair, expressed in his recently aired "Song of Impotent Rage," arguing that it is precisely this affective performance that the world, and his network, need less of. It's also worth pointing out that Gimlet's biggest shows featured white male hosts, and that their first big hit, *The Mystery Show*, hosted by *TAL*-alumna Starlee Kine, was abruptly canceled after its first season. Kine took to social media to call out the network's sexism.[51]

---

[49]Shane, "Towards a Third Podcasting."
[50]See Jesse Signal, "The *Reply All* Implosion Shows What a Toxic, Circular-Firing-Squad Mess Media is Right Now," *Signal-Minded*, Blog, February 26, 2021, https://jessesingal.substack.com/.
[51]Dustin Rowles, "'Mystery Show' Host Starlee Kine Tried to Tell Us About Gimlet Media," *Pajiba*, March 3, 2021, https://www.pajiba.com/podcasts_1/mystery-show-host-starlee-kine-tried-to-tell-us-about-gimlet-media-.php.

Meanwhile, on the sub-Reddit dedicated to the show, longtime fans debated the show's shifting focus, most of them longing for the good old days before Dzotsi, before "The Test Kitchen," before everything became "more hostile and fraught."[52] For his part, Emmanuel Dzotsi navigates his new position as if keenly aware that sounding like he belongs will require equanimity from him as well as different types of listeners and listening. In a September episode on feelings during the pandemic, Dzotsi confessed to feeling "hella moya moya," which he explained, with an explanatory comma, means frustrated and uneasy in Japanese. The news shared by Goldman and Dzotsi in May 2022, that the show was coming to an end, was greeted by online fan communities with relief as much as with sadness.[53] It was clear that "sounding like *Reply All*" had become impossible.

Within a year, Spotify had absorbed Gimlet entirely into its audio platform, which led to the eventual cancellation of almost all of its original programing. Alex Blumberg, Gimlet's founder and producer of *TAL*'s "Testosterone" (2002) episode, in which he sought "a way to, as a man, talk without getting in trouble," had found a successful formula for two decades. As much as anyone, Blumberg, from his early work on *TAL* to his foundational role on *Planet Money*'s defense of free markets, to his creation of the for-profit Gimlet that nurtured dozens of empathy-driven shows, was a chief architect of the public radio structure of feeling.[54]

The 2019 purchase of Gimlet Media by Spotify accelerated the ambient mood of confrontation and discomfort already underway, bringing its roster of empathy-forward shows into an unapologetically corporate streaming ecosystem with little interest in Gimlet's branding around the humanistic values of its content. The "Spotification" of podcasting was met with mixed feelings by critics, scholars, and listeners alike, including the concern that

---

[52] https://www.reddit.com/r/replyallpodcast/comments/pgeum1/178_i_am_not_a_bot/.
[53] Nicholas Quah, "This Era of *Reply All* Is Ending," *New York Magazine*, May 18, 2022.
[54] Eric Johnson, "Alex Blumberg and Matt Lieber explain why they sold Gimlet to Spotify," *Vox*, February 7, 2019, https://www.vox.com/2019/2/7/18214941/alex-blumberg-matt-lieber-gimlet-spotify-deal-acquisition-peter-kafka-media-podcast-audio-interview.

corporate platforms might offer greater visibility for the format, but that it might also "undermine some of the format's earliest promises of accessibility and diversity of voices."[55]

The next year, Spotify shelled out $250 million for exclusive streaming rights to *The Joe Rogan Experience (JRE)*, a podcast juggernaut that occasionally trafficked in conspiracy theories and odd displays of toxic masculinity. The outcry among artists who didn't want their work streamed alongside *JRE* was swift, calling into question the relationship between platforms, artists, content, and the divisive politics of public health and identity politics.[56] Wendy Zukerman, of *Science Vs*, a podcast Gimlet acquired from the Australian Broadcast Company in 2015, that Vulture described as "a fun, peppy, nerd-empowerment program,"[57] chose not to quit, but instead to transform her show into a point-by-point counter-programming of all of Rogan's specious claims with a side helping of critical investigations into Spotify's complicity. She was moved to do so by Rogan's notorious December 2021 interview with Dr. Robert Morgan, a vaccine conspiracy theorist. She responded first to that one episode with a forensic takedown of each of Morgan's antivax arguments presented on *JRE*.[58] In one of his final hosting gigs at *Reply All*, Dzotsi invited Zukerman on to share that episode.[59]

Dzotsi's experience wasn't unique. Jazmine Green's account of being a contract producer for *The Nod* (2017–20), Gimlet's promising but short-lived show that sought to "gleefully explore

---

[55] Jeremy Wade Morris, "The Spotification of Podcasting," in *Saving New Sounds: Podcast Preservation and Historiography*, eds. Jeremy Wade Morris and Eric Hoyt (Ann Arbor: University of Michigan Press, 2021), 209, http://www.jstor.org/stable/10.3998/mpub.11435021.16.
[56] Mary Biekert, "Here's a List of Artists Who Are Boycotting Spotify Because of Joe Rogan," *Time*, February 2, 2022, https://time.com/6144634/artists-boycott-spotify-joe-rogan/.
[57] Nicholas Quah, "*Science Vs* Does Joe Rogan," *Vulture*, February 10, 2022, https://www.vulture.com/2022/02/new-podcasts-joe-rogan-science-vs-trojan-horse.html.
[58] Wendy Zukerman, "The Joe Rogan: The Malone Interview," *Science Vs*, Gimlet, Podcast, February 4, 2022, https://gimletmedia.com/shows/science-vs/49hngng/joe-rogan-the-malone-interview.
[59] Linda Qui, "Fact Checking Joe Rogan's Interview with Robert Malone that Caused a Stir," *The New York Times*, February 22, 2022, https://www.nytimes.com/2022/02/08/arts/music/fact-check-joe-rogan-robert-malone.html.

all the beautiful, complicated dimensions of Black life," paints a damning picture of stark racial hierarchies and hypocritical pretensions to equality.[60] Stefanie Fu, a producer for a short time at *TAL*, described work as "a space where I was forced to think about white supremacy and violence against people of color all day every day—all while facing prejudice and abuse from my own management."[61]

The departure of high-profile women of color from key positions at NPR and other podcasting networks in the following year added to the nagging questions that Dzotsi had posed about who sounds like they belong. "We're hemorrhaging hosts from marginalized backgrounds," tweeted *All Things Considered* host, Ari Shapiro.[62] Independent audio and print investigative reporter Amy Westervelt tweeted her alarm at the "wave of bipoc journalists announcing their quitting."[63] David Folkenflik, reporting on his own network, conceded the currency of "a belief that NPR has proven incapable of doing the right thing when race is a factor and is willfully or carelessly driving away its future stars, even as it aspires to attract more Black and Latino listeners."[64]

Another blow to diversity in the industry was WNYC's 2023 cancellation of *The Takeaway*, a long-running daily programming with a history of controversies, that had reached what many hoped was an even keel with Melissa Harris Perry as the host. Shortly thereafter, NPR canceled four programs hosted by people of color—*Invisibilia, Louder than a Riot, Rough Translation*, and *Everyone and Their Moms*. By this time, public radio veteran and gadfly Celeste Headlee's manifesto, "An Anti-Racist Future: A Vision and Plan for the Transformation of Public Media," had already been circulating for nearly two years. It called out on-air and off-air racism and discrimination, linking it to the larger crisis

---

[60] Jazmine Greene, "Glass Walls," *JTGreene*, Weblog, 2022, https://www.jtgreen.me/now.
[61] Stephanie Fu, *What My Bones Know* (New York: Ballantine Books, 2022), 64.
[62] Ari Shapiro, Tweet, January 4, 2022.
[63] Amy Westervelt, Tweet, January 12, 2022.
[64] David Folkenflik, "Hosts' departures fuel questions over race. The full story is complex," *NPR*, 11 January, 2022.

in public media and calling for actionable change.⁶⁵ Media coverage of this document, which had the backing of scores of colleagues across the industry, was relatively muted compared to the wall-to-wall coverage garnered by a backlash manifesto published by another NPR veteran, Uri Berliner, the following year that claimed that the network had lost its way in the pursuit of diversity and "wokeness."⁶⁶

The larger collision of music-streaming platforms like Spotify and Apple, with the growing but still comparatively miniscule podcasting market, made for lopsided battles such as this one. Podcasting, which had been driven by "a proleptic imaginary" for nearly a decade, seemed to have hit some kind of plateau in the popular discourse surrounding it by the early 2020s.⁶⁷ Talk of innovative creative practices disappeared into complaints about an ocean of undifferentiated and overvalued "content."⁶⁸ The journalistic discourse around podcasting shifted from a sunny technological utopianism to disappointment, overwhelm, and cynicism.⁶⁹ "Gimlet's story was always going to end this way," chimed *The Defector* in a 2023 piece published as Spotify laid off hundreds of Gimlet and Parcast staff, consolidating the two production units into its own in-house shop.⁷⁰

---

⁶⁵Celeste Headlee, "An Anti-Racist Future: A Vision and Plan for the Transformation of Public Media," *Medium*, January 18, 2021, https://celesteheadlee.medium.com/an-anti-racist-future-a-vision-and-plan-for-the-transformation-of-public-media-224149ab37e6.

⁶⁶Uri Berliner, "I've Been at NPR for 25 Years. Here's How We Lost America's Trust," *The Free Press*, April 9, 2024, https://www.thefp.com/p/npr-editor-how-npr-lost-americas-trust.

⁶⁷Verma, *Podcasting in an Age of Obsession*, 8–10.

⁶⁸https://pacific-content.com/feeling-overwhelmed-by-the-sheer-number-of-podcasts-its-not-just-you-d46b37a8fdcc/; Alex Sujong Laughlin, "If you love podcasts, dump Spotify," *Defector*, Blog, April 12, 2024, https://defector.com/if-you-love-podcasts-dump-spotify.

⁶⁹Jonathan Sterne, J. Morris, M. B. Baker, and A. M. Freire, "The politics of podcasting," *The Fibreculture Journal* 13 (2008), https://thirteen.fibreculturejournal.org/fcj-087-the-politics-of-podcasting/.

⁷⁰Alex Sujong Laughlin, "Gimlet Media's Story Was Always Going To End Like This," *Defector*, June 12, 2023, https://defector.com/gimlet-medias-story-was-always-going-to-end-like-this.

Such a conclusion was hard to resist; even Blumberg's business partner Matt Lieber admitted "returning a nice profit for our investors" was a key goal of the entire project, upon the sale to Spotify.[71] The political economy of platformization had produced conditions in which creativity had been transformed into "fundamentally contingent cultural commodities... increasingly modular in design and continuously reworked and repackaged."[72] Such a process seemed at odds with the intimacy and specificity of the storytelling impulse in public radio's structure of feeling. Colin Campbell, Gimlet's former head of new show development, described the feeling in his last year there as a "rudderless, opportunistic... folly." Campbell, part of the public radio diaspora who built Gimlet, was by early 2024 back at NPR, heading up their podcast division and happy to be working with people who "work with a sense of mission."[73]

The 2020 purchase of Serial Productions by *The New York Times* was met with considerably less controversy. At the time, the acquisition made sense given the larger industrial trends in legacy news media. *The Times*' ambition to become a major player in the audio content industry followed the logic of the past twenty years of legacy print media embracing digital platforms. Nonfiction podcasts focused on investigative journalism and benefitted from the resources and reputation of major newspapers like *The Times*, which meant, among other things, a brisker schedule of seasons thanks to a larger staff.[74] As mentioned above, it was an extension of *TAL*'s franchising efforts with for-profit content providers. It seemed an obvious collaboration and was hailed as a win-win, bringing together two simpatico content producers joining forces in an era when a few massive corporate streamers were gobbling up audio.

---

[71]Johnson, "Why Gimlet Sold to Spotify."
[72]D. B. Nieborg and T. Poell, "The platformization of cultural production: Theorizing the contingent cultural commodity," *New Media & Society* 20, no. 11 (2018): 4275–92, https://doi.org/10.1177/1461444818769694.
[73]Ariel Shapiro, "What NPR's new podcast chief learned from the mess at Gimlet," *The Verge*, January 23, 2024, https://www.theverge.com/2024/1/23/24048118/npr-gimlet-collin-campbell-spotify-throughline-israel-story.
[74]Rachel Abrams, "New York Times to Buy Production Company Behind 'Serial' Podcast," *The New York Times*, July 22, 2020, https://www.nytimes.com/2020/07/22/business/media/new-york-times-serial.html.

Podcast listeners, for their part, had taken quickly to consuming news on multiple channels and platforms, a contrast to the more vexed relationship between podcasting and streaming music that complicated Spotify's aggressive acquisition of podcasting assets, antagonizing podcast listeners.[75]

For these reasons, *Serial*'s 2022 offering, "The Trojan Horse Affair," represented one of the most anticipated shows of that year. At one level, it was like previous *Serial* podcasts, an exemplary piece of investigative journalism taking aim at structural problems of inequality told with a focus on characters, careful plotting, and compelling thematic attention to an unfolding mystery and the play of symbolic antitheses. It was also a savvy exploration of how hateful conspiracy theories circulate through mass media and local communities, a topic that had become increasingly important to investigative journalism in general and podcasting in particular.

The season proceeded as a methodical dissection of the origin of a nasty Islamophobic hoax in 2014, the so-called Trojan Horse Letter, that had massive consequences in the UK educational system. The anonymous letter, sent to Birmingham City Council officials, falsely claimed to have been written by a local schoolteacher and alleged an Islamist plot to radicalize South Asian students in East Birmingham. However, the biggest dramatic moments occur between the two hosts, one a white veteran journalist, Brian Reed, a longtime *TAL* producer who had also hosted *Serial*'s Peabody Award–winning "S-Town," a literary and ethically fraught story of a troubled autodidact in a small Southern town; the other was Hamza Syed, a doctor turned journalist, a child of Pakistani immigrants, and a Muslim. Reed finds himself uncomfortable with Syed's "unprofessional" anger during interviews. Syed is impatient with Reed's cool distance from the real-life implications of the hoax and their project. Reed wants to "tell a good-ass story"; Syed wants to have an impact. Syed opens the series with unresolved anger that this story, his first as a journalist, will also likely be his last.

The show won praise for its "metajournalistic transparency" and for how it challenges and renegotiates the construction of journalist

---

[75] Ellis Jones and Jeremy Morris, "Competing Sounds? Podcasting and Popular Music," *Radio Journal: International Studies in Broadcast and Audio Media* 20, no. 1 (2022): 3–15; Cory Doctorow, "Podcasts are hearteningly enshittification resistant," *Pluralistic*, Blog, January 27, 2023, https://pluralistic.net/2023/01/27/enshittification-resistance/#ummauerter-garten-nein.

authority and credibility.⁷⁶ Reviewers and academics saw it as a timely intervention in a moment of crisis for public trust. The open conflict between Reed's traditional journalistic objectivity and Syed's impassioned search for truth and consequences offered "a model, perhaps *the* model for salvaging contemporary journalism, due to its radical honesty at a time of deepening suspicion of the news media."⁷⁷ Amidst the creeping Spotification of podcasting and the simmering violence of the culture wars in the United States and the United Kingdom, this "twisty," "punchy" show was welcomed as a vigorous reassertion of the form's artistic and social power.⁷⁸ "The Trojan Horse Affair" offered proof, argued the BBC's Deb Grayson, that "complex issues" like Islamophobia and journalistic credulity in the face of misinformation "needs a long-form format like an eight-hour podcast if it is to be really grappled with."⁷⁹

The podcast seemed perfectly timed to address a newfound urgency among those who saw journalism's role in the preservation of liberal democracy in dangerous decline. Nicholas Quah praised it for the way its unresolved conflicts and contradictions demonstrated "how a commitment to 'objectivity' and a distrust of activism can speak to the fundamental whiteness of the institution—how journalists are expected to conduct themselves as if insulated from the consequences of their reporting."⁸⁰ It was precisely its failure to comfort that made "The Trojan Horse Affair" successful.

In the aftermath, however, there was some uncertainty about what this discomfort had accomplished. Syed himself seemed to experience it most acutely. Throughout the podcast and afterwards,

---

⁷⁶Spencer Jones, "Transgressive (Un)Scripting: Metajournalistic Discourse and 'Sonic Friendship' in The Trojan Horse Affair," *Radio Journal: International Broadcasting in Broadcast and Audio Media* 22, no. 1 (2024): 25–41.
⁷⁷Jones, "Transgressive (Un)Scripting."
⁷⁸Nicholas Quah, "The Trojan Horse Affair is a Twisty Thrill," *New York Magazine*, 2022; "Trojan Horse rides again; Islam in Britain," *The Economist*, March 5, 2022, 29. *Gale Academic OneFile* (accessed July 12, 2024), https://link-gale-com.proxy-bc.researchport.umd.edu/apps/doc/A695577174/AONE?u=umd_umbc&sid=bookmark-AONE&xid=0d4ee1b0.
⁷⁹Manuwi C. Tokai, Jannat Hossain, and Debs Grayson, "Review of *The Trojan Horse Affair* by Brian Reed and Hamza Syed," *Soundings: A Journal of Politics and Culture* 81 (2022): 106–9.
⁸⁰Quah, "The Trojan Horse Affair is a Twisty Thrill."

he indicated that the process of "becoming a character in the show" was "deeply, deeply uncomfortable." The editing process was one in which Syed, a neophyte who was only "days into my quote-unquote 'professional career'" was fighting Reed, a Peabody Award–winning veteran, over how much of Syed's emotional turmoil to include. "I was like, 'No, no, no. Let's take it out. Let's take it out. We don't need this. Let's focus on the story.'" And while he ultimately acknowledged that this approach had value, he doesn't seem completely sold on the idea that it was worth it. "I was reluctantly doing it."[81]

Reed, for his part, seems, like Koenig, to understand the dramatic impact of their conflict. "I love how mad you are," he tells Syed after a particularly frustrating interview. Syed's doubt and anger, on the other hand, are continuous on and off the microphone, in part because he doubts, almost from the start, that the podcast will have any actual impact, something he says in the first episode. Even Spencer Jones, the show's most enthusiastic critic, acknowledges that the impact of the podcast on actual policy in the United Kingdom hasn't yet been seen.[82] Sarah Larson, writing in *The New Yorker*, suggests the impact was more of a whimper than a bang, stating "the series ends in lightly poetic writing and investigative frustration."[83] Syed says he "was deeply, deeply disappointed by the project's inconclusive ending. I felt a sense of failure." For Reed, the creative goal of "telling a good-ass story" was accomplished. It is unclear whose feelings are more pertinent to the state of nonfiction podcasting's immediate future but the show's formal success suggests that feeling uncomfortable had become another rippling affect in the medium's traffic in feelings.

The industry of podcasting itself, in its explosive growth in the years after 2016, had been a source of discomfort. On Spotify alone, the number of programs grew from 700,000 at the end of 2019 to more than 4.7 million in September 2022.[84] Articles began

---

[81]Ismail Aymann, "*Trojan Horse Affair*'s Hamza Syed Has Some Regrets," *Slate*, March 9, 2022.
[82]Jones, "Transgressive (Un)Scripting."
[83]Sarah Larson, "'The Trojan Horse Affair' Works Best When Studying Itself," *The New Yorker*, March 20, 2022.
[84]Tyler Aquilina, "Podcast Exclusivity Is Quickly Becoming an Outdated Strategy," *Variety*, January 20, 2023.

to address the podcast glut as a source of "overwhelm" and "brain-melt," leading some listeners to quit cold turkey in the name of mental health.[85] Even Ira Glass confessed to feeling overwhelmed: "Why does everybody need to do a podcast now?" he asked a journalist in 2016.[86]

NPR hosts like Leila Fadel and Sam Sanders were drafted to pitch the network's news offerings as an antidote to the "exhausting" "plague of misinformation" coming from other media.[87] A 2022 animated short touted *All Things Considered*'s "calm curation, as a balm to the larger media environment, featuring a fantasia in gentle pastels, where passengers on a train pass through a tunnel in which affects circulate freely." During "those brief minutes on the bus or subway, your story is connected to everyone else's riding with you," explains the filmmaker.[88] This post-pandemic remediation of the Driveway Moment as the paradigmatic ideal of reception, replaces isolation with shared sociality and temporary rest with forward motion. Finally, it nicely adapts Ahmed's notion of the economic circulation of affects, moving over and through passengers and listeners.

At the same time, there was a growing unease at the unsustainable economics of podcasting. After the pandemic of 2020 sent daily listening routines along with the larger economy spinning into chaos, uncertainty grew about how listening habits and the advertising dollars that chased them would rebound. Spotify complained of "fairly dismal" margins in 2021 and even *JRE* saw its growth curtailed after it joined Spotify's walled garden of exclusive content.

---

[85]Dan Misener, "Overwhelmed by the Sheer Number of Podcasts? It's Not Just You," *Pacific Content*, January 11, 2019, https://pacific-content.com/feeling-overwhelmed-by-the-sheer-number-of-podcasts-its-not-just-you-d46b37a8fdcc/; Marcus Gilmer, "I'm giving up podcasts to save my brain and soul from overload," *Medium*, March 2, 2019, https://mashable.com/article/quitting-podcasts-information-overload. https://mashable.com/article/quitting-podcasts-information-overload.

[86]Steph Harmon, "Ira Glass: 'I Feel Like I'm Actually Sort of Scared All the Time,'" *The Guardian*, May 6, 2016, https://www.theguardian.com/tv-and-radio/2016/may/07/ira-glass-i-feel-like-im-actually-sort-of-scared-all-the-time.

[87]Leila Fadel, "Fundraising appeal," *NPR*, September 8, 2022.

[88]Sommer Hill, Sergio Romano, Cara Tallow, and Yu Zhang, "Behind the Frames of *All Things Considered*'s New Ad Campaign," *NPR*, August 5, 2022, https://www.npr.org/sections/npr-extra/2022/08/05/1114288282/behind-the-frames-of-all-things-considereds-new-ad-campaign.

"Decelerating growth" and a mismatch between ad revenue and the number of shows contributed to a sense of unsustainability.[89] By the early 2020s, it was clear that podcasting was confronting a boom-bust cycle familiar to other tech-based sectors of the economy. Spotify's decision to make its podcasts exclusive seemed by 2023 to have been a disaster, "a self-inflicted wound."[90] Scores of audio producers and technical staff took to Twitter to complain about peremptory layoffs and to ask for help in finding new jobs.

The discourse around podcasts (business, ethical, social) began to sound like an echo of the larger discourse of chaos, conflict, unsustainability, and uncertainty of the larger timeline rather than an appreciation of the stories that provided an escape, a way to "feel less crazy and separate." Metaphors for this decline were easy to find, like the rise of Better Help, an online therapy service, to the very top of all ad purchases in US podcasting for much of the early 2020s.[91] Square Space, Mail Chimp, and any number of online mattress vendors epitomized the ad space imaginary of podcasting's golden year of 2014, evoking a world of entrepreneurs, e-commerce, and a restful night's sleep in a Brooklyn walkup. The landscape of early 2020s podcasting on the other hand, with ads for Better Help and Talkspace, suggested anxiety and isolation.

---

[89]Aquilina, "Podcast Exclusivity."
[90]Doctorow, "Podcasts are hearteningly enshittification resistant."
[91]Brad Hill, "Podcast Advertisers in November," *Radio and Internet News*, December 17, 2021, https://rainnews.com-magellan-november-2021/.

Conclusion: Maybe It's a Feeling? How *This American Life* Long Endured

When asked "what is podcasting?" radio historian Jennifer Hyland Wang responded with this deceptively simple answer: "Maybe it's a feeling?" This was itself a question, a tentative framing appropriate to trying to solidify something still very much in solution even as late as 2020. She had in mind both the sheer variety of content and the expanding range of platforms, beyond RSS feeds and including YouTube videos of women's knitting collectives and video gameplay sessions, that were popularly referred to as podcasts. But she was also pointing, I think, to a specific fellow feeling that members of small, marginalized podcast communities shared.[1] I borrow this line because it has been a helpful one as I formulated the arguments presented in this book, especially as I tried to understand the role of empathy as a dominant ethos in the public radio structure of feeling.

Since coming across William Siemering's 1970 declaration that public radio should provide "an affective education," I have been interested in how this idea and his related injunction "people should feel," applied to literary, ideological, and civic interpretations of nonfiction audio media that blossomed in public radio and then

---

[1] Jennifer Hyland Wang, "What is Podcasting," webinar, *Radio Scholarly Interest Group, Society for Cinema and Media Studies*, August 20, 2000.

in podcasting. Raymond Williams's notion of a structure of feeling helped me understand the contradictions I heard on my radio, not as absurdities or lies, but instead as different possible futures vying to be heard and to cohere into something irrefutable.

Affect theory in its application to hate speech (Ahmed) and workplace indoctrination (Richard and Rudnyckyj) helped me to understand some of the powerful affordances of media as empathy machines. Barthes (via Glass himself) helped me to understand the emotional delights to be found in the play of formal narrative elements as they ripple through a story. Feminism and queer theory helped me to understand the powerfully gendered currents of attraction and disidentification circulating through public radio's gendered vocal performances. Theorists of the public, and its imbrication with the intimate, helped me grapple with the contradictions, even impossibility, of an American public while remembering it is a concept we cannot do without.[2] Theorists of neoliberalism, along with Barbara Ehrenreich and Lauren Berlant, have helped me sort out the hard mixture of empathy and market cruelty that have poured out of my radio speakers and earbuds for the last thirty years. All of it takes me back to Wang's question about the centrality of feeling and the rippling of affects through and across subjects, making sense before they even take on their final form.

Thirty Years of *TAL*

These many ways of thinking about the role of feeling have been particularly useful as I consider the main question I want to address in this conclusion: How has *This American Life* persisted and adapted through three decades of technological, industrial, and social changes in nonfiction narrative audio? More acutely, how did it weather the last decade of reckonings (economic, political, and social) that have wiped out or severely diminished so many other radio shows and podcasts? In the fraught years since 2016, it has racked up two Peabody awards, three DuPont-Columbia

---

[2] Bruce Robbins, *The Phantom Public Sphere* (Minneapolis: The University of Minnesota Press, 1993), xxi.

CONCLUSION

prizes, and a Pulitzer for stories that leaned into the thorniest and most difficult issues of the era: abortion rights, Black Lives Matter and racialized violence, immigration and asylum, undocumented workers, and rape.

In 2025, the year in which public funding for public media was clawed back and zeroed out, *TAL* boasted more than three million weekly listeners, almost evenly split between podcast downloads and tuning in to radio. It aired on the radio in Canada and Australia.[3] It ranked in the top ten on Apple's list of "Society and Culture" podcasts in the US, South Korea, Singapore, Germany, China, and was #3 overall in Japan, to name a few examples.[4] It employed 36 full-time employees, making it a massive operation by contemporary podcasting standards. Perhaps most importantly, the show continued to innovate in its stories, distribution scheme, its partnerships with new seasons of *Serial*, and in live road shows. In what follows, I suggest some answers as tentatively as Wang's and with the humble recognition that the news may have changed by the time you read this.

The first temptation when trying to account for its remarkable staying power is to locate *TAL* as *radio* (i.e., durable, endlessly remediated, and solid), the "greatest generation" of media. Podcasts, like CB radios on the other hand, can be thought of as epiphenomenal, at least until they celebrate their first centenary. In a recent email exchange with Glass, I noticed that he continues to refer to *TAL* as simply a "radio show." But such an argument requires ignoring lots of problems facing terrestrial radio, in general, and public radio in particular. The decimation of public funding followed already massive budget cuts and layoffs at NPR. Relentless demographic forces aided and abetted by the pandemic and new technology in phones and car dashboards had already reduced audiences for public radio. This argument also requires paying short shrift to the painful and necessary reckonings for the sexual abuse and harassment scandals that have rocked NPR,

---

[3] *This American Life*, "About," website, https://www.thisamericanlife.org/about#:~:text=Our%20show%20reaches%20more%20than,radio%20in%20Canada%20and%20Australia.

[4] *This American Life Podcast* rankings, 2024, Chartable, https://chartable.com/podcasts/this-american-life/charts.

WNYC, and public radio outlets in recent years,[5] to say nothing of the larger chaos of the political environment in which "the media" have been identified as the enemy.[6]

Most importantly, this explanation fails to account for the fact that *TAL* is every bit as much a podcast as it is radio show. Worst of all, this answer steers us into debates about podcast versus radio that move us away from the particular to the general. Arguments about radio's exceptional place in media history ignore that it has always been part of transmedia cultural practices, from its earliest remediation of the theater and the concert hall to its assemblage with the automobile, to its endless technological advances from audion to satellite.

Podcasting and Broadcasting: Feeling Media

On the other side, it's easy and frankly irresistible for the radio historian to pick apart the podcast exceptionalist arguments. Portability, really?[7] Intimate? Ahem.[8] Arguments about time-shifting stand on more solid ground, but members of Generation X have been quick to point out that the practice of taping off the radio, though clunky and impractical, was ubiquitous. On my first day of college in 1983, an apple-cheeked Midwesterner on my residential hall, when asked about what records she brought, proudly presented a small suitcase filled with taped cassette tapes of *Prairie Home Companion* (APM, 1974–87, 1992–2016) shows taped directly off of WSUI-FM, Iowa City's NPR affiliate. Such

---

[5]David Folkenflik, "NPR's Sexual Harassment Scandal," *NPR*, November 5, 2017, https://www.npr.org/2017/11/05/562188679/nprs-sexual-harassment-scandal.
[6]Tyler Falk, "Ted Cruz questions NPR about foundation funding in letter to CEO," *Current*, July 24, 2024, https://current.org/2024/07/ted-cruz-questions-npr-about-foundation-funding-in-letter-to-ceo/.
[7]Jason Loviglio, "The Car-Radio Assemblage," in *The Routledge Companion to Radio and Podcast Studies*, eds. Mia Lindgren and Jason Loviglio (London and New York: Routledge, 2022), 226–36.
[8]Jason Loviglio, *Radio's Intimate Public: Network Broadcasting and the Mass-Mediated Public* (Minneapolis: University of Minnesota Press, 2005).

practices have recently begun to receive scholarly attention for their role in reception, musical subcultures, and musical innovations.[9]

One utility of examining claims of podcast exceptionalism is what they reveal about how podcasters themselves felt about the new medium, what they saw as its greatest affordance. This may help us to better understand *TAL*'s success in straddling this debate and in riding out the tumultuous fortunes of both forms of audio. Podcasting has been described and celebrated for two decades in terms of *freedom* for producers, hosts, and listeners alike. This is often framed in terms of freedom *from* the strictures associated with radio. Spinelli and Dann identify podcasting's "greater freedom" from the "timing and scheduling constraints of broadcast media" which, they note, are burdened as well with high production costs and narrow formats. Low technical and capital barriers to entry meant talented amateurs could "do it without having to deal with all of the bullshit of a station NPR or whatever," declared Roman Mars, one of the many podcast titans who got started in public radio.[10]

Jad Abumrad, whose *Radiolab* began as a live WNYC program, embraced similar language when describing his show's transition to its podcast-forward version, "we're not beholden to radio forms." Journalism scholar David Dowling also notes that podcast journalists are "free to develop far richer and more dense content." Echoing Spinelli and Dann, he also notes that "podcasts liberate hosts from" radio's scheduling and production limitations.[11] As a born-digital medium that arrived as sound files, listeners found new temporal freedoms, listening and re-listening, whenever they wanted.

---

[9] Eleanor Patterson, *Bootlegging the Airwaves: Alternative Histories of Radio and Television Distribution* (Champaign: University of Illinois Press, 2024); Zita Joyce, "Taping Radio: Recording Memories," in *The Routledge Companion to Radio and Podcast Studies*, eds. Mia Lindgren and Jason Loviglio (London and New York: Routledge, 2022), 389–98; Pacey Foster and Wayne Marshall, "Tales of the Tape: Cassette Culture, Community Radio, and the Birth of Rap Music in Boston," *Creative Industries Journal* 8, no. 2 (2025): 164–76, doi:10.1080/17510694.2015.1090229.

[10] Martin Spinelli and Lance Dann, *Podcasting: The New Audio Media Revolution* (London: Bloomsbury, 2019), 88; Eric Steuer, "Roman Mars: The Man Who's Building a Podcast Empire," *Wired*, January 28, 2015, https://www.wired.com/2015/01/podcaster-roman-mars/.

[11] David Dowling, *Podcast Journalism: The Promise and Perils of Audio Reporting* (New York: Columbia University Press, 2024), 88.

Even Cory Doctorow celebrated podcasting as "hearteningly resistant to enshittification." Like all new electronic media over the last century, podcasting has been framed by the rhetoric of the technological sublime. "Revolution" appears in the titles of scholarly books, articles, and journalists' headlines, and the largest professional conference calls itself a "movement."[12] "These liberties, by their nature," observe Nee and Santana, speaking of the new affordances of podcast journalists, "including setting aside long-held institutional norms in the news dissemination."[13]

There are traces in this discourse of the language of neoliberalism, which David Harvey and Wendy Brown tell us, is often sold in terms of freedom of economic actors from constraints of governmental regulations, including those having to do with pesky questions of the public interest as a matter not fully addressable by market forces.[14] Neoliberal reforms are often hailed in the language of "revolution," as mentioned above, a keyword in the emerging lexicon of podcast studies and promotion and one that is closely associated in the American imaginary, with the soaring rhetoric of the Enlightenment (e.g., freedom, liberty, and independence). Neoliberal economic reforms and revolution have been paired explicitly in the work of economists like Friedrich Hayek, Milton Friedman, and others associated with the Chicago School.[15]

Gimlet's privatization of public radio talent and its rapid enshittification by selling out to Spotify, followed by massive layoffs, and the datafication of listeners has been likened to a chapter right out of the neoliberal playbook.[16] Keeping one foot in the world of public radio, *TAL*, still a production of Chicago's WBEZ, may have

---

[12]Spinelli and Dann, *Podcasting*; Arran Bee, "Podcasting: The Audio Revolution," in *The Radio Handbook*, eds. John Collins and Arran Bee (Oxfordshire: Taylor and Francis, 2019); Swish Goswami, "The Podcasting Revolution," *Forbes*, December 20, 2019, https://www.forbes.com/sites/forbestechcouncil/2019/12/20/the-podcasting-revolution/.
[13]Rebecca C. Nee and Arthur D. Santana, "Podcasting the Pandemic: Exploring Storytelling Formats and Shifting Journalistic Norms in News Podcasts Related to the Coronavirus," *Journalism Practice* 16, no. 8 (2021): 1559–77.
[14]David Harvey, *A Brief History of Neoliberalism* (London: Oxford University Press, 2007); Wendy Brown, *Undoing the Demos: Neoliberalism's Stealth Revolution* (Princeton, NJ: Princeton University Press, 2017).
[15]Naomi Klein, *The Shock Doctrine: The Rise of Disaster Capitalism* (New York: Picador, 2007).
[16]John Sullivan, *Podcasting in a Platform Age: From an Amateur to a Professional Medium* (London: Bloomsbury, 2024).

avoided the worst predations of this boom-bust cycle. But Glass has by no means avoided and has in fact very publicly embraced "capitalism" with partnerships and sell-offs with several corporate media entities. As I've tried to show in this book, especially Chapter 4, the entire history of public radio has been captive to the political economy of neoliberalism and has, in some ways, been an expression of its cultural logic.

The Show as Transmedia Form

Instead of finding an explanation for *TAL*'s longevity in the success or failure of either platform, it makes sense to look at the very specific ways it has combined questions of form and content to make a show that is never any one thing in particular, but always in motion between things. This has been less a story about freedom than one about a long-term commitment to the endless play of meanings and circulation of affects. *TAL* has endured because it has never entirely given up its perch between the two media along with forays into others, such as film, TV, and live events. As a hybrid of media forms, it has always been, first and foremost, a "show."

Glass and other on-air producers are careful to refer to it that way, referring to a "podcast version" of the show or a family friendly version with "curse words beeped out" and directing to "the version of the show on our website." Thus, there is no single original text, a feature that Spinelli and Dann observe is unique to podcasts; for *TAL*, these multiple versions afford a kind of freedom from any single medium as well as the ability to edit content after publication. The promiscuous play of possible meanings circulating around the word "show" is, I think, part of its appeal for Glass, along with its nostalgic historical resonances. Glass says of his childhood career as a magician, which I explored in Chapter 3, "there was something about [putting on] shows, that got me into media, and that was what got me into radio."[17]

The first use of the word "show" to describe an exhibition of any kind dates back over 800 years; its association with entertaining a crowd was first used in the 1560s. By 1760, a show could refer to

---

[17]Claudia Dreifus, "To Get Things More Real: An Interview with Ira Glass," *The New York Review of Books*, August 8, 2019, https://www.nybooks.com/online/2019/08/08/to-get-things-more-real-an-interview-with-ira-glass/.

"an exhibition of strange objects, trivial performances, etc." and for the last 200 years, it has meant "any kind of public display or gathering."[18] Radio programs have been called "shows" since at least 1932, by which time "the show business" had begun to migrate from live theatrical events to the emergent transmedia phenomenon that joined broadcasting, motion pictures, recorded music, and live events into a coherent cultural force. Over the last century, "the show" has gathered from these multiple connotations its own cultural logic and even an ethos. The phrase, "the show must go on," dating from the 1890s, exceeds purely commercial imperatives to embrace an emerging sense of professionalism, workplace solidarity, and an ambient sense of higher service to an audience, a public.[19] The stoicism implicit in that saying is leavened by "show's" connotations of playfulness and benign deceit, a trick played on willing victims.

TAL, which began as *Your Radio Playhouse*, has always leaned into these historical traces, combining a keen, if often ironic, attachment to older forms of talkshow, as in the interview with veteran talkshow host Joe Franklin in the debut broadcast (see Chapter 3). Combining the show's connotations of event, "exhibition of strange objects and trivial performances," "public display," and the twentith-century notion of a transmedia "show business," it has borrowed from these resonances a status beyond radio or podcast. With its penchant for moving audiences between empathy and irony, constructing narrative surprises and moments of amusement, it conjures an atmosphere of improvisation through painstaking editing. Steeped in a tradition with benevolent trickery at its core, like a magic trick, *TAL* is first and foremost, a show.

To the extent that *TAL* embodies the qualities of a show, it has not been closely attached to any particular topic, trend, or theme. This is a distinction that cannot be said of many podcasts that engage listeners in ways that are specific to a moment in time or a particular social problem, cultural preoccupation, or format. Indeed, its central structuring principle is its presentation of novel

---

[18]"Show," *Etymology Online*, 2024, https://www.etymonline.com/word/show#:~:text= Middle%20English%20sheuen%2C%20from%20Old,%2DGermanic%20root%20 *skau%2D%20%22.
[19]"Show."

themes in each episode, often juxtaposing wildly different takes on that theme, an emphasis perhaps on the elasticity and recombinant playfulness of its commitment to that theme. It is the movement between and among narrative threads and the sleight-of-hand storytelling maneuvers to bring them together that provide a sense of continuity.

Meeting the Moment

It could be countered that the show's title refers to the theme of "American life," which in an age of global digital networks and geographically unbounded communities of interest, could certainly be counted as a limit, if not a liability. However, the national specificity of the show, as I've tried to show in Chapters 3 and 4, was haphazard from the beginning. The term "American" evoked its broadcast ambitions circa 1996 and a sense of universality rather than a narrow sense of an America distinct from the rest of the world. Even in the post-9/11 era, when the focus on national identity was explicit and acute, topics and guest speakers were often Americans *in motion*—recently arrived or recently departed, like Hyder Akbar who left the United States as a teen for Afghanistan.[20]

One theme running through the previous chapter was the difficulty public radio and podcast organizations and programs had in navigating new and urgent diversity, equity and inclusion demands in the workplace while maintaining the same tonal and emotional register of their shows and the same affective bonds with their listeners. These listeners had, by and large, been inculcated into a public radio structure of feeling in which socially powerful storytellers demonstrated empathy for and ironic distance from socially marginal others. The collapse of *Reply All*, the demise of *Invisibilia*, and the other racial and gender reckonings rippling through public radio and podcasting seemed to pose the Spivakian question, "can the subaltern feel?"[21] To be more precise, these

---

[20] Ira Glass, "Come Back to Afghanistan," *TAL*, episode #230. January 31, 2003.
[21] Gayatri Chakravorty Spivak, "Can the Subaltern Speak?" in *Marxism and the Interpretation of Culture*, eds. Cary Nelson and Lawrence Grossberg (London: Macmillan, 1988), 271–313.

changes posed the challenge: can podcasting's traffic in feelings accommodate the inclusion of socially marginal people as subjects and not merely as objects?

Emmanuel Dzotsi's awkward and doomed stint as a co-host for *Reply All* suggested not. He was brought in only a couple of months before the larger contradictions of the public radio structure of feeling brought the show to an abrupt and emotionally fraught conclusion. *TAL* suggests a potentially different conclusion. It has managed to make steady and dramatic changes to its personnel in ways that managed this diversity with relative ease and to significantly impact the sound of the show to evolve with the changing traffic in feelings circulating through the narrative audio landscape that *TAL* had done so much to define. That these changes have not resulted in complaints from listeners, as was the case at *Reply All*, speaks, I think, to *TAL*'s central and abiding commitment to putting on a show with all of the connotations of display, amusement, strangeness, and triviality that come with it. This commitment, with its implied acknowledgment of the importance of timing, narrative enchantment, emotional stakes, and delight in sleight-of-hand movements of one thing into something else meant that Glass and company did their best work when things were in motion.

We can trace some of this movement through the example of Emanuele Berry, hired as a producer/editor in 2019 and promoted to executive editor in 2021. Berry began her career at a small public radio station in Lansing, Michigan, before shifting to Gimlet where she edited and produced for shows like *The Nod*, *Undone*, and *Startup*. A young African American woman from the Midwest, Berry briefly attended a community college, won a Fulbright to Macau, and covered the uprising in Ferguson, Missouri after the 2014 police killing of Michael Brown. She brought very different experiences to the show's management along with a compatible approach to the central role of feelings in storytelling.

In a decade that saw a massive outflow of public radio talent to Gimlet and other for-profit networks, a privatization of the commons and a major first step in the enshittification process, *TAL* responded by cheerfully poaching back from these corporate walled gardens, hiring Berry as well as Sarah Abdurrahman, a Muslim woman of Libyan descent, as its new managing editor in 2020. Abdurrahman was another public radio veteran poached by Gimlet before *TAL* hired her. Editor Phia Bennin and Producer Diane Wu followed a similar trajectory. The movement of audio producers from public radio to for-profit networks and back again speaks to the show's preference for circulation and play over stasis. Rather than dwell on public radio's failures to develop, promote, and retain talented staff from diverse backgrounds, *TAL* simply hired these folks from Gimlet and other for-profit networks with an alacrity and acumen not often associated with public media.

These hires took place with little fanfare in the years after 2016. The organizational chart for the show reveals that six of the eleven top management positions directly reporting to Glass belong to women; three of which are women of color. Ten of the thirty-five full-time positions are held by people of color; and two thirds are women or non-binary. These transformations happened during a period of tumult and growth in podcasting and political turmoil in the country about questions of race and gender equity and representation from Hollywood to Washington, DC.[22] In a 2025 interview, Glass mused that it became easy to hire talented staff once his show and his style had become dominant—"they all knew how to do it and I didn't have to teach them."[23]

Berry first guest-hosted an episode in 2022 titled "Talking While Black," which was dedicated to the specificity of Black American experience and feelings at that precise historical moment, two years after the summer of George Floyd's murder and the massive uprising it sparked and well into a period of sustained backlash against "Critical Race Theory" (CRT) curricula in schools and diversity, equity, and inclusion (DEI) initiatives in workplaces of all kinds.[24] It is hard to imagine a white host managing the particular challenges of moving between context, anecdote, and takeaway without some uneasiness. And, because of the show's relentless focus on emotional presence, that uneasiness (and its whiteness) would have become a central character. The episode won a DuPont-Columbia Award, the

---

[22]"This American Life," *The Org*, https://theorg.com/org/this-american-life; This American Life, Staff page, https://www.thisamericanlife.org/about/staff.
[23]PJ Vogt, "Is it OK to Work All the Time?" *Search Engine*, January 10, 2025.
[24]Rashawn Ray and Alexandra Gibbons, "Why Are States Banning Critical Race Theory," *Brookings: Commentary*, November, 2021, https://www.brookings.edu/articles/why-are-states-banning-critical-race-theory/; Eesha Penharkar, "These Were the Most Banned Books in 2023," *Education Week*, April 24, 2023, https://www.edweek.org/teaching-learning/these-were-the-most-banned-books-in-2022/2023/04.

show's sixth. Berry centers her own feelings right off the bat but does so in a way that is tonally aligned with Glass's signature play of diffidence and emotional presence. She describes, with sardonic forbearance, the ubiquitous and very public virtue signaling in the summer of 2020 around issues of race in the aftermath of the police killing of George Floyd.

> Emanuele Berry: Remember when everyone on Instagram posted black squares for a day to show solidarity with the Black community? I'd started to roll my eyes at all the MLK and Baldwin quotes. The murder of George Floyd had forced the country into another racial awakening.

There is, of course, another element in this particular distancing move that distinguishes it, I think, from Glass's 1990s hipster's disdain for the popular, the over-hyped. The complicated stew of emotions that many Americans of color expressed during periodic spasms of public concern over police violence against them, much of it symbolic and performative and emotional, is hard to capture in a simple anecdote.[25] The word "another" to modify "racial awakening" and the choice of "awakening," with its spiritual, vaguely New Age-y connotations, is doing a particular kind of emotional work that is both subtle but also hard to miss if you're tuned in. That's partly why the episode's theme, "talking while Black," manages to be evocative, layered, and challenging all at once. There's much to say, but the conditions for saying it in a public forum like *TAL*, which has been explicitly coded as white for so long, is tricky, requiring narrative dexterity. It requires what Raymond Williams calls "characteristic elements of impulse, restraint, and tone."[26]

The first anecdote, shared in the introduction, sets the scene for the entire episode, just as it would in a Glass-hosted show. A Black principal in a town near Dallas, Texas, Dr. James Whitfield, is

---

[25] Gene Demby, Shereen Marisol Meraji, Leah Donnella, Steve Drummond, Brianna Scott, and Alyssa Jeong Perry, "The Racial Reckoning That Wasn't," *Code Switch*, NPR, June 9, 2021.

[26] Raymond Williams, *Marxism and Literature* (Oxford: Oxford University Press, 1977), 132.

caught up in the whipsaw of racial politics in the US when a public statement he wrote in 2020 at the height of the racial awakening, which thanks his "white brothers and sisters" for their solidarity and pledges to do what he can to "disrupt systemic racism and eradicate it," came under scrutiny during the backlash of 2021 when an "anti-CRT" law passed in Texas. A predictable protest by white residents of the town ensued—another public scene at the school board! Whitfield, the school's first Black principal, was let go. The story ends with him describing what it's like to drop his daughter off at the nearby elementary school and the feeling of seeing the high school building he once ran but is now not allowed to enter.

> Whitfield: And so, yeah, it's really hard to pass by a place that you love and know that there's staff members that you love in there… And you know that's where you're supposed to be. But you're not allowed to be in there, right? It's like, what kind of person is not allowed to be in there? You know, it's disheartening.

Centering the emotional stakes locates this small story as part of a bigger one while still making it, essentially, a story about feelings, linking Berry's sardonic tone when describing the racial awakening of 2020, to the brutal wakeup call of Whitfield's encounter with racial backlash in 2021. It is a story that Glass could have done, but it hits different with Berry as host and narrator. Berry builds the episode's thematic trajectory in the house style: anecdote followed by grand takeaway: "The line of what's acceptable to say about race and racism in America, it moved… And that shift has left many Black people exposed and vulnerable and living with those consequences."

In addition to tone and structure, the opening anecdote also hews closely to the journalistic details that elucidate the theme. At the next school board meeting, Whitfield is given exactly one minute to defend himself and his record and to address the shifting contexts in which his words have been interpreted. Berry plays the audio so we can hear him racing through his argument, trying to make a complicated case about a complex issue in a dynamic era, in very few words. This impractical and unfair time limit, the episode implies, is part of the constraints of "talking while Black" that set the stage for the rest of the episode. Berry confirms that a minute is not quite enough time, setting up the rest of the hour: "what to say in this moment of backlash. Stay with us."

Solving for a Different X

One thing that has shifted in the years since *TAL*'s early years, however, is the role of surprise. Covering the political often means focusing on patterns and acknowledging the force of layers of historical residue that accumulate in the lived experience of hierarchy and struggle. Berry makes this explicit, as if making sure that her sardonic reference to "another racial awakening" has not been lost on anyone. "This backlash, it's not surprising." She's taking an editorial stance that Glass has often eschewed. "This is what America does, Reconstruction, then Jim Crow, the Civil Rights movement, to the war on drugs, Obama to Trump." Berry pivots, as if to acknowledge that surprises may not be the stuff of politics, but they are the stuff of emotions: "But what I am surprised by is the way people have been caught up and tangled in [the backlash politics]."

Berry follows this thread of backlash and its surprising affective currents as it ripples through Traverse City, Michigan, a small mostly white town that, in 2021, had its own moment of awkward public meetings and heated rhetoric. The piece begins with audio of white parents angrily denying the existence of racism in their town or in the nation. Berry sets up the backstory that led to this scene. A group of white high school students set up a Snapchat group text titled "Slave Auction," in which participants could bid on purchasing their Black classmates. It also included such messages as "All Blacks should die. Let's have another Holocaust."

What starts as a story of a town in turmoil zeroes in on the impact it has on a single Black girl, Naveah, and how she tries to navigate her friendships with white students, some of whom were on the group chat. The story follows the emotional currents common to other kinds of high school drama. Apologies are awkwardly tended and awkwardly accepted; friends speak on behalf of other friends; support ripples toward Naveah at first then wanes when some begin to feel like the controversy has taken up too much attention and who want to "move past this incident."

CONCLUSION

Talking while Black, for Neveah, is explored in this story as part of a universal story of fitting into high school's maze-like social structures and friend groups, even as you're developing into an adult. It's a John Hughes movie and Spike Lee Joint in one. Through it all, Neveah tries to figure out how to feel as she navigates the aftermath, and her story follows a kind of hermeneutic code logic of uncovering a mystery. Berry submits a Freedom of Information Act (FOIA) request to get the full transcript of the group texts, because Neveah cannot reconcile her friendships with not knowing who said what. The stacks of paper she gets are heavily redacted with thick black lines blocking out most of what was said, a frustrating bureaucratic layer of meaning framing the larger backlash dynamic in which the 2020 summer of accountability fades to black. The final scene, Neveah sitting on her bed, surrounded by piles of FOIA documents, trying to sort them into some kind of sense, is affecting in its lack of closure. Berry leaves her there, "sifting through the clues, still puzzling, still trying to find something to help her make sense of it all."

The second act tracks the backlash and its impacts as it circulates through the banning of a specific book, *New Kid* written by Jerry Kraft, a Black children's book author. Berry reminds us of the vertiginous rise in interest in Black-authored books in the summer of 2020 and the equally fierce round of book bans in schools and libraries around the country in the following year. Kraft and his book, which is about "finding your way in middle school," get caught in the crossfire. Chana Joffe-Walt reports the story as a set of dueling arguments about whose feelings should be prioritized. She makes it clear that the book is "about being uncomfortable," a theme that is of course universal to the middle school experience and which took on new currency in the post-2016 era, as I discussed in the previous chapter.

Because Kraft's protagonist, the eponymous *New Kid*, a boy named Jordan Banks, is Black, his discomfort becomes a target. A white parent in Katy, Texas, Bonnie Anderson, who'd run unsuccessfully for the school board on an anti-mask platform, lodged a complaint about a scheduled book event Kraft had been invited to at a local school, citing a new Texas anti-CRT bill that explicitly forbids books that cause students to feel "discomfort, guilt, or anguish." Such laws have taken effect in twenty states, Joffe-Walt tells us. "These laws almost never list specific books you can't teach. They talk about feelings." But her story centers almost entirely on how Kraft tries to make sense of the controversy. Like Nevaeh, he's bemused, struggling to understand.

Joffe-Walt describes a scene in another of Kraft's books, the sequel to *New Kid*, when Jordan's white teacher admonishes him for the "angry" drawings she finds in his sketchbook. Just as Jordan tries to exonerate himself for depicting his own experiences of race and fitting in, Kraft is at pains to locate his character's experiences in his own biography. "We are exactly the same," Kraft says of his protagonist. The story ends with another school board scene, impassioned arguments, and uncomfortable parents. Joffe-Walt observes that "Jerry is already uncomfortable... He never had any laws protecting him from discomfort... What he had was the ability to write it down, to talk about it." In an interview about the episode at the DuPont-Columbia awards ceremony, Berry said she thought it was "silly" to imagine telling such a story and not bringing her feelings as a Black woman into it. The interview, like "Talking While Black," are utterances that simultaneously reaffirm *TAL*'s innate deference to emotional presence while subtly shifting the terms of what can be felt and who can do the feeling and talking, a narrative sleight of hand worthy of her boss, Ira Glass.

The Playhouse

Longevity in show business was a preoccupation in *TAL*'s first episode, back when it was called *Your Radio Playhouse*. Glass muses on the show's early creative period after only a few minutes, a humorous acknowledgment of the rapidity of the modern media lifecycle. In an interview that balances courtly deference and subtle mockery, Glass also interviewed talkshow veteran and self-proclaimed pioneer, Joe Franklin. Glass wants to know the secret to Franklin's long tenure in broadcasting, a medium he both loves and hopes to revolutionize. Thirty years later, it is interesting to consider Glass as a media figure whose longevity is now strikingly impressive and whose ability to find creativity in each historical era represents a counterpoint to the boom-bust cycle in much of nonfiction podcasting.

In the introduction to this book, I wrote that Glass, as his name suggests, is best appreciated for his ability to reveal and reflect: a window into radio's historical talkshow format, a mirror for sharing images of contemporary life, and a looking glass through which the imagination could find freer rein than in more staid nonfiction work.[27] It has just as often been a funhouse mirror, celebrating the absurdity in the everyday. If there's a throughline, from *Your Radio Playhouse* to the version of *TAL* that 1 million radio listeners and 1.6 million podcast listeners tune into in the mid-2020s, it is perhaps the endless play of reflections, refractions, and funhouse surprises that this metaphor suggests.

A 2023 episode titled "Say it to My Face" begins with a confession from Glass that is both hilarious and awkward. It strikes me as an example of the show's ability to age gracefully. Glass begins, "A few months ago, somebody said something to me that I found so humiliating that I have never talked about it." He reasons that while he can't tell friends and family, he can say it on the show. "We have a different relationship, right." He sets the scene, an awkward public one. He's biking home from work in a suit, coat, reflective vest. His folding bike, he says, is "poncy" and he confides, "I have gray hair at that point, a neatly trimmed beard. I was wearing a helmet. All in all, I was a picture of fastidious care." A young woman, "full of cheerful loose energy," sees him and remarks, "do you fart out the front?" He laughs as he recalls this, even as he struggles to decode the image literally. But he takes her point: "It's kind of like, you fussy, foppish, overcareful old man."

Anecdote shared, he follows up with a trademark reflection: telling strangers something is easier than telling our intimates. "With our friends, with our loved ones, it can be so much more difficult to be honest, sometimes. Right?" What strikes me about this introduction is the way he layers the central emotional insight; being honest to strangers is different, over and over in a tight little anecdote. It's what gives him license to share something embarrassing on the radio/podcast. It's how he makes sense of the woman's comment, her candidness, an extension of her "loose energy." But it's also a way to frame the episode: a pair of stories about the difficulties involved in hard conversations with friends about old hurts and unspoken feelings and the way that putting on a show somehow makes it all easier.

---

[27] Wayne Munson, *All Talk: The Talkshow in Media Culture* (Philadelphia: Temple University Press, 1993).

The first and longer of the two acts tells the story of a young man named Gabe who deals with his pain and anger about a friend's betrayal by putting on a one-man-show about it rather than confront his friend directly. *TAL* producer Aviva de Kornfeld not only tells this story, but she also actively intervenes, inviting Gabe's estranged friend Tim to the show, bringing him to town, and walking him to the theater. After the show, she mediates between the two, eventually asking Tim to apologize to Gabe. In another narrative layer, de Kornfeld is at the same time also delaying an awkward encounter of her own with a former friend who dumped her. Gabe is unstinting in his honesty in his show and in his interviews with de Kornfeld for her show. Throughout this episode, and really across the entire *oeuvre*, the emotional power of putting on a show persists. It is in its dedication to the notion of a show that *TAL* has managed to find powerful moments of truth through play. It helps to explain the show's penchant for distilling feelings into cinematic "scenes." And in its self-conscious layering of reflections, refractions, and fun-house grotesques, *TAL* retains, thirty years and over 850 episodes on, a radio playhouse.

In these pages I have attempted to map the play of affects as they've rippled through *TAL*'s long history as a radio show, a podcast, and a transmedia brand. I started with its inspiration in the radio show's century-old legacy of evoking and regulating feelings as they brought listeners together "in a state." I situated its early preoccupation with empathy within public radio's "make-them-feel" history of liberalism, understood not as a political philosophy but as a bundle of affects circulating through and constituted by audio narratives. And while empathy was at first a way to "feel good about feeling bad"—stories to make privileged public radio listeners feel for "others," the uses of empathy shifted after September 11, 2001. In a new era of crisis, terror, and starker lines between right and wrong, *TAL* sought to evoke empathy for the storytellers themselves, who suddenly found safe narrative alcoves and artful stories about universal interiority harder to come by. Feeling liberal in the early 2000s could best be conveyed by an uneasy longing for middle ground and the pleasurable stasis epitomized in the Driveway Moment. The literary pleasures of narrative became both a refuge from and way into the wars abroad and new divisions at home.

CONCLUSION

The subprime mortgage crisis and ensuing Great Recession produced another challenge to public radio's liberal structure of feeling, pitting brutal material and conceptual contradictions against a house style that prized nuance, mystery, and the play of in-between-ness. Formidable journalistic resources were brought to bear on the crisis, along with a empathy for the storytellers, mediators, translators whose struggle for middle ground became increasingly difficult. These stories resonated in part because public radio journalism had, by the 2000s, successfully marketed itself as a source for inspirational, "moments" of feeling, rather than for sustained and urgent calls to action that the decade seemed to require of a liberalism based in "active constructive participation," per NPR's founding "Purposes."

It is impossible to understand public radio's appeal to liberal sensibilities without an understanding of the role that gender ideologies and vocal performances have played in the creation of its unique sound. *TAL* perfected the sound effect of the soft male voice as a cipher for a kind of post-feminist liberalism, the perfect sonic alibi for avoiding confrontation with the neoliberal usurpation of some powerful feminist and queer signifiers. Further, the show made explicit the centrality of masculinity and its queer relations to feminism, liberalism, and to the men whose power to speak without getting in trouble depended upon its freedom from political and aesthetic constraints. *TAL*'s men, straight and queer, discovered in the bind of masculinity another privileged alcove from which the play of opposites could be seen and talked and talked about.

If *TAL* can be said to have invented a certain kind of audio narrative style, its impact was most strongly felt in the explosion of high-quality journalistic podcasts of the 2010s. There the pleasures of audio storytelling spread across myriad platforms, many of them for-profit, and across many topics, including those that experimented with a bolder, muckraking style of journalism. Calling out the failures of liberalism, with newfound attention to "implicit bias," and longstanding tolerance for stubborn inequities, these new shows pushed the boundaries of what feeling liberal might sound like. It was inevitable that the contradictions and hypocrisies of the liberal empathy machine would begin to founder in an era of urgency, impatience, and reckonings. Public scenes in which liberal ideals met with failures, impasses, and humiliations represented a dramatic way to "turn it into a feeling." Narrators, mediators, and journalists once again pleaded for empathy as they sought to find balance between irreconcilable opposites.

Almost as soon as podcasting seemed to beckon towards a vast new world of audio journalistic possibility, the form itself seemed to stagger under the weight of its own cringe-y shortcomings, as illustrated by *Reply All*'s seeming hypocrisy in ferreting out race- and class-based hierarchies in other organizations while reproducing the same dynamics itself. Market saturation, shifting ad revenue calculus, and the platformization of audio content represented additional challenges to the form as it exited the pandemic in the early 2020s. *TAL*'s success through each of these eras represents a singular achievement. But by late 2024 Ira Glass struck an elegiac tone when discussing the show, as if merely surviving to continue to translate the impossible American affective landscape into narrative sound was no longer an imperative, but instead perhaps, just a job he'd gotten very good at.[28]

As radio, podcasting, and streaming audio industries continue to navigate changing technological possibilities and the attendant financialization and enshittification of creative output, *This American Life* has provided an interesting model for centering change. As the culture wars continue to rage through fictional and nonfictional representations of American life, it has modeled grace under pressure; like Glass on his bike in a suit and reflective safety gear, grace doesn't always look graceful. Now that podcasting's first golden age has passed, I'm optimistic that we can move beyond the medium-specific analysis of nonfiction audio narrative. Instead, I hope we can refocus on the ancient affordances of oral storytelling, which Walter Benjamin reminds us, both depends upon a certain temporality while imposing its own.[29]

At the same time, I hope that this book has helped us think about the modern legacy of the show as an historical form for shaping modern experiences of shared presence and the traffic in feelings across shifting distances, modalities, and reception practices. Finally, I hope that this book has helped to shed some light on the historical linkages between the limits of empathy—

---

[28] Ira Glass, "Nancy's Deep Cuts, Part 2," *TAL*, podcast. October 30, 2024.
[29] Walter Benjamin, "The Storyteller: Reflections on the Works of Nikolai Leskov," in *Illuminations* (San Diego, CA: Harcourt Brace Jovanovich, [1936] 1968).

partial, fleeting, passive—and the enduring contradictions inherent in liberal capitalism. If affect is the medium in which subjects are formed, and stories are our most affecting forms of creativity, we need to listen in to the stories that we are telling each other about our daily experience. We need to listen back to the stories we've told to get us to this strange point in history. And we need to continue to listen out for new voices and new stories, even when they make us a bit uncomfortable.

Afterword

The election of 2024 brought Donald Trump back to the White House with renewed attacks on the very underpinnings of liberal democracy—civil liberties, constitutional protections, pluralism, even the rule of law. It followed a presidential term by a Democratic centrist whose unpopularity was due in some part to a "vibecession," an affective current of unhappiness running counter to many economic and social indicators. The start of the second Trump term seemed designed to maximize the political power of shock as a way to further demoralize a liberal opposition whose life force seemed to have already been scattered across an affective landscape of confusion, anger, and despair. Complicating the work of understanding and resisting the assaults on democratic norms and civil and human rights was the steady barrage of spectacular scenes of far-right stagecraft: Nazi salutes, ICE raids, and outlandish proclamations of a revivified US manifest destiny stretching from the Gulf of Mexico to Gaza. Cancelling federal support for public media appeared, in the context of these onslaughts, a relatively minor shock.

Naomi Klein has made clear the power of the shock doctrine in right-wing coups of the past, providing insights into the limitations of the press, the people, and the institutions of liberal democracy to cope with spectacular shows of force and violence. But what proved most challenging in early 2025 was the supplemental bastions of seemingly sober and reasoned discourse on right-wing media channels, patiently arguing in favor of absurd propositions like "Elon Musk wasn't making a Nazi salute; he was just waving." Much of this discourse took place on podcasts, which had become one of the most potent bulwarks of the right-wing attack on liberal democracy in the years following the attempted coup of January 6, 2021.

The year 2024 was, by some measures, the first podcast election. Much was made in 2024 about the role of podcasting in the campaign strategies of Trump and his last-minute opponent, Vice President Kamala Harris. If the political currency of nonfiction journalistic audio had waned from its mid-2010s peak, the chatcast/interview podcast was still riding high. Trump and his vice presidential candidate J. D. Vance spent hours on shows like *The Joe Rogan Experience* (*JRE*). Such shows represented the triumph of AM talk radio's legacy of right-wing conspiracy-mongering and anti-government animus. But in its podcast avatar, the talk format seemed to lose its connotations with the lowbrow atavism of the AM band. *JRE* and other shows assumed a mantle of (relative) intellectual prestige, dedicating hours to deep-dive explorations of science, policy, and often fringe theories. Eschewing the broadcast clock and associated limitations of radio formats, long-form interview shows reinforced claims to seriousness of purpose, intellectual curiosity, and the exchange of opposing views. Such shows, with their unhurried conversations, wide array of political opinions, and moderated tones, could be said to have inherited the mantle of the liberal media.[30]

"Just asking questions" became, in the talkshow podverse, a way to insinuate conspiracy theories alongside mainstream arguments, providing a rhetorical symmetry to the views and the appearance of a robust, even virile liberal toleration. John Durham Peters warned about the dangers of liberalism's "abyss artists," who in "warming themselves by the fires of hell," courted the enemies of liberal democracy. It was in these very terms that politically savvy cynics began to shift their alignments towards the inevitability of Trump redux. Witness Jeff Bezos's invocation of journalistic objectivity in refusing to let his *Washington Post* editorial board endorse a candidate. Or Mark Zuckerberg's appeal to free speech in his abrupt decision in January 2025 to drop Facebook's fact-checking bot and weaken its anti-hate filters. Even TikTok embraced the

---

[30]Kim Fox, "The Turn to Podcasts as a Mass Campaign Medium," *The Journal of Radio and Audio Media*, forthcoming; Nivaldo Ferraz and Daniel Gambaro, "Podcasting and the Language of Destruction: Far-Right Discourse for a Parallel 'New' Brazil," in *The Palgrave Handbook on Right-Wing Populism and Otherness in Global Perspective*, eds. Rui Alexandre Novais and Rogério Christofoletti, Global Political Sociology (London: Palgrave Macmillan, 2025), https://doi.org/10.1007/978-3-031-77868-1_8.

incoming president as a free-speech champion. Such risible claims echo the language of skepticism and freedom of choice while also embracing the paleo conservatism of Christian nationalism and the crypto-finance movement on the libertarian fringe. In other words, liberal rhetoric in its virtuosic plasticity, has managed to animate the current of affects rippling across a perilously illiberal moment in US history.

As the landscape of podcasting moves beyond the RSS feeds, amateurish enthusiasm, and proleptic perspective of the 2010s, we might want to consider audio media more broadly as a collection of sonic elements and social practices that now put pressure on the idea of a discrete "podcasting space." Meanwhile, podcasts increasingly include visual elements, as on the chat shows featured on YouTube, which often feature little more than a perfunctory webcam image of two or more people chatting. Podcasting, Jonathan Sterne et al., pointed out in 2008, has always been best understood as first, a social practice.[31] And modern sound media have, for more than a century, been inextricably linked in industrial practice and consumption habits with the media of visual spectacle like films and theater and concerts. In short, podcasting is best understood as always already part of a transmedia process shaped by and in social practices.

But the success of right-wing podcast chat shows suggests that the specific affordances of the long-form sound media centering the human voice merit an extra layer of scrutiny for how they manage to evoke affective currents of identification and hate, precisely because of their proximity to forms of liberal discourse perfected by radio and its podcast progeny. The shouting "shock-jock" of AM radio's golden age of the 1970s to the 1990s, has left a lasting legacy on the political identity of the American soul. But the podcasts of the early 2020s that embraced Trumpism offered a kinder, gentler form of rightist philosophy. In this way, they provide a fresh horizon of analysis and struggle for those fighting not just for liberalism's sound effects but for liberation itself.

---

[31] Jonathan Sterne, Jeremy Morris, Michael Brendan Baker, and Ariana Moscote Freire, "The politics of podcasting," *Fibreculture* 13 Article: FCJ–087 (2008), http://thirteen.fibreculturejournal.org/fcj-087-the-politics-of-podcasting/. Accessed December 2015.

INDEX

Abdurrahman, Sarah 316
Abel, Jessica 105, 113, 131, 132, 133
Abumrad, Jad 214, 311
Adams, Noah 131
adminstrative perspective 2
Adventures in Sound 125
affective economy 4, 19
Afghanistan 162–4
African American voices 249–51
Ahmed, Sara
    affective economy 4, 19, 105, 174, 261, 268
    audience research literature 78
    emotions 31, 123
    hate speech 308
    political speech 121
    racial animus 83–4
Akbar, Hyder 162–4, 166, 315
Alexander, Erica 291
Allison, Jay 57, 100–1
All Things Considered
    coverage of human experience 43–4
    debut of 38, 40
    double remediation 7
    music on 244–6
Ally Bank 204–5
Anderson, Arlie 51–2
Anderson, Bonnie 321–2
Anwyl, Jeremy 201
The Argument 6
audience participation program 125–7

audience research 62–9, 69–81
Audience 88 study 75–7
Audience 98 study 77–9
Audience 2010 study 82

Baker, Courtney R. 289
Baker, Edwin 51
Bannon, Steve 281
Baran, Madeleine 272
Barber, Benjamin 146
Barnett, Charlie 57
Barthes, Roland 20, 145, 148–9, 150–2, 251–3, 308
bathos 134–40
Battles, Kathleen 30
Beardsley, Eleanor 219
Beckman, Adam 226, 227
"Before and After" (2001) 141–2, 151
Bellow, Saul 55
Benjamen Walker's Theory of Everything 92
Benjamin, Walter 326
Bennin, Phia 316
Bergdahl, Bowe 269, 271
Berlant, Lauren 9, 168, 170–4, 308
Berliner, Uri 210–11, 299
Berry, Emanuele 316–22
Bezos, Jeff 328
Biewen, John 103–4, 132, 242
Birdsall, Carolyn 21
Black, Baxter 50
Blaine, David 109
Block, Melissa 218

INDEX

Bloom, Paul 289
Blumberg, Alex
    foreign economics 202
    Gimlet 207, 296
    Goldman, Alex 295
    Hansbury interview 227–33, 235–7
    mathematical formulas 132
    Obama health care bill 200–1, 202
    Planet Money 184–6, 188
    storytelling 182–3
    ticket scalper interview 157
    voice 242
Bogart, Humphrey 252
Bon Appetit 293
Bottomley, Andrew 104
Boudreau, Don 193–4
Bourdieu, Pierre 102–3, 215, 252
"Brigadoon" 118
Brini Maxwell's Hints for Gracious Living 92
Brown, Michael 316
Brown, Wendy 66, 223–4, 312
Burton, Susan 163
Bush, George W. 160
Butler, Judith 30, 235–7

Campbell, Colin 300
Candow, David 67, 222
Cantril, Hadley 31
Capote, Truman 247–9
Capra, Frank 169
Carey, James 38
Carnegie Commission on Educational Television 69
car radio 35
Carrier, Scott 115, 118, 119, 227
Car Talk 139
Carter, Jimmy 42, 50
Cash, Johnny 248
Chace, Zoe 123, 124, 259
Chadwick, Alex 48

Chang, Ailsa 219
Charles, Ray 251
"Cheap 90" study 78
Cheever, John 145
Cher 248, 254–5
Chion, Michael 20–1
Christo, Jane 74
Church, Tom 71
Clifford, Theresa R. 76
Cline, Patsy 247–248
Clowney, Peter 12
Code Switch (2016–) 6, 291–2
Codrescu, Andrei 51
Coleman, Korva 218
Conaway, Laura 189, 193–4
Conflicted 6
Conners, John 238
Cornish, Audie 218
cringe 279–80
Croce, Jim 251
cruel optimism 9
Cwynar, Christopher 182
Czitrom, Daniel 8

The Daily 214
Daisey, Mike 136
Daniel, Drew 236, 251
Dann, Lance 153, 311, 313
Dark, Allison Elliot 145, 153
David, Larry 279
Davidson, Adam
    driveway moment 57
    foreign economics 202
    New York Times Sunday Magazine 207
    Planet Money 184, 186, 188–9, 191–3
    US invasion of Iraq 160–2, 169
    Warren interview 205–6
"The Deepest, Darkest Open Secret" (2014) 164–7
Deforest, Lee 35
del Barco, Mandalit 219

Delgaudio, Derek 109–10, 111, 113
Demby, Eugene 6, 291–2
DeRose, Jason 147
Dewey, John 268
"Didn't Ask to Be Born" (2002) 122
Doctorow, Cory 273, 312
"Do-Gooders" (1999) 106
double remediation 7–8
Douglas, Susan 16, 21, 34, 117, 230, 233
Dow, Whitney 291
Dowling, David 284, 311
Doyle, Irene 258
driveway moments 54–8
Duggan, Lisa 66, 224, 235
"Dutiful Aggregators" 84–7
DXers and DXing 61, 117
Dzotsi, Emmanuel 293–8, 316

Earplay 97
"Eat the Rich" 292
Ebert, Roger 3
Eby, Tim 146
economic crisis (2008) 183–94
Eddings, Eric 293
Edwards, Bob 221
Edwards, Cliff 248
Eggers, Dave 129
Ehrenreich, Barbara 47, 68–9, 70, 308
Ehrick, Christine 21, 219, 251
Ehrlichman, John 50
empathy 1–2
empathy machines 1–2, 13
"The End of Empathy" (2019) 286–90
"Enemy Camp" (2003)162
Engels, Friedrich 18
"Entanglement" 284–5
explanatory comma 6–7

Fadel, Leila 219, 304
The Fall of the City 30–1
family theme 128–31
Ferguson, Niall 194
fiascos 106
Fierstein, Harvey 248, 254
Fine, Cordelia 230
Finkel, Jim 185
fireside chats 28–9, 34, 55
Fitzgerald, Ella 246
Florida, Richard 204
Floyd, George 317–18
Folkenflik, David 223, 298
Fox, Nicols 39
Francis, Richard 230
Frank, Joe 97, 244
Franklin, Joe 95–6, 97, 314, 322
Freddie Mac 192
Freeman, Morgan 247, 249–250
"The Friendly Man" (1995) 115
Frier, Betsy 53
Fu, Stefanie 298
Fukuyama, Francis 101

Gaines, Alisha 289
Gans, Herbert 72
Garfield, Bob 260
gay voices 216, 253–5, 259
Geiger, Jack 113–14
Geithner, Timothy 204
gender and sexual identities 225–42
General Motors Acceptance Corporation (GMAC) 204–5
Gill, Rosalind 230
Gimlet 207–8, 293, 295–300
Giovannoni, David 73–5, 78, 79–80, 82
Gitlin, Todd 2
Glass, Barry 130

INDEX

Glass, Ira
   autobiographic influence on *This American Life* 99–105
   capitalism 178–9
   Chadwick as role model 48
   driving and radio 54
   early years with Stamberg 53
   feelings and emotions in radio programs 11, 12, 15
   fund-raising 80–1
   impact of Barthes on 148–9, 150–1
   longevity 322–3
   magician job 105, 107, 108–9
   Malatia gag 134–7, 140
   masculine cultural authority 224
   mission 19
   natural, conversational voice 223
   recession 188
   Serial 267
   soft masculinity 234–5
   storytelling about strangers 112–14, 116–17, 120–2
   This American Life as "show" 313–15
   voice 242
   voices of NPR 49
   "Why We Fight" 169
   word-centered narrative guidance 131–3
   Your Radio Playhouse 95–8
Glazer, Ilana 279
Glover, Donald 279
Goebel, John 201–2
Go Figure 89–90
Golden Age of radio 219–20
Goldman, Alex 207–8, 292, 294–6
Goldstein, Jonathan 207, 240–2
gothic family mysteries 128–31
Grandin, Temple 57
Grayson, Deb 302
Green, Jazmine 297

Greene, David 223
Grosvenor, Vertamae 50–1
Gulf War 39

Hagood, Mack 21
Hannah-Jones, Nikole 6, 147, 274
Hansbury, Griffin 227–33, 235–7
Harris, Evan 98
Harris, Kamala 328
Harvey, David 175, 312
Hayek, Friedrich 203, 312
Hayes, Chris 198
Headlee, Celeste 298
health insurance 200–1
Helgren, Jamie 89–90
Hendy, David 11, 29
Hernandez, Imee 278
heterosexuality 236
Hilmes, Michele 4, 219, 249, 251
Hindenburg 28
Hit, Jack 118
homonormativity 235–6
homophobia 238–42
Hussein, Saddam 161

Ilchik, Bob 53
Inskeep, Steve 221
"Interesting People, Interesting Radio" (campaign) 88–9
Invisibilia (2015–23) 6, 126–7, 138–9, 214, 282–90, 292
Isay, Dave 281
"Is This Working?" (2014) 275

Jackson, Samuel L. 251
Jacobsen, Abbi 279
Jahad, Shirley 141–2, 151
Jamieson, Kathleen Hall 219–21
Joffe-Walt, Chana
   conversational style 188
   "make them care" 147

"Make Them Care: Crafting Narratives About Entrenched Social Problems" 6
New Kid 321–2
"Nice White Parents" (2020) 275–7
Obama health care bill 200–1, 204
Social Security benefits 208
story framing 112
Toxie (bond) 196–200
John, Nicholas 59
Johns, Bill 202
Johnston, Windsor 218
Jones, Norah 247
Jones, Spencer 303
Josephson, Larry 71–2
Juman, Jillian 278
Justice Center (Cleveland, OH) 265–6, 269

Kamen, Jeff 38
Kaplan, Lori 84, 90
Karpf, Anne 220, 252
Kasell, Carl 221
Kavanaugh, Brett 288–9
Kern, Jonathan 57
Kestenbaum, David 196, 198–9, 201, 203
"The Kindness of Strangers" (1998) 112–14
Kine, Starlee 207–8, 295–6
Kipling, Rudyard 257
Kitchen Sisters 57, 144
Klein, Naomi 43, 206, 327
Kling, Bill 180
Klosterman, Chuck 98, 100
Koenig, Richard 198–200
Koenig, Sarah 164–7, 215, 265–7, 269–70
Kornfeld, Aviva de 324
Kraft, Jerry 321–2
Kroc, Joan 58–9, 93, 146
Krulwich, Robert 57, 214

Kumanyika, Chenjerai 262–3
Kuralt, Charles 43–4, 49–50

Lacey, Kate 16, 17, 18, 32, 219
Laham, Nasser 162, 173
Larson, Sarah 303
LaRusso, Glen 187
Lee, Canada 113–14
Lee, Hae Min 167, 267
Letson, Al 272
liberalism 3, 9–10, 40, 68–9
liberal media 3
libraries 123–4
Lieber, Matt 139, 300
Lindbergh baby kidnapping and murder trial 27–8
Lindgren, Mia 207
listening back 17, 18
listening in 16
listening out 16
"The Lives of Others" 123
Llinares, Dario 242
"Lost in Translation" (2003) 161, 173
Love+Radio 92
Low, Tobin 103
"Low Opportunity Categories" 84–7
Luther, William J. 203

magic and *This American Life* 105–11
"Magic Show" (2017) 108
Magliozzi, Ray 139
Magliozzi, Tom 139
Malatia, Tory 134–7, 140, 241
Marcus, Greil 258
Marisol-Meraji, Shereen 6
Marketplace 180–1, 291
Maron, Marc 215
Mars, Roman 215
Marx, Karl 18
Matho, Lauren 53
McCauley, Michael 59, 67, 70, 80
McChesney, Robert 63

INDEX

McConnachie, Brian 246–7
McCourt, Tom 37
McDonald's 58–9, 93, 146
McGovern, George 40
McLuhan, Marshall 66
men's voices 221–2, 224, 234–5
Meraji, Shereen Marisol 291
Mesiti-Miller, Pat 132
"Middlemen" (2002) 154–7
Miller, Edward 30
Miller, Lulu 6, 127, 282, 284
Minow, Newt 36, 64
Misitzis, Lina 286–90
Mitchell, Jack 40, 46, 51, 64, 67
mobile privatization 5, 29
Monroe, Marilyn 247
Montaigne, Renee 218
Montopoli, Brian 258
Morgan, Robert 297
Morning Edition 38–9
Morrison, Herbert 28, 29
mortgage crisis 185–6
Moryl, Rebecca 203–4
The Moth 126
Murray, Matthew 241
Murrow, Edward R. 100
music on shows 243–5
Musk, Elon 327

Nakamura, Lisa 122, 289
narrative nonfiction podcasts 268
Nathan, Clarence 185–6
National Public Radio (NPR)
    audience research 62–9, 69–81
    branding 58–9
    and crisis 38–43
    driveway moments 54–8
    early years 36–8
    funding 41–3, 55
    "listeners should feel" 44–53
    podcasts 92–3
    premeditated elitism 69–81
    psychographic studies 81–7
    sound quality 47–9
    voices of 48–53

Natisse, Kia Miakka 290, 292
Navarro, Lulu Garcia 219
Nelson, Soraya Sarhaddi 219
neoliberalism 68, 175, 177, 192
New Kid 321–2
Newton, Wayne 255
New York Times 300
Next Door Stranger (2018) 126
"Nice White Parents" (2020) 275–7
Nixon, Richard 40
No Compromise 6
No Feeling Is Final (2018) 283
Noguchi, Yuki 219
Norris, Michele 218–19
nostalgia 32–5
Not Past It 6
NPR Listens 84
NPR's Playhouse 97

Obama, Barack 200–201, 280
Odetta 247, 249
Office of Communication Research 71
Ohmann, Richard 111
One Small Step 126
Ong, Aihwa 191
Overby, Peter 221, 223
Overholser, Geneva 44

Patterson, Eleanor 41, 180, 215
Pazarbasioglu, Ceyla 188
Peabody, Charles 193
Perry, Melissa Harris 298
Personal People Meters (PPMs) 62
Peters, John Durham 40, 152, 153, 168, 328
Peterson, Jack 286–89
Peterson, Wallace D. 52
Phillips, Lisa 50
Pinnamaneni, Sruthi 293
Planet Money
    coverage of underwriters 204–8

economic crisis (2008) 183–94
fans of the show 203–4
foreign economics 202–3
health insurance 200–1
origin of 175–78
structure of feeling 208–12
Toxie (bond) 195–200
podcasts
double remediation 7–8
empathy machines 4
as feeling media 310–13
Podcast Movement 214–15
Poggioli, Sylvia 256
politics of grievance 2–3
Porter, Jeff 97, 220, 252
A Prairie Home Companion 180
Prakash, Snigdha 218
premeditated elitism 69–81
Presley, Elvis 248
Princiatta, Sal 154–6, 166
psychographics 65
psychographic studies 81–7
Puar, Jasbir 235
Public Broadcasting Act 36–7, 63
public radio, origins 63–4
public scene 268

Quah, Nicholas 302
queerness 227, 229, 234, 239–42
Quist-Arcton, Ofeabia 219, 256
"Quitting" (1995) 98

radio
crises ("in a state") 27–32
double remediation 7–8
empathy machines 4
history 5
listeners 61
mobile privatization 5
nostalgia 32–5
"Radio" (1998) 116—20
Radio Diaries 125
Radio Juventud 92
Radiolab 214

Rae, Issa 279
Ragusea, Adam 213, 214
Rakoff, David 224, 227, 232–4, 242
Reader's Digest 103
Reed, Brian 301–3
Reinvigorating Public Radio's Public Service & Public Support (study) 81–4
Reparations: The Big Payback (2021) 291
Reply All 139, 293, 295, 315–16
Resistance 6
"Retraction" (2012) 136
Richard, Analiese 22, 23, 207, 273, 308
Richman, Joe 125, 132
Robbins, Bruce 177
Roberts, Cokie 67, 217
Robeson, Paul 250
Rogan, Joe 297, 328
Ronson, Jon 242, 244
Roosevelt, Franklin D. 28–9, 30, 33–4, 55
Rosenstein, Rob 132
Rosin, Hanna 6, 19, 127, 282–4, 286–90
Royko, Ben 135–6
Royko, Dave 136
Royko, Jake 135–6
Rudnyckyj, Daromir 22, 23, 207, 273, 308
Rusesabagina, Paul 106
Russo, Alex 195
Rwanda 106
Ryssdal, Kai 291

Sallie Mae 192
Sanders, Sam 304
"The Santaland Diaries" 237–8
Sarnoff, David 27
Savage, Dan 224, 227, 239
"Say It to My Face" (2023) 323
Scam, Iggy 118

INDEX

Schafer, R. Murray 20
Schiller, Vivian 42
school inequalities 274–80
Schorr, Daniel 221
Schudson, Michael 30
Schumacher-Matos, Edward 206, 210–11
Schwartz, Tony 125
Second World War, the 32
Sedaris, David
  books by 244
  family stories 129
  Frank, Joe 97
  gayness 237–9
  Malatia gag 134
  masculinity 224
  queerness 227
  Santaland Diaries 57
  stranger stories 119
  unusual voice 242
September 11, 2001, terrorist attacks 39, 141–6
Serial 165, 265–72, 301
Seward, Senta 21
Shane, Jess 292, 295
Shapiro, Ari 259, 298
Sharon, Tzlil 59
Shaw, Yowei 290
Shaywitz, David 204
Shearer, Harry 97
Shepard, Alicia 205
Shepperd, Josh 37
Sherman, Scott 258
Shipley, Sharon 52
Short, Bobby 248, 254
"Shouting Across the Divide" (2006) 170–3
Shuster, Mike 133
Siegel, Robert 40, 67–8, 144–5, 221
Siemering, William
  celebrating human experience 43
  curiosity over authority 168
  nonfiction storytelling 101
  NPR's purpose 44, 45, 72, 220
  public radio and affective education 307
  radio and public radio as feeling medium 11, 23, 93
Singh, Lakshmi 218
sissies 238–42
Smith, Robert 22, 185
SmithGeiger 84–7
Snow, Tony 257
Snyder, Julie 6, 106, 139, 142–3, 167, 267
Solberg, Wit 196–7
soundwork 4
Spiegel, Alix 58, 127, 170–1, 282, 286–7
Spiegel, Elise 6
Spinelli, Martin 153, 311, 313
Spivak, Gayatri Chakravorty 30
Spotify 296–7, 299–300, 303
Stamberg, Susan
  feminism 222
  NPR and voices 217, 258
  public broadcasting's role 37
  1980s 47
  voice 50, 51–3, 168
  warmth of 214
Stavitsky, Alan 71, 72, 76
Sterne, Jonathan 329
Sterner, Ashley 243
Stevens, Shay 218
Stewart, John 139
Stoever, Jennifer 21, 249
StoryCorps 125
Storycorps 11
"S-Town" 301–3
strangers 111–27
Strangers 126, 283–4
structure of feeling 12–13, 17–18, 54–5
"The Strugglers" 84–5

Sullivan, Lily 123
Syed, Adnan 167, 269
Syed, Hamza 301–3

Talbot, Keith 244
"Talking While Black" (2022) 317–18
Tarantino, Quentin 251
Taylor, Ann 218
"Team Captains" 84–7
"The Test Kitchen" (2021) 293–6
"Testosterone" (2002) 227
Thau, Lea 126, 283–4
"The Cheap 90" 73–4
"The Giant Pool of Money" (2008) 184–6
This American Life
   bathos 133
   branding 59
   diffidence and wonder 99–105
   double remediation 7
   driveway moment 54
   elasticity of gender and sexual identities 225–42
   empathy machine 15, 16
   focus on strangers 111–27
   "The Giant Pool of Money" (2008) 184–6
   gothic family mysteries 128–31
   impact of 325–7
   inequality in schools 274–80
   magic 105–11
   post-9/11 shows 141–3, 149–52
   prominence of 10–13, 41
   thirty years of 308–10
   as transmedia form 313–15
   word-centered narrative guidance 131–3
   Your Radio Playhouse as original name 95–8
This I Believe 100–1
Thomas, J. 76
Thomas, S. 76

"Three Miles" (2015) 275
time spent listening (TSL) 74
Titanic 27
"To Be Real" (2017) 109–10
Torbati, Yeganeh 124
Totenberg, Nina 40, 217, 255–6
Toxie (bond) 195–200
"The Traditionalists" 84–5
traffic in feelings 19–20
transgender 227–28, 229
translators 159–60, 162, 173
Traverse City, Michigan 320
"The Trojan Horse Affair" (2002) 301–2
Trump, Donald 1–2, 40, 271, 281, 327–8
"Two Steps Back" (2004) 274

"Vacations" (1995) 102
VALS (Values and Lifestyles) methodology 72, 75, 76
Vance, J. D. 328
Van Der Kolk, Nick 92
Verma, Neil 184, 208, 275
vocal fry 259–60
Vocal Impressions 246–7, 252–6
vocal performances 215–25, 242–6, 246–56
Vogt, P. J. 207, 292, 293–5
"Voracious Voyagers" 84–7
Vowell, Sarah 119, 242, 244, 257

Wang, Jennifer Hyland 307–9
War of the Worlds 29, 30–1
Warren, Elizabeth 205–6
Washington, Glyn 214
Webb, Tom 202
Weiss, Ellen 205
Welles, Orson 30–1
Wells, H. G. 34
Wertheimer, Linda 46, 67, 217
West, Mae 248
Westervelt, Amy 298
Wheeler, Soren 132, 133

INDEX

White, Barry 250, 253
Whitfield, James 318–19
"Why We Fight" (2002) 159–60, 169
The Wild Room 97
Williams, Cynthia 53
Williams, Flawn 245
Williams, Juan 90–1
Williams, Kim 51
Williams, Raymond
    impulse, restraint, and tone 318
    listening back 18
    mobile privatization 5–6, 29
    structure of feeling 12, 54, 99, 129, 159, 177, 308

Wiltenburg, Mary 105–6
Winchell, Walter 27, 29
wireless transmissions 27
women's voices 217–220, 233, 257–58
The Wonder Years 129
Wu, Diane 316

Xaykaothao, Doualy 219

Your Radio Playhouse 95–8

Zarroli, Jim 223
Zuckerberg, Mark 58, 328
Zukerman, Wendy 297
Zwerdling, Daniel 242